The Anzin Coal Company, 1800–1833

The Anzin Coal Company 1800–1833

Big Business in the Early Stages of the French Industrial Revolution

Reed G. Geiger

University of Delaware Press

University of Delaware Press,
Newark 19711
 Published 1974
Printed in the United States of
America

International Standard Book
Number: 0-87722-100-6
Library of Congress Catalog Card
Number: 74-77774

Printed and distributed for the
University of Delaware Press by
Temple University Press,
Philadelphia, Pennsylvania 19122

Contents

Illustrations

Maps

Figures

Tables

Preface

Traditionally, historians have regarded nineteenth-century France as a curious example of backwardness or underdevelopment in a world where Great Britain, Germany, and the United States provided the normal pattern of a comparatively sudden takeoff into rapid, sustained economic growth based on the development of industrial capitalism and modern machine and energy technology. Scholars have proposed an almost limitless list of explanations for the supposedly eccentric or unique character of French industrialization ranging from inept government policy to sluggish entrepreneurship to poor coal resources. In the last decade, however, some general comparative studies and studies of the French economy as a whole have tended to blunt the view that France's development was particularly retarded or stunted. Further elucidation of the problem will require more specialized monographs for key industries, regions, and firms like Jean Vial's study of the iron industry or Jean Bouvier's work on the Crédit Lyonnais.

In this context, the history of the Anzin Company is a particularly interesting case study from several points of view. First, Anzin was almost certainly the largest single, concentrated industrial firm in France during the early nineteenth century, with a symbolic reputation for the French much like that of General Motors or Standard Oil for Americans today. Second, it produced about one-fourth of the French output of coal at a time when coal was becoming the crucial raw material for industrialization. Third, the company was located in the Nord department, which was just beginning its development into the most important of France's modern centers of industry. Finally, because of Anzin's size and the nature of its product, its history touched most of the key issues concerning economic growth and development during an age of transition from a traditional to a modern industrialized economy: technological innovation, especially the application of steam power to industry; the revolution in transport resulting from the improvement of internal waterways; the increased width and depth of the market that resulted from better transport and the growth of coal-using industries; foreign competition and tariff protection; the need for a disciplined labor force and trained, responsive managerial cadres; capital formation in a rapidly expanding firm; the relationship of government and business (Anzin's chief leader between 1821 and 1832 was first minister of France in 1831–32); the strategy, structure, and control of the giant firm; and the strengths and weaknesses of French entrepreneurship. Drawing mainly on company records and government documents, I have tried to explain the interplay of technological innovation, market expansion, and entrepreneurial skill that revitalized Anzin after the disasters of the Revolution and made it a prime example of industrial power and profitability in the early stages of the French Industrial Revolution.

A great many people have contributed in one way or another to the completion of this study. My former advisor, Professor Paul W. Bamford of the University of Minnesota, first suggested that the coal industry was an important part of early nineteenth-century French industrial history that deserved more thorough exploration. The United States Educational Commission of the Department of State made it possible for me to spend the year 1959–60 in France under the provisions of the Fulbright Act, and leave granted by the University of Delaware in 1970 allowed me to complete my research. Among those who have helped an innocent abroad are François Crouzet of the University of Nanterre and Marcel Gillet and Firman Lentacker of the University of Lille. The library staffs of the Universities of Minnesota, Delaware, and Lille as well as those of the Archives Départementales of the Nord, the Archives Nationales, and the Bibliothèque Nationale have been uniformly courteous and helpful. Of the many scholars to whom thanks are due, I must mention especially Paul Bamford and Marcel Gillet, both of whom read my manuscript at an early stage, and William Parker of Yale University and Stephen Salsbury of Delaware, whose comments during the process of revision were most useful. I should also like to express my appreciation to Robert Wolcott of the College of Wooster and the late Herbert Heaton of the University of Minnesota. Both contributed more than they knew (or perhaps would wish to acknowledge) to the direction my work has taken since they first stirred my interest in the economic life of the past. Finally, I gratefully acknowledge the patient help and encouragement of my wife during the lengthy preparation of this study.

The Anzin Coal Company, 1800–1833

Introduction

As recently as 1959, Charles de Gaulle could proclaim that coal was an "élément toujours essential, mais aujourd'hui capital de l'économie française."[1] Although even then petroleum was rapidly replacing coal as the prime source of energy in France, the fate of its industrial economy still seemed indissolubly linked to the strength of the European coal industry. The bonds have slackened still further today, but a time when coal has lost its preeminent place as the fuel of modern societies is perhaps the appropriate moment to look back through its history to the original energy crisis of the modern age, the period when the first crucial links between coal and the process of modern economic growth were forged.[2] Wisely, one historian has recently warned against the danger of overrating the industrial importance of coal in France during the first half of the nineteenth century.[3] Nonetheless, the period witnessed significant attempts on the part of some French industrialists to adapt themselves to a source of energy that was cheaper, more abundant, and more efficient than wood or water. Although the typical nineteenth-century linkage of coal with iron, steam, and the railroad did not culminate until well after 1850 in France, the first decades of the century saw the beginnings of modern, coal-based industry.

This study attempts to explain how the leaders of the Anzin Company, far and away the largest French mining firm of its day,[4] overcame a series of handicaps to realize the opportunities created by the rising demand for coal in the early nineteenth century. Located in the Nord department near Valenciennes on the Belgian border, Anzin mined about one-quarter of all French coal output by

1830 and held a virtual monopoly on production in northern France. With its five thousand employees it was probably the largest centralized industrial establishment in the country. Its owners included some of the greatest names in French and Belgian politics, society, and business—Cambacérès, arch-chancellor of the Empire, the Duke of Arenberg, Rémusat, Walckiers, and above all Casimir Perier. Perier, first minister and restorer of order after the July Revolution of 1830, was the leading figure in the modernization and expansion of the company during the twenties. The size of Anzin and the importance of its owners made it a center of public controversy. To its enemies, Anzin seemed to epitomize monopoly and the power of vested interest. Yet friends and enemies alike regarded it as an example of what good management and up-to-date technique could do for French industry in its struggle to shorten the lead of the English. Anzin was a symbol of the new industrial capitalism, a prime example of large-scale, highly capitalized private enterprise in early nineteenth-century France. From a variety of points of view, then, Anzin merits the attention of historians.

Considerable research has been done and many of the sources have been published on the Anzin Company before 1800.[5] The story of Anzin after 1833 has attracted interest as part of the history of the rapidly expanding basin of northern France.[6] But the period 1800 to 1833, which has been briefly dismissed by most scholars as one of stagnation, remains to be investigated in detail.[7] The history of the French coal industry has suffered the same neglect for this period,[8] although many historians have conceded, indeed insisted, that a shortage of good, cheap coal was a crucial limiting factor in French industrialization.[9]

This neglect is all the more surprising in view of the recent interest and discoveries in French economic history of the early nineteenth century.[10] New investigations indicate that the annual rate of growth of French industry during these years compares favorably with that of most later periods.[11] It is true that no Rostovian takeoff occurred then or later, and rates were moderate throughout the century as compared to the more spectacular phases of British, German, or American industrialization. Furthermore, the relatively rapid growth of the textile, metallurgical, chemical, and equipment industries after 1800 was in part due to the fact that the figures for output at the beginning of the period were so low.[12] But that is precisely the point:

la grande industrie, in our modern sense, got its start in France in the first third of the century. The phrase itself was ceasing to be used, as it had been in the eighteenth century, to designate the great state workshops or far-flung putting-out empires and was coming to mean mechanized factory production on a large scale.

The groundbreaking work of Charles Ballot fifty years ago pointed to the beginnings of textile mechanization and industrial chemistry, and Henri Sée was aware of an "indubitable industrial development" under Napoleon.[13] Yet only recently have historians managed to escape the trap of writing of industrial development from 1800 to 1815 as a branch of Napoleonic politics.[14] When, in contrast, one focuses on the beginnings of large-scale mechanization, especially in textiles, and the resulting concentrations of capital, workers, and plant, or on the rise of a new class of capitalist industrialists made rich by war-stimulated demand and a new class of private bankers made rich by currency speculation and war loans, the achievements of Napoleonic industry appear less transitory and political, more fundamental and constructive. The Empire saw at least the vigorous stirrings of a new industrial way of life. After the legal, political, and monetary chaos of the Revolution, Napoleon provided a stable political and juridical ground for the development of new ideas and discoveries in science, technology, administration, and education germinated by revolution.[15] Granting that no industrial revolution had yet taken place in France and that French industry in 1815 lagged even farther behind English industry than it had in 1789, some knowing men, nonetheless, could see soon after Waterloo the advent of a new industrial order and could justly claim that French industry was second only to that of England.[16]

It is well known that the Restoration of 1815 restored far less than a complete Old Regime; the Napoleonic institutions, laws, and bureaucracy were preserved almost intact under the last Bourbons.[17] Moreover, the transition was quick and peaceful. There was no general upheaval of the foundations of economic life comparable to that after 1789. The most drastic change was the reversion to prerevolutionary boundaries with the loss of many continental markets now protected by tariffs. But French industrialists found some compensation for the loss in the reopening of overseas markets and the institution of French tariffs aimed at securing the home market for domestic industry. The high tariff policy of the Restoration may have been shortsighted, but it succeeded temporarily in keeping out most cheap English manufactured products and in closing off much of a lucrative market to the growing industry of the area that was to become Belgium in 1830.* The protectionist policies of the Restoration were to benefit Anzin greatly in the struggle against its Mons rivals across the restored frontier.

The aristocratic reaction hardly jarred economic activities. The parliamentary clashes of liberals and conservatives had mainly to do with politics, social status, and religion, not with economics. Arguments about economic questions of the day—tariffs, internal improvements, and the budget—revealed no clear division and much agreement between merchant, or industrial bourgeois, and landowning aristocrat on goals if not on means of resolving these bread-and-butter issues. Like Napoleon, the Bourbons refused to renew the more galling restrictions of an outmoded mercantilism while encouraging industrialists to use the power of the state wherever it might help them. Industrialists were free to pursue prosperity and expansion in much the same stable legal framework that had existed since 1800.[18]

Coal played an increasingly important role in this expansion. Although the growth rate of coal lagged slightly behind that for French industry as a whole under the Empire, the rate during the Restoration and early years of the July Monarchy clearly surpassed the industrial average.[19] While still far from being a "leading sector,"[20] coal had nevertheless become one of the faster growing industries, an *industrie vedette* ("star industry") of the twenties and thirties.[21]

About fifty "coalfields" were thought worthy of inclusion in a map of French coal production for 1814.[22] By that time, however, several fields had begun to be worked on a relatively large scale and to acquire a more than local importance. Most of these were located in southern and central France, where they abutted on the borders of the *Massif Central.* Among them, Alais, Decazeville-Aubin, Carmaux-Albi, and Le Creuzot-Blanzy shed their rural aspect during the early nineteenth century and attracted much attention and capital, but because they lacked really large reserves of good

*For the sake of brevity, the term "Belgium" is often used throughout to refer to the territory known by that name only since 1830. See note 26.

quality coal they were never to become much more than suppliers of local industries that grew up in their shadow.

Chief among the central French fields were Rive-de-Gier and Saint Etienne, known together as the Loire field. The oldest in France, this field served the manufacturers of various small metal goods in and around Saint Etienne, a sort of minor Birmingham, and was chief supplier of the Paris workshops by way of a long and difficult network of rivers and canals. Dozens of *sociétés*, composed mainly of local men interested in the metal-goods industry, worked the few hundred pits. Production per pit and per société was incredibly small; what was apparently the largest company in 1826 produced only 36,000 tons in 1812.* Total production for the field rose rapidly from about 145,000 tons in 1789 to 229,000 tons in 1812 and 683,000 in 1830.[23] Although production more than tripled from 1800 to 1830, techniques remained primitive, with manpower predominating even for drainage and haulage, and the unit of production remained small, with little investment in capital equipment.

Except for a few very small mines on the lower Loire and in western Normandy, the only coalfield in western or northern France was the one nearly monopolized by Anzin near Valenciennes on the Belgian border of the Nord. The Nord had produced about 40 percent of all French coal and about two times as much as the Loire in 1790. Subject to fewer disturbances during the Revolution and Empire, production in the Loire expanded fairly rapidly and surpassed the declining production of the Nord by 1812. In 1830 the Loire output was twice that of the Nord, but the latter was once more expanding at a rapid rate, and together these two fields supplied well over one-half of all French production, the proportion rising from 58.4 percent in 1812 to 67.4 percent in 1830.[24] Except for the Loire, one or a very few companies controlled most of the larger French fields, but only in the Nord were production and profits large enough to make the monopoly significant. Anzin was distinguished from all other French mining companies by its sheer size, whether measured by production, capital invested, profits, or technical sophistication.

Anzin exploited only the small western wing of an almost continuous coalfield stretching to the Ruhr and beyond, with larger

*Throughout, except with reference to England, tonnages are expressed in metric tons (ca. 2,200 pounds).

breaks, into Saxony and Silesia. Then as now, this field concentrated by far the greater part of continental production, but the relative position of the various basins was singularly different from their present ranking. The southern Belgian basins around Mons, Namur-Charleroi, and Liège were one of the chief reasons why "Belgium was . . . the one country which kept pace industrially with England in the first half of the nineteenth century."[25] On these basins was the sole industrial area on the continent where coal was of basic importance before 1830. Belgian miners had a tradition of skill reaching back to the Middle Ages and were much sought after all over Europe, although their clear technical superiority in 1800 was being challenged by the 1830s. Despite some areas where companies of working miners predominated, Belgian mining companies were large by early nineteenth-century standards, especially around Mons. Production in the Liège and Namur-Charleroi basins was comparable to that of Anzin, but the Mons basin was by far the largest on the continent. It seems certain that the Mons production alone in 1810 amounted to 800,000 tons or slightly more than that of all the rest of France.[26] By the end of the Empire, production for Belgium as a whole had reached 1,500,000 tons, much of it exported to France, whose total production at this time amounted to only about 900,000 tons.[27] By 1829 production for Belgium had risen to 2,513,000 tons,[28] a rate of expansion somewhat slower than that of France, but sufficient to keep Belgium the leading producer on the continent, although far behind England, which had reached 21,000,000 tons in 1826.[29]

The only other important continental fields before 1830 were in Germany, mainly Prussia. Of the chief centers—Aachen, the Ruhr, Saxony, Upper and Lower Silesia, and the Saar—all except the Saar constituted part of the eastward continuation of the great northern field of which Belgium and Anzin formed the western wing. Saxony, "longtemps la terre classique des mineurs," was now in relative decline, but in all the other German basins serious exploitation did not begin until the late eighteenth century and made great strides between 1800 and 1830.[30] In the Saar, production rose from about 50,000 tons in 1804 to about 200,000 in 1830. Even more impressive results were achieved in Upper Silesia. In technique and scale of operation, the Upper Silesian mines were second to none, and production rose extremely rapidly from almost nothing in 1800 to

311,000 tons in 1815 and nearly 4,000,000 tons in 1850. In Lower Silesia, production was already 47,000 tons in 1790 and reached 200,000 in 1830, after which the basin suffered a relative decline because, like Saxony, it lacked large reserves. Aachen's development was also slow; in 1828 output was only 150,000 tons.[31]

As for the Ruhr, "at the beginning of the nineteenth century the diminutive coalfield of Ibbenburen attracted fully as much attention . . . and for the first half of the century the Ruhr was regarded as inferior in every respect to the coalfield of Upper Silesia."[32] Production around 1804 was less than 400,000 tons and did not increase significantly until the decade of the twenties, when production reached 571,000 tons. Among the larger European fields, it ranked with the Loire near the bottom in scale of production and technical sophistication. In 1821 a mine regarded as big for the Ruhr had forty-four miners and produced only 13,782 tons of coal per year.[33]

Several conclusions can be drawn from this brief survey of the continental coal industry in the early nineteenth century. First, production increased considerably almost everywhere, especially after 1820. Second, almost all the increase took place on the few fields that had large reserves. Most of the numerous surface workings of the eighteenth century stagnated or were shut down. Third, the growing fields, most of them first developed in the eighteenth century, were on the whole the same ones that dominate the industry today. Fourth, on these few fields large reserves and large output both permitted and required rapid technological advance. In fact, there tended to be positive correlations between rate of technical advance, rate of growth of output, and even the recentness of intensive development of output. Fifth, technological progress and increased production demanded large initial capital investments, larger units of production, and new organizational forms tending toward the modern corporation and away from the company of working miners. There was a tendency for coal mining to move away from its agrarian and craft heritage and become a truly industrial undertaking. It is evident that mining companies in most of the major continental fields faced reasonably favorable conditions for growth. The demand for coal was increasing and the supply of coal could be expanded. Yet there were serious stumbling blocks on both sides of the supply-demand equation. Obviously, the situation called for an intelligent, aggressive entrepreneurial response.

The Anzin Company was potentially supplied with strong leadership. The peculiar circumstances of its founding in 1757 had placed complete control of company policy in the hands of the six self-perpetuating members of the board of directors, and the catastrophic events of the Revolution had compelled the inclusion on the board of some of the most powerful and enterprising Parisian businessmen of the period. Unfortunately, the new owners chose to defer to the presumed wisdom of the founding families, their local experience, and their prerevolutionary record of success. It was not until 1821, when the company's very existence seemed threatened, and two deaths created an opening, that Casimir Perier emerged as the strong man of the board. Until his death in 1832, he drove Anzin along a path of rapid expansion and daring innovation by the force of his domineering, aggressive personality. Even he, however, was unable single-handedly to maintain absolute control over a company the size of Anzin. A clash between innovation and tradition led to the dismissal of two general managers and a great strike before Perier's success and Anzin's future were assured.

Just as the wise policy of Casimir Perier was only slowly translated into effective action, so too the obvious opportunity presented by a growing demand for coal did not automatically call forth an equivalent increase in output. Anzin was faced with difficult geological problems. Its coal seams were rich and yielded a wide variety of types of coal, but they were deep, thin, wet, and gassy. Thus Anzin was forced to use a complex and costly system of extraction. In comparison with smaller, less affluent companies who benefited from more favorable geological conditions, Anzin was inclined to adopt improvements in mining techniques rather rapidly. By 1800 Anzin was already a model in the use of advanced technique, and so there was little need for radical changes in methods of extraction during the early nineteenth century. The company did, however, make some important modifications in the basic system during the next thirty-three years. In the first decade of the century Anzin became one of the first mining companies on the continent to use steam engines for hoisting coal to the surface. In the 1820s it placed rails underground to expedite movement of the coal from the working face to the bottom of the mine shaft. With the aid of the safety lamp, improved ventilation, and better pumping practices, it conquered seams previously unworkable because of gas or water.

By themselves, the improvements in mining techniques at existing pits would not have been sufficient to increase production. To meet the rising demand for coal, it was necessary to discover, acquire, and reach new seams. Anzin made intense efforts to explore its old concessions, acquire new concessions containing rich coal deposits, and win its battle with the *Torrent*, a mass of water which overlaid some of the company's best coal deposits and which proved impenetrable until the 1820s.

The introduction of new techniques and the ability to exploit new seams affected output, costs, and productivity. Between 1800 and 1833, output rose about 250 percent, with most of the increase coming in the last half of the period. The cost of production fell only about 10 percent, and productivity rose by about the same amount. Yet these apparently slight changes were something of a victory. As shafts went deeper and expenditures for new pits and equipment increased, productivity would have fallen and costs risen without the rationalization of the techniques of production.

The problems involved in marketing coal were much less amenable to the company's control than those concerning production. Only where water transport was available could coal compete in distant markets with older fuels like wood. Although the waterways in the Anzin market area were relatively good, the high cost of transport was one of the major brakes on the expansion of coal consumption. Toward the end of the 1820s, a series of improvements on most waterways, in which Anzin sometimes participated, began to lower the cost of transport and allowed Anzin to reach new markets such as Paris. No complete breakthrough occurred, however, until about 1850, when further improvements on waterways and the competition of the railroad drastically lowered freight rates.

Just as Anzin usually benefited automatically from the improvement of the waterways, so too it gained new markets as a result of the demand created by the birth of modern industry in its region. Of course, both the waterways and industrialization were in part a response to the existence of Anzin and its growing ability to feed them with coal. Many waterways were improved or built for the purpose of carrying Anzin or Mons coal. And the conversion in the twenties of established industries from charcoal or wood to coal, the development of new industries using coal, and the beginnings of a switch from horse and water power to steam clearly depended in part on the availability of Anzin coal at a price competitive with other fuels. It was no accident that industrialization proceeded much faster and farther in the Anzin market area than in most other parts of France where coal was higher-priced or less available. Nevertheless, it would be easy to overstress the effects of industrialization on the Anzin market during the early nineteenth century. Until the very end of the period studied, much, perhaps most, of the growth in the company's market derived from the slow growth of population and of the traditional coal-burning industries.

In part because the Anzin Company lacked the power to change either the composition or the area of its market, it followed a traditional and unenterprising marketing policy. The company continued to sell most of its output to a network of wholesale dealers rather than to the ultimate consumers of its coal. The one example of an attempt to open a new market by direct action, at Paris, depended more for its ultimate success on improved waterways than on Anzin salesmanship. Yet the shelter provided by transport costs and a substantial tariff allowed the company to set prices competitive with its Mons rivals and still achieve a rate of profit on sales of about 40 percent.

Around 1833, an era was drawing to a close with the death of Casimir Perier, Anzin's first great strike, the threat of a reduction of tariff protection, and the beginnings of competition from rival companies in the Nord itself. Despite this cluster of ominous events, Anzin faced the future with justifiable hopes for continued prosperity because the previous thirty-three years had witnessed momentous changes in company policy and practice. Did those changes constitute a turning point in the company's history or were they merely transitional adjustments? Turning point or transition, the period 1800 to 1833 produced changes in the company environment that in turn forced or encouraged the Anzin Company to change. The old markets were beginning to recede in importance as new markets developed toward Paris and among industries using the new technology of iron and steam. Although some production processes such as the actual cutting of coal from the seam remained essentially unchanged, others such as winding and hauling underwent radical alteration. Finally, there was a drastic revolution in entrepreneurial style as the eighteenth-century local owners gave way to a more aggressive, more innovative group of Parisian businessmen.

This last change was crucial, but it would be a mistake to regard this study as a celebration of the virtues of "un-French" entrepreneurship.[34] The men who ruled Anzin were not absolute monarchs. They were restricted at every turn by the hard facts of geology, the petty pace of technological innovation, the recalcitrance of the working force, the resistance of government bureaucrats and of public opinion, the competition of foreign and domestic rivals, the high cost of transport, and the modest rate of growth of French population and industry. Even for Casimir Perier, the inherent difficulty of controlling effectively from Paris an organization of five thousand men scattered over dozens of square miles in dozens of separate work locations meant he was bound to suffer resistance and occasional setbacks. Grand strategy was captive to inadequate structure;[35] the triumphs of enterprise had their limits.

1 Owners, Managers, and Miners: The Exercise of Control

In 1757 three infant mining firms were fused into the Anzin Company. A self-coopting board of directors was established to exercise almost total control over company affairs. Under the leadership on the board of the founding fathers, a handful of local nobles and bourgeois, and their descendents, Anzin on the eve of the Revolution had become far and away the largest and most prosperous mining company in France.

The Revolution, however, destroyed the company's commercial prosperity and broke up its ownership; emigration and confiscation gave the government over one-half of the twenty-four *sols* ("shares") that represented ownership in the firm. Out of the welter of personal and revolutionary politics at Paris and in the Nord, a reconstructed Anzin emerged, infused with new money and talent by a group of Paris bankers. Nevertheless, until the early years of the Restoration, control of policy continued to rest with the remaining old owners, especially Stanislaus Desandrouin, son of the company's chief founder in the eighteenth century. Despite the fact that they soon owned a majority of the firm's twenty-four sols and held three of the six seats on the board, the new owners deferred to the supposed expertise of Desandrouin and regarded their sols first as a financial speculation during the Revolution and then as a passive source of income under the Empire. Only because of the obvious lack of capacity of Desandrouin by 1816–17 and the equally obvious lack of a successor among the other original owners' descendents did the company turn to the Perier brothers, representatives of the interlocked world of Paris finance, business, and government.

Even then the Periers' ascendency over the board and the owners did not convey effective control of company operations and policies. Casimir Perier was forced to beat down the passive resistance of the company's managers in the Nord, who, like the owners, constituted a small, closed circle of interrelated families, sure of their indispensability, proud of their experience, and impervious to change and direction from Paris. One is tempted to compare Perier's problem to that of the feudal king whose vassals ignore him while struggling among themselves for control of the provinces. As company overlord, Perier's greatest contribution to Anzin's success—and the prerequisite for all the others—was his ruthless campaign to reorganize the managerial hierarchy and subordinate the men of the Nord to Paris. It was a battle for authority and order he would fight again as first minister of France after the Revolution of 1830.

Significantly, it was only after Perier's death that Anzin experienced its first major strike, in 1833, a spontaneous explosion expressing the growing discontent of the mining population with his attempts to extend company control over its labor force and exploit it more efficiently. While rhetorically supporting the patriarchal spirit and practices of the old Anzin leaders toward their workers, Perier had in fact tightened work discipline and more or less covertly cut wages. When the dam burst in 1833, Perier's successors were forced to give ground before the combined condemnation, moral or self-interested, of most of the company's workers and the local population, the business communities of the Nord and Paris, and some sections of the government. A return to a true paternalism was impossible, but, softened somewhat and administered in a more humane and equitable spirit, Perier's rationalizing policies could serve as the basis for restored peace between management and labor.

The Background to 1800

In 1700 coal already played a small but important role in the economy of the parts of the north European plain located near the then known portions of today's great northern coalfield. Miners had long extracted coal in small quantities at outcroppings along the river valleys.[1] Since the Middle Ages, the area had been remarkable for a profusion of small-scale industries of all kinds, for an intensive agriculture on the flat, well-watered plains, and for a dense concentration of busy towns. Lacking extensive forest reserves, the coalfield areas experienced a shortage of wood for both industrial and domestic use long before most of Europe and consequently led in the substitution of coal for wood wherever possible.[2]

By 1700 the area around Valenciennes was already buying an increasing number of cartloads of coal from the Mons district, the Borinage, just a few miles to the east and across the frontier.[3] The man who could discover coal near Valenciennes would find himself the master of a market protected from Mons competition by high transport costs. The known seams to the east disappeared underground near Charleroi, passed below Mons, and disappeared again at a considerable depth on the western edge of the Borinage. The Hardinghen mines just discovered near the Calais coast appeared to be the missing link between the deposits of England and those of the continent, and local opinion held that the seams extended deep underground along the Mons-Hardinghen axis.[4]

The immediate cause of the discovery of the Valenciennes basin, however, was compounded of high international politics and individual determination. The treaty of Utrecht in 1713 handed over most of the old Spanish Netherlands to Austria and split the two western provinces of Flanders and Hainaut between France and Austria. The inhabitants and industrialists of Lille and Valenciennes were now at the mercy of the "foreigners" of Mons for their coal supply, and the modest but galling entry tariff established by the Austrians might be raised at will. That was believed to be the prospect faced by these new inhabitants of the Kingdom of France, some of whom were already part-owners of Mons and Charleroi mines.[5]

The Viscount Jean-Jacques Desandrouin, *bailli* of Charleroi, *seigneur* of Lodelinsart near Charleroi, but a resident of Fresnes in French Hainaut, had gained a considerable fortune as owner of a coal mine, forges, and glassworks on his *seignurie* of Lodelinsart.[6] Strongly convinced of the existence of coal in French Hainaut, he placed his fortune, a considerable knowledge of mining, and a phenomenal doggedness at the service of its discovery, and until his death in 1761 was the dominant figure in the development of the Valenciennes basin. Joined by a brother, Pierre, and several local bourgeois, he brought Jacques Mathieu, the director of his Lodelinsart works, and twenty of its miners to Fresnes in 1716 to begin the search for coal.[7]

From the beginning, mining in the Valenciennes basin was marked off from most contemporary experience on the continent by the depth at which the coal beds lay, the enormous quantity of water stored in the intervening rock layers, and the consequent necessity of expending large amounts of capital before a single ton could be brought to the surface. Armed in 1717 with a special exploration permit from the French government, an exclusive mining privilege for fifteen years, and a subsidy of 5,000 florins,* Desandrouin's first company reached coal on its seventh attempt in February 1720. People flocked from miles around to witness the discovery, and "full of joy carried pieces of coal home with them."[8] A grateful government extended the company's privilege to twenty years and granted a new subsidy of 35,000 livres. In November of the same year, the company reached a workable vein, but almost immediately a subterranean flood forced abandonment of the pit and a loss of 20,000 livres. The company was obliged to reorganize. After yet another reorganization, it reached a new vein at Fresnes in 1723. The coal, however, proved to be anthracitic, fit mainly for baking bricks and roasting limestone. Because the company was unable to supply the domestic hearth or blacksmith's forge, production remained low. Between 1725 and 1732, this third company spent perhaps 163,400 florins searching for a bituminous deposit. But in 1734, nearing the end of his financial rope, popularly regarded as a maniac, an *illuminé* in the grip of an *idée fixe*,[9] Desandrouin was saved and the future of French Hainaut revolutionized by the discovery of the vein *Maugretout* ("in spite of everything") in the main bituminous deposits of the basin at the village of Anzin near the gates of Valenciennes.

On the strength of this discovery, the government extended the company's privilege to 1760 and granted an extension of territory to the north. Exploration of the concession and the sinking of new pits in the Anzin area led to a steady expansion of production and sales. By 1756 the company employed fifteen hundred workers and extracted about one hundred thousand tons from sixteen pits. Desandrouin's company was already part of la grand industrie by virtue of its size alone.[10] More significantly, it foreshadowed the giants of the Industrial Revolution in its dependence on advanced technology

*Four florins equaled one livre. The franc replaced the livre at a ratio of 80:81.

and a liberal use of private capital. The discovery of 1734 had hinged on the installation of the first Newcomen mining engine in France. By constant pumping in specially constructed shafts invented by Mathieu, the company had met the technical problems of overcoming flooding. Although these techniques proved successful only at enormous costs, the company showed a neat profit while competing successfully with Mons producers by lowering prices on the local market by one-third. "Heating was no longer a subject of worry for the poor and most factories [*usines*] took advantage of [lower prices] by increasing their production."[11] The company's very success, however, led to a legal struggle for existence with the local nobility.

The French crown had long claimed rights to the products of the subsoil. A royal decree of 1744 had strongly reaffirmed these rights by specifically requiring both landowners and *seigneurs haut-justiciers* to obtain permission from the King's Council to open mines on their lands and by implication asserting its power to refuse such authorization and grant permission to someone else, as it did with its concessions to Desandrouin and his associates. The royal edict met determined resistance in the new provinces of the north where the haut-justiciers claimed possession of the subsoil by provincial custom and asserted that the treaties of 1713–15 guaranteed the customs and charters of Hainaut.[12]

This legal question had been of no importance while the Desandrouin company was struggling to establish itself. Most of the seigneurs haut-justiciers had agreed to permit the company's operations on payment of a fee. Such payments, like the one agreed upon with the Duke of Croy in 1737 for his haut-justice of Fresnes, were, of course, a recognition of the haut-justicier and his rights. The company made a similar arrangement with Croy in 1741 for the seigneurie of Vieux-Condé, but it was Croy himself who obtained the royal concession for Vieux-Condé in 1749, although he permitted Desandrouin and his associates to continue extracting coal there for a fee. Not content with the increasingly dominant position he was gaining in the company's affairs, he obtained a decree from the council in 1756 by which Fresnes would be conceded to him directly upon expiration of Desandrouin's grant in 1760. Croy, one of the great nobles of the kingdom, prince of the Empire, and future marshal of France, was on the verge of becoming himself an important mining entrepreneur by stealing part of the fruits of Desandrouin's lifework.

An even more serious challenge arose from the Marquis of Cernay, haut-justicier of Raismes, within the Desandrouin company's original and principal concession of Anzin. In 1754 the intendant of Hainaut granted him the sole right to exploit Raismes. He bought up the rights to Saint Vaast-la-Haut from its seigneur and obtained permission to exploit the seigneuries within a three-mile radius of his pits at Raismes. He was casting a hungry eye on Desandrouin's entire Anzin concession, which, like that of Fresnes, also expired in 1760. Soon both companies were working pits on territory claimed by the other. Desandrouin, Cernay, and Croy, far from performing their supposed aristocratic role as economic parasites, had shown themselves as eager capitalists and industrialists as their English counterparts.[13]

Desandrouin took his case against Cernay to the local intendant, who upheld Cernay. The next step would be an appeal to the King's Council, which would have an extremely ticklish legal and political decision to render. Croy had apparently been maneuvering matters toward just such a crisis. Now he used his soldier's bluffness, noble courtesy, and good business sense to calm the personal bitterness between Desandrouin and Cernay and avert a costly legal battle, whose highly uncertain outcome might have ruined any or all of them.[14] His solution was the merger of the three rivals' interests in a single company, "la Compagnie d'Anzin."

The union was consummated on 19 November 1757 by the "Acte de Société et du Réunion," which was to govern the company in nearly its original form until the late nineteenth century.[15] The pits, equipment, concessions, and other titles, as well as the capital invested—all the assets which each of the old companies had brought to the union—were sunk without any future claim by their original owners into the new company's assets. Possession of a given number of the twenty-four sols conveyed ownership in the new company. The sols did not represent a fixed capital value or even a market price, since the Acte de Société closely restricted their transfer in practice to the families of the original owners. Their possession merely conveyed a proportionate ownership in the company's assets and a proportionate right to share in any profits. The initial parceling out of the sols was, of course, a matter of diplomatic juggling, since only Desandrouin and his associates had important tangible assets. Cernay's company had only a few unproductive pits, but he and Croy together obtained a half-ownership in the new company on the same grounds which had enabled them to force a union in the first place: their possession of government concessions. In the division, Desandrouin's company received twelve sols, Cernay's eight, and Croy's four. Anzin, Raismes, Fresnes, and Vieux-Condé, totaling about fifty thousand acres, remained distinctly separate concessions held individually by grant from the French crown, but each was now in the hands of the united Anzin Company (see map 1). The Acte de Société placed complete control of the company in the hands of the members of the board of directors, selected from among the largest and most influential owners. Anzin was in essence a multiple family partnership in perpetuity, ruled by the heads of three local noble families.

The subtitle of the Acte de Société stated that it "concerns all the coal pits in Hainaut." Anzin's monopoly of the French portion of the northern coalfield was continually but unsuccessfully threatened during the eighteenth century. Anzin "attempted to concentrate in its hands the concession of all land that could possibly be considered useful. It made every effort to repulse competition. Guided by the experiences of the Desandrouin Company, it made every attempt to obtain both the royal and seigneurial concessions of the land it wished to exploit."[16] Anzin obtained several new concessions between 1757 and 1789, primarily to prevent other companies from receiving them, but abandoned most without any real effort at exploitation or even exploration. It bought up the rights of the remaining haut-justiciers within its concessions where it could and fought the others through the courts to prevent them from setting to work on their own or coming to terms with its rivals. In all these more or less unsavory operations, the political power and social prestige of Croy and the other directors at Paris and in Hainaut were usually triumphant. The companies composed exclusively of bourgeois of such towns as Lille, Douai, and Cambrai were easy prey; those headed by other nobles were more tenacious. The Aniche Company of the Marquis of Trainel and his associates obtained a huge concession west of Anzin in 1773. It was the only rival actually producing coal by 1789, but its production was never more than a tiny fraction of

Map 1

Anzin Company and Its Neighbors, 1833

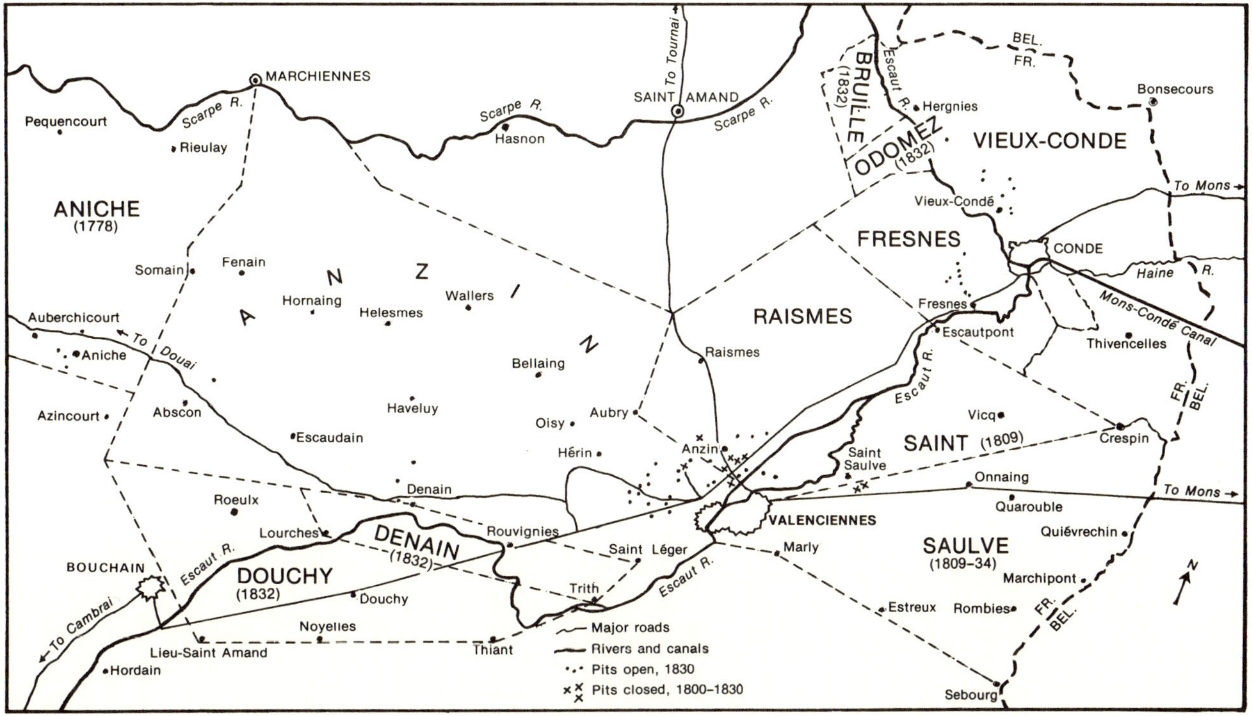

Anzin's, and it encountered continuous financial and technical difficulties.[17]

If its rivals were in a condition "très peu prospère," all observers of Anzin agreed that "on the eve of the Revolution, this immense establishment had attained a very high degree of splendor."[18] Expansion in all phases of operations was steady after 1757. Production rose from about one hundred thousand to three hundred thousand tons, easily the greatest of any French field and incomparably greater than that of any other single company. The technical and organizational aspects of production were handled "with the greatest intelligence." Men like Croy, Desandrouin and his son, and the Mathieu family exercised a close personal supervision over the details of day-to-day production and a paternal watch over the efficiency and welfare of the company's four thousand workers. Twenty-eight of the thirty-seven pits were used for extraction, the others for ventilation and drainage. The deepest pit reached three hundred meters at a time when no shaft in Belgium extended below two hundred meters. Although horse-powered winches were still used to extract the coal, twelve Newcomen engines were required for drainage. In a few areas, such as ventilation, Anzin's methods were superior even to the best English practices.[19]

Anzin's prosperity was reflected in high profits and a sharp increase in the value of the sol, already valued at 15,000 livres by 1771. In 1781 a *denier*, one-twelfth of a sol, sold for 33,250 livres; a sol was thus worth 399,000 livres, and the company's total market value was about 9,576,000 livres. Even more impressive was the relationship of profits to income. On sales of about 3,750,000 francs in 1789, costs totaled perhaps 2,400,000 francs, leaving a gross profit of about 1,350,000 francs or over 50 percent.[20] Although the company plowed back some of this profit in capital improvements, the enviable financial position of Anzin's owners on the eve of the Revolution helps explain the hatred and vindictiveness they and the company faced when the storm broke.

During the first years of the Revolution Anzin maintained its position.[21] Production remained high through most of 1792. The company bought over one million francs' worth of *biens nationaux* ("nationalized property"), especially church lands located within its concessions. With the aid of an impassioned speech by Mirabeau, the advocates of exploitation by concession won a partial victory over the advocates of exploitation by the surface owners. The latter were at last checked by the mining law of 1791, which declared, though equivocally, that the mines belonged to the nation and were to be disposed of in the public interest. Mirabeau had been briefed, perhaps bribed, by the Anzin owners, and he used the example of Anzin, "une des plus belles [mines] de l'Europe," to clinch the argument that rational exploitation on a large, highly capitalistic basis required state control over a system of concessions as a means of reconciling the principle of private property, so dear to the heart of the National Assembly, with the equally dear general interest.[22]

Then late in 1792 catastrophe struck. The Austrians invaded France. The armies of the Allies swept over the Anzin Company pits; Valenciennes was beseiged. After the respite of the French victory at Valmy (20 September 1792), the Allies again invaded. Exposed to enemy fire, eleven of the fourteen pits near the village of Anzin lay idle. Three-fourths of the miners left the pits for the army or the more lucrative jobs of supplying the rival armies with food or transport. Anzin production in 1793 was only 80,000 tons against 291,000 in 1791 and 275,000 in 1792. In 1794 it fell even lower to 65,000 tons. The local government established a system of requisitions only partly respected by the company. The maximum set on prices and wages failed to hold the line. Favoritism, black-market operations, and evasions of local and national attempts at regulation were rampant.[23]

In August 1793 Valenciennes had fallen and Anzin, although never occupied, was subjected to repeated pillaging. Enemy patrols prowled the countryside and sacked or burned the company's warehouses and surface installations. They smashed the great Newcomen engines, causing much flooding in the pits and consequent damage to the system of underground timbering, pit linings, and the coal seams themselves. On the eve of the victory at Fleurus (26 June 1794), which marked the final expulsion of the Allies from French soil, the mines acclaimed by Mirabeau three years earlier as the finest in Europe lay in nearly complete abandon.[24]

The depredations of war were serious enough, but they were to prove more temporary and reparable than the company had dared hope. In the long run, the increasingly radical governments of the internal Revolution were a greater threat than the external enemy. The breakdown of the old institutions, the war and its attendant chaos and fear, the sudden rise to power of new men with old scores to settle, put all the institutions of the Old Regime in danger, the more so as they had been powerful, wealthy, or identified with privilege. Anzin was eminently powerful, wealthy, and privileged and for all these reasons had long endured the cordial detestation of the bulk of the local population, which, since the union of 1757, had come to regard it as a monopoly artificially maintaining inflated prices.

In addition to the general envy of success, the local *patriotes* in control of the district and departmental councils included a good number of industrialists who had suffered as members of ill-fated rival companies. Their efforts to find and mine coal had been scotched, at least in their minds, more because of Anzin's machinations than by their own ineptitude or bad luck. Then too, among Anzin's owners were members of the nobility who were suspect as such and whose protestations of loyalty to the new regime rang more and more hollow. Finally, Anzin had cooperated only half-heartedly with the zealous wartime regulations of the councils dominated by the patriotes.

The result of this fund of local hatred was a double-pronged attack against Anzin and its owners. First, intimidation by the successively more radical local and national governments led the owners of over fourteen of the company's twenty-four sols to join the emigration from France. The second or third generation of the Desandrouin and Taffin families and a few others remained, but the heirs of Croy and Cernay swelled the ranks of the *émigrés* as did eventually many of the top-level employees, including the works director, Léonard Mathieu. The revolutionary government now held as biens nationaux over 50 percent of Anzin, confiscated from the absent owners.

The district and departmental councils launched a second attack at the remaining owners by pressing the national government for complete nationalization of Anzin on the grounds that the mines belonged to the nation anyway and that Anzin had long shown its abuse of the trust vested in it, most recently by its attitude and failures in the war crises. In Paris, the Commission of Arms and Powder was less concerned with satisfying local greed and envy than with enabling Anzin to aid the war effort. Vested with majority control of the company, it sent a series of government mining engineers to take charge of stepping up production. The engineers, with training in only theory, proved unable to cope with the enormous practical problems: local governmental pressures; resistance on the part of the remaining owners; severe shortages of workers, horses, wood, and other necessary supplies; war damage; and their own lack of practical experience and lack of knowledge of local men, customs, and mining conditions. Production was very slow to recover even after victory at Fleurus in June 1794 freed the pits from the threat of enemy patrols.

In September of that year, Stanislaus Desandrouin and the Taffin family, the chief remaining owners, assessed the situation. Management was in the hands of blundering governmental appointees and the company was losing money. The immediate commercial outlook was highly uncertain, and the future financial outlook, with the government as dominant partner, was totally confused. Agreeing that their position was untenable, they decided to offer to sell their remaining interest to the government at its prewar value. Their petition to the National Convention stated that, "being unable by themselves to turn to account their property, now administered by the agents of the Republic, the petitioners limit themselves to asking the just indemnity due them for the losses they have suffered from the invasion and the presence of the enemy, and to offer to the Republic, . . . the ownership of the shares belonging to them in the company stock." They claimed that the interest of the nation would be better served by a single administration.

The Commission of Arms and Powder, to which the petition was referred, recommended its denial on the grounds that the experience of the old company was necessary to help the government engineers reestablish production. The situation was thoroughly snarled. Desandrouin and the Taffins were in an impossible position; they could neither operate nor abandon their interests in Anzin. The district of Valenciennes was pressing for nationalization and the remaining owners were ready to capitulate, but the national government refused the gift pressed on it from both sides.

At this crucial point in Anzin's history, a law was passed on 7 December 1794 authorizing "citizens having an interest in commercial or manufacturing establishments where one or several associates have been struck by confiscation" to purchase the confiscated part from the government if they would assume all outstanding obligations and undertake to keep the establishment in operation.

There was some question in the minds of Desandrouin and the Taffins whether or not this law applied to mining properties. As an alternative to repurchase, they proposed the division of the Anzin properties, with Fresnes and Vieux-Condé to be ceded to them as constituting one-third of the total value of the four concessions and Anzin and Raismes to be the exclusive property of the state. Nevertheless, they also proceeded under the new law to obtain the necessary evaluation of the net worth of the company. The evaluation of 4,105,327 francs meant that the repurchase price of the government's fourteen sols would be 2,418,505 francs. The entire project met continuous harassment and delay from the district government, and only the timely purge of the departmental, district, and communal councils by a commissioner of the National Convention in May 1795 made the purchase possible. On 11 June the district made final cession to the reconstituted Anzin Company.

The purchase price was, of course, far beyond the shrunken means of the remaining owners. In order to pay the government its 2,418,505 francs, Desandrouin was obliged to borrow the sum from a group of Parisian private bankers. In fact, the "loan" was merely a legal device by which the bankers acquired the government's share of Anzin. Desandrouin immediately turned over eleven sols, four deniers to the Parisians as repayment for the loan. Desandrouin had preserved intact the Anzin Company and his share in it at the price of handing over a near-majority of the twenty-four sols to a group of absentee owners.

The chief figures among the new owners were Claude Perier, Guillaume Sabatier, and Lecouteulx de Canteleu. All three were private bankers and businessmen who were in the process of pyramiding their modest prerevolutionary fortunes by seizing on dangerous but lucrative opportunities for speculation in *assignats*, government contracts, and biens nationaux. Their investment in Anzin, then, appears as merely one speculation among many. It differed only in being larger than most and hence requiring a pooling of resources. They had no experience or interest in operating a mine, nor was any a local figure from the Nord. Although they took three of the six seats on the board of directors, Desandrouin, also resident at Paris, continued to direct the company's operations and set out broad policy, subject only to the Parisians' approval.

The company that had seemed such a dubious venture in 1795, when it was losing money and almost all operations were at a standstill in the broken and deserted pits, slowly reestablished itself as a highly profitable investment. The outsiders had saved the company. They had saved it from total disintegration at the lowest point in its history by furnishing money in an amount which Desandrouin could not have found outside their own circle of private bankers at Paris. Their dependence on Desandrouin's experience with mining and the local situation could not hide his dependence on their financial support. Soon a majority of the sols of Anzin was no longer in the hands of the company's founding families, men of lifelong acquaintance with each other, the company, its traditions, and its employees. It was held by absentee owners who looked on Anzin not as a way of life or the foundation of the family fortune, but as simply one of many investments, valuable as a steady producer of income. It was only after 1817, however, that this new group made the crucial decision to move to active management in an effort to maximize that income.

The first fruit of absentee ownership was a radical shakeup in administrative personnel and organization. The general intent was to tighten, clarify, and regularize duties and procedures loosely conceived at a time when owners and employees knew each other well and issued or carried out directives informally, without sacrifice of close supervision and efficiency. The new owners insisted on spelling out a hierarchy of officials, with individual lists of duties, connected by clear lines of authority. In addition to theoretical considerations of efficiency, they also hoped to safeguard their interests against the staff employees whom they found indispensable but also feared as local men loyal to the old ways and old owners. Among the figures produced by the new regime, the most important was Jacques Renard, who had received a modest share in the reconstituted company for his role in the purchase as front man for the Parisian group. He became "general agent," and his post combined for the first time the duties of day-to-day technical and business supervisor. He retained the position until forced to retire in 1825.[25]

Although the reconstitution of the company in private hands laid the basis for commercial recovery in the future, the immediate outlook remained bleak. Production doubled in 1795 but was still well under half the prerevolutionary level, and the government reserved part of that production for its exclusive purchase. The continued depreciation of the assignats, which the law forced the company to accept in payment for its coal, led to a classic price-wage spiral.[26] The company attempted to meet the demands of its workers by paying them in grain purchased with the depreciated assignats, but labor unrest increased with the cost of living. Strikes broke out in August and September 1795 led by *révoltés* who claimed the right to choose their own foremen. The cost of living reached its peak at the end of that year. The directors observed "that it was no longer possible to find any kind of supplies without silver money; that the workers threaten to quit the works if we continue to pay them in assignats, because they cannot buy their necessities with this money."[27] The company finally, after much petitioning, received permission from the government to sell to its customers in silver or paper at the customer's choice. It probably used this concession as a means to force payment in metal, although it innocently claimed that the reason sales remained low was that its customers preferred to pay in silver! In November 1795 the board of directors reset the price of coal and the workers' wages in terms of silver; the price of a load of top-grade coal fell from 250 to 3 francs. The vicious spiral broke and thereafter prices and wages remained fairly stable.[28]

Once the company could again pay its workers a living wage and establish some order in its commercial operations, it turned to the problem of raising production. By the end of this same year, 1795, it had stopped flooding in the pits and reconstructed most of the Newcomen engines. It repaired two pits that had collapsed and bought five hundred horses and other supplies. The mines were operating at one-third efficiency. All that was lacking was "men, money, and tranquillity."[29] Tranquillity was easiest to find; henceforth the government's policy toward Anzin alternated between active and passive benevolence. The necessary funds were rounded up by Renard from various merchants, preferably old customers of the company. The appeal to outside capital was at all times a last resort for Anzin, but its owners desperately needed "funds to nourish the enterprise."[30] They borrowed at short term some amount up to 25,000 francs in Lille and Arras. Men were harder to come by. Barely half the prerevolutionary work force was at hand in the winter of 1795–96, and it grew only slowly as the government turned a deaf ear to the company's pleas for draft exemptions and permitted only a trickle of emigré employees to return.

Governmental and company sources for the next four years are meagre, but the impression is one of rapid restoration. By the spring of 1796 the worst was over, and the company undertook a variety of new projects. It began a trial shaft for a new pit, purchased a boiler for a Newcomen engine, erected a "horse hospital," built surface roads, pierced connecting galleries between pits, and started explorations for new seams.[31] Production reached 248,000 tons in 1799, a figure not reached again until 1808. By 1800 Anzin had almost recovered from the loss of equipment, labor, owners, and tariff protection due to the Austrian army, the French army, the flight of the emigrés, and the annexation of the Austrian Netherlands. The basic structure of the company survived the Revolution intact, and materially at least the company was within reach of its old days of splendor.

The Company and Its Owners

Although not unique among mining companies, Anzin's legal structure had a "caractère particulier incontestable."[32] When Napoleon's new Commercial Code went into effect in 1807, the company petitioned the authorities that it be recognized under the code as a *société anonyme* ("corporation"). The minister of the interior pointed out that the 1757 Acte de Société provided for a board of directors consisting of six self-perpetuating members but not for any general assembly of all the owners; that the sols representing ownership had no par value and their transfer was restricted; and that each owner's liability for the debts of the company was unlimited.[33] On the other hand, the company had a legal persona separable from that of its owners. This, together with the character of its concessions giving it the right to exploit certain seams of coal in virtual perpetuity, especially after the mining law of 1810, clearly separated it from the various forms of French partnership known as *sociétés en commandite*, either with or without transferable shares. In fact, like other mining companies in the area, Anzin was, in the eyes of the law, not a *société civile* falling under the Commercial Code but real property, an exploitation of the soil like any agricultural holding.[34]

Anzin may have been a union of "persons" rather than "capital," but most of its owners had no share in the government of the company. Anzin's ownership descended in three tiers of privilege and power. Alone at the pinnacle were the six members of the board of directors. Whenever a vacancy occurred on the board, the old members were free to chose their new colleague from among the entire roster of Anzin owners. In practice, they restricted their choice to the oldest, most able, or most eminent son, brother, or son-in-law of the previous member. Only when a director died or retired, and left no obvious heir to his seat did the board look beyond his family to the head of the most important family group as yet unrepresented.[35] In addition, it maintained a constant balance of power on the board by assuring that half the members belonged to the prerevolutionary ownership and the other three to the Parisian group that had bought into the company in 1795.

According to the Acte de Société, this tight little self-perpetuating board had "to the exclusion of all others," absolute control over all the company's affairs.[36] Meeting three and then two times a year in the company's headquarters at Saint Vaast near Anzin, the board was not obliged to render any account of its proceedings or decisions except for a bald statement, at an owner's request, of the total annual sum of expenses and receipts.[37] It controlled access to ownership by demanding prior notice from any owner who wished to sell or transfer his sols and by refusing to allow any sale or transfer without first renouncing its own right to buy up the proffered sols, in the name of the company, at the going price.[38]

A second level of owners consisted of those recognized by the board as *chefs de famille*.[39] They alone were eligible to receive for their respective family groups the dividends voted by the board each year. Including the directors, there were thirty chefs de famille, who represented in turn the more than one hundred fifty men, women, and children who by 1830 were enrolled as owners. At this third level of ownership, each individual had only a mediated relationship to the board through the chef de famille he had chosen to represent his interests in the company.

This insistence of treating owners as so many fragments of a permanent cadre of families reflected the fact that most owners regarded their interest in Anzin as a patrimony, an endowment to be preserved and passed on within the family. Transfers of ownership averaged perhaps ten a year between 1800 and 1833. Even where the transfer represented an outright sale, the parties involved were almost always connected by family, social, or business ties. Most transfers, however, were the products of death or marriage. With the passing of each generation, the number of individual owners grew as the ever more complicated fractions of Anzin's sols and deniers provided dowries, settlements, and inheritances for wives, widows, divorcées, legitimate heirs, and even the genteel spinster daughters of Desandrouin's mistress.

Despite its inevitable fragmentation, ownership in Anzin continued to signify more than the mere possession of an impersonal stock certificate.[40] By controlling access to ownership and accepting few bids from outsiders, by insisting on grouping owners under a few chefs de famille, and by themselves representing six of the most important family groupings, the board members reinforced a natural tendency for the owners to remain a small and intricately interrelated group.

As important chefs de famille, the six directors almost always controlled close to a majority of the twenty-four sols. On the other hand, no individual, whether in his own right or as chef de famille, ever approached a position of majority control. Most owners held one denier or less and only two chefs de famille ever controlled more than two and one-quarter sols. Desandrouin's holding of more than four and one-half sols was the largest of any single owner in 1800. After 1806 his increasing financial problems led him to borrow on his holdings by temporarily "selling" them piecemeal to the company with the right of repurchase. He never recovered his financial health, and at his death in 1821 the company was in effective possession of his entire interest in Anzin.[41]

By that time the Perier family had become the largest block of owners. Beginning with the two and one-quarter sols bought by Claude in 1795 and divided "fairly equally" among his numerous heirs, the total family holdings by 1826 probably amounted to about three and one-half sols, of which about one sol apiece was owned directly by Casimir and his brother Scipion's widow.[42] As company bankers, Casimir and Scipion were frequently employed as Anzin's agents to help finance the extended dealings over Desandrouin's sols or to purchase those bought up by the company to indemnify the former emigré owners. [43] Strategically placed to increase their holdings and those of their family, they were unwilling or unable to acquire as much as 15 percent of total ownership. Clearly, the Periers' leadership on the board after 1817 stemmed from recognition of the company's need for business ability rather than any financial control.

Not surprisingly, Anzin drew its owners exclusively from the middle and upper strata of French society.[44] In 1833 over one-fourth of the twenty-four sols were in the hands of persons possessing noble titles. Some, like the Duke of Arenberg, were descendents of the great eighteenth-century aristocrats who had founded the company. Others, like the Viscount Hocquart, had been ennobled under the Empire for their loyalty and service to Napoleon's government. Except for Desandrouin himself, however, the nobility played a minor role in Anzin's operations, although both Arenberg and Hocquart served as directors in the twenties.

More than one-third of the company's sols were in the hands of members of *la grande bourgeoisie*, especially those great Paris bankers and their relatives and descendents—the Periers, the Lecouteulxs, and the Sabatiers—who had entered the company ranks in 1795. Noted men of law like Berryer, Anzin's house lawyer, and high government officials with noble titles like Cambacérès, archchancellor of the Empire, and Baron Louis, minister of finance in the first years of the Restoration, were part of the same milieu. It was this group, with its preponderance of financial and administrative abilities and its influential connections with the upper ranks of business and government, that gained the upper hand at Anzin after Desandrouin's fall in 1817.

Most of Anzin's other owners, the numerical majority, were of more local eminence—lesser bankers, merchants or industrialists, lawyers, government officials or army officers, or propertied gentlemen of Paris, Belgium, or the Nord region. Typical members of this group were the branches of the Taffin family, who collectively owned over two sols of Anzin and were scattered over the Pas-de-Calais, the Nord, and western Belgium. Grandchildren of Pierre Taffin, an important legal official under the Old Regime and the original Desandrouin's most important partner, they lived quietly on their estates and served their country in a variety of respectable local positions. At Douai in the Nord, Taffins were in the municipal council, the district council, the electoral college, and at various levels of the court hierarchy. Taffin-Sorel, the only Anzin director regularly resident in the Nord during the twenties, served on the General Council of the Nord department almost continuously across two regimes from 1813 until his death in 1825.[45]

Other equally typical members of the family were content with less worldly success. Taffin de Givenchy followed his father into exile during the Revolution but returned in 1800 as an officer in Napoleon's armies. His father's death in 1810 led him to resign his commission, marry, and settle down on the family estate near Saint Omer. There he lapsed into peaceful and somewhat penurious obscurity except for a brief flash after Napoleon's return from Elba when he took up his sword again, this time for Louis XVIII. The turning point in his fortunes occurred after 1820 when six deniers of his father's confiscated holding in Anzin were restored to him under the company policy of indemnifying its emigré owners and their descendents. Then he could allow "free rein to his taste for books."[46] He gathered a large library, edited or wrote several works on the early centuries of French history in his capacity as permanent secretary of a local society of antiquarians, and gave generously of his fortune to the local archives, library, museums, and churches. An "homme d'intelligence et de coeur, plein de droiture et de modestie, serviable, généreux, loyal, modèle de bon ton," he contributed nothing to the steady increase in the income from his six deniers that made his comfortable and benevolent existence possible.[47]

It could be argued that such a man's life does not belong in a study of the Anzin Company. But for most of the company's owners, whether *rentiers* like Taffin de Givenchy or modestly successful lawyers, merchants, or bureaucrats, the lucky accident of finding themselves possessing even a fractional holding in Anzin must have made a crucial difference in their economic and social status. The value of a denier rose from 30,000 francs in 1808 to 45,000 francs in 1821, 60,000 francs in 1823, 90,000 francs in 1828, and 120,000 francs in 1832. Dividends rose almost proportionately.[48] Thus, Anzin's increasing prosperity was the foundation of several noble fortunes and played no small part in the success of even those owners who were members of a "bourgeois dynasty," but it was still more significant for those owners who were drawn from the middle levels of French society. Because they were deprived of any voice in the affairs of the company by its legal structure, their well-being was profoundly affected by the decisions of those at the top.

The Decision Makers and the Managers

The Anzin Company was essentially a simple organization. It produced one commodity and sold directly to consumers and wholesalers at its pits or the loading areas on the nearby Escaut River. It needed no elaborate marketing system, and expansion of production merely involved the replication of existing facilities. Administratively, it required a single hierarchy of control that would transmit orders downward, assure a flow of information to the top, and maintain discipline at all levels. Compared to the larger and more modern textile firms or ironworking establishments of the time, where a variety of processes, materials, skills, and equipment were beginning to be vertically integrated into complex, delicately articulated systems, Anzin's administrative needs appeared rudimentary.[49]

On the other hand, Anzin's size alone created serious problems. The producing units, the pits, lay geographically dispersed, separated from each other by as much as thirty-five kilometers. The product, bulky coal, moved in three stages from the coal face to the bottom of the pit shaft, up the shaft, and along the surface to the banks of the Escaut. Vast supplies of wood and other materials were bought from scattered dealers, stored in huge sheds, processed in surface workshops, and parceled out to the producing units in the requisite amounts at the appropriate times. Accurate and complete record keeping of the material inputs of supplies and labor and outputs of coal as well as the financial flows they created put a severe strain on the company's capabilities. It was still more difficult to combine a multitude of such records into a coherent picture of the company's operations that could serve as the basis for determining entrepreneurial strategy and managerial tactics.

Human weaknesses at every level of the hierarchy compounded the problems arising from size. At the top, the board lacked any driving entrepreneurial leader so long as Desandrouin was its chief figure. Its most important members lived far from the Nord, in Paris, where they were immersed in other, more immediate concerns than Anzin's welfare. Finally, until the appointment of Mark Jennings as general agent in 1828, they found no one who could effectively carry out their policies on the spot and exercise responsible local leadership.

Moreover, it is clear that the company's middle- and lower-level managers constantly resisted the orders of the board and its general agents. The *porions*, who supervised *chantiers* ("working areas" in each pit), the master porions, who directed the operations of the *cantons* (groups of pits), the small band of general production supervisors (directors, subdirectors, and inspectors), and the receivers and controllers with their staffs of assistants and clerks who recorded and directed the streams of cash and materials that flowed through the company's operations—all were drawn from some fifteen or twenty interrelated local families.[50] Clearly distinguishable from the bulk of the mining population, some of these families had served Anzin for several generations.[51] In effect, their members claimed the same nearly hereditary right to fill Anzin's managerial ranks as did the leading chefs de famille to positions on the board. Their long practical experience made them suspicious of theory. Their local roots made them hostile to outsiders. Self-interest led them to view innovation as dangerous. And their cohesiveness allowed them to present a solid front of resistance to change. So long as Desandrouin dominated the board, they would be the real decision makers for Anzin.

Unlike his father, who had thrown heart, soul, and fortune into the success of Anzin, Stanislaus Desandrouin may have regarded his company holding as a source of income rather than an opportunity for active entrepreneurship. Under the Old Regime, Léonard Mathieu had provided the necessary leadership from his post as works director. Owner of a small interest in Anzin and one of the most famous mining engineers in France, Mathieu was forced to emigrate in 1794.[52] The company held his position vacant in expectation of his early return. Meanwhile, Desandrouin and the new owners appointed Jacques Renard, a Parisian lawyer and businessman, as general agent in 1795, intending to limit his functions to those of a glorified office manager. He was to oversee the company's local commercial, financial, and legal operations and to serve as the board's eyes and ears in the Nord.[53] When it became evident that Mathieu would not return, George Boisseau, an old employee, assumed some of the functions of works director without the title, and

Renard gradually assumed a general supervision over all company operations, technical as well as commercial.[54] On Boisseau's retirement in 1804, the company still further weakened its control over mining operations by dividing his functions between two works directors, one for Anzin and Raismes, the other for the pits on the Fresnes and Vieux-Condé concessions.[55]

Ignorant of the art of mining and overburdened by duties as correspondent with the company's customers, lawyers, and directors, Renard depended on the locally trained staff of managers for technical advice. Although these men were skilled enough in carrying out the routine direction of day-to-day operations, they lacked any theoretical knowledge or a broad vision of the company's needs as a whole. Resistant to new ideas or new leadership, they preferred to follow the traditional ways that had proved so successful in the past. Weakness in the Nord was thus joined to weakness at Paris. Desandrouin, who was already sixty-two years old in 1800 and embroiled in ever more serious financial difficulties after 1806, trusted the experience and loyalty of local men trained in Anzin's service and acquiesced in their routine responses to such basic problems as the increasing exhaustion of the company's best seams of soft coal. He lacked the will to provide or enforce the new policies that would have aroused Anzin from its somnolence.[56]

Nor did the new owners of 1795 provide the needed spur. Absentees like Desandrouin, they too tended to regard their holdings in Anzin primarily as a source of income and concentrated their talents on business, social, or political affairs at Paris. Only Claude Perier and two of his sons, Casimir and Scipion, paid much attention to Anzin in the years of Desandrouin's ascendency. On Claude's death in 1801, the two brothers became partners in a wide variety of business enterprises whose core was their private bank in Paris.[57] The most important client of the bank was the Anzin Company, in which Scipion became assistant director in 1801 and full director in 1805, when he named Casimir as his assistant.[58] A young but experienced chemist with a broad knowledge of applied industrial technology, Scipion knew personally most of the leading scientists, technicians, and manufacturers of Paris.[59] It was almost certainly his technical knowledge, personal contacts, and initiative that led Anzin to adopt Watt rotary steam engines after 1802 for hoisting coal to the surface. Through Casimir, the Perier bank helped finance the purchase of the three sols used to indemnify the former emigré owners for a part of their prerevolutionary holdings. Anzin deposited its growing reserve fund with the bank, and Casimir did his unsuccessful best to help the company open new markets in Holland and Paris. Nevertheless, the Periers seem generally to have acquiesced in Desandrouin's policies and local control under the Empire.

After 1816 it became increasingly evident that Desandrouin was not the man to seize the opportunity presented by the separation of the Mons area from France, the protective coal tariff of 1816, and the upsurge in the demand for coal. Once Anzin's largest single owner, he had pledged his entire holding as security to the company for loans arranged through the Perier bank. Furthermore, he was old and weary and in failing health until his death in 1821 at the age of eighty-three. Control of basic policy gradually slipped into the hands of the Periers. They were not only Desandrouin's largest creditors and representatives of the largest single family holding in Anzin but also eager advocates of a thoroughgoing policy of renovation and expansion. Scipion in particular emerged during the period 1816 to 1821 as the strong man of the board. It was he who defended Anzin's interests against attacks on the coal tariff, instigated the replacement of the Watt engines with the improved machines of the Periers' new associate, Humphrey Edwards, and pressed for the sinking of new pits in hitherto unexploited portions of the company's concessions. Then his sudden death in 1821, coupled with that of Stanislaus Desandrouin, cleared the way for the ascendency of his brother.

Casimir succeeded Scipion as a director, assumed control of the board, and ruled Anzin as an autocrat until his death in 1832. Unlike his brother, who despite his ability and devotion to his calling was subject to fits of melancholy and mysticism,[60] Casimir was a leader of men, an *homme fort* determined to rejuvenate Anzin just as he was determined to consolidate the July Monarchy as first minister in 1831–32. Tall and handsome, exceptionally hard-working and vigorous, he had the reputation with friends and enemies alike of a man avid for power. He combined stubbornness and a hot temper with a cold, precise insistence on "order and method" and attention to detail. He demanded loyalty from his family and associates and absolute obedience from subordinates.[61] His mind was strongly practical in its bent. Although he had been educated by priests, he showed only a "respectful indifference" to religion, and as a dabbler in Saint-Simonianism, he was drawn to its severely utilitarian rather than its utopian and mystical elements.[62] Between such a man and the entrenched staff of Anzin, accustomed to the easy-going and traditional ways of a Desandrouin, conflict was inevitable.

The battle was soon joined. Perier believed the company's position was critical; to survive Anzin must expand. He launched a thoroughgoing overhaul of every aspect of the company's operations that would, he hoped, increase sales, decrease costs, and build a solid foundation for future growth. His program of renovation and improvement of equipment and techniques and rapid expansion of output, and his determined search for new seams are fully discussed in later chapters. Here we must consider his shake-ups of the managerial hierarchy and personnel, his attempts to create a smoothly functioning administrative machine immediately responsive to his demands at every level of the hierarchy.

The chief source of weakness was at the very top where the "slackness" of the general agent, Jacques Renard, allowed his subordinates to "get into the detestable habit of doing only what suited them."[63] Perier first attempted to reduce the pressure on the overworked and aging general agent by appointing a new staff of expert technicians under Renard's supervision. He reorganized the warehouses, surface workshops, and loading areas as separate staff divisions with directors hired from outside the company. To coordinate the work of these divisions with the line functions of the inspectors and master porions, Perier appointed a general director, Jules Lachapelle, another outsider, who was to supervise all surface and mining operations.[64] While continuing to handle commercial operations directly, Renard found his role in production reduced to transmitting and enforcing Perier's orders.

Enforcement sometimes proved too great a task for the general agent. Anzin's old hands hated and resisted the new men appointed by Perier. Lachapelle and his enemies were agreed on only one issue; they all contented themselves with shufflng paper reports in their offices and refused to see for themselves what was happening in the pits.[65] Even a threat entered in the Registres that "when an order emanating from the general agent" as a result of a decision formally taken by the board "is not executed in the prescribed time, the negligent employee will be fined one-sixth of his salary" was not sufficient to overcome the "slackness and inertia" that continued to triumph over "order, regularity, and economy."[66]

Thus, on 11 April 1825 the board accepted Renard's resignation with thanks for his "eminent services."[67] His successor was Christian Paulze d'Ivoy, who had been a close friend of Scipion Perier and was temporarily unoccupied after being discharged for political reasons from his post as a departmental prefect.[68] Casimir Perier clearly expected the new general agent to serve him loyally as a sort of glorified clerk while exercising a firm administrative control over the fractious staff in the Nord. Once again, he failed to take account of local susceptibilities and the intricacies of the mining business. Paulze's zeal and administrative abilities were not enough to make up for his tactlessness and ignorance of coal mining. He clashed immediately not only with Lachapelle and Perier's other new appointees but also with their enemies among the local managers. Within a year, Paulze was at loggerheads with Perier himself. The former prefect could not adjust to the role of faithful servant played so well by Renard or accept the imperative tone in which Casimir issued his almost daily stream of detailed orders from Paris.[69] No thanks accompanied the board's acceptance of this general agent's resignation barely two years after Renard had stepped down.[70]

At last, on his third attempt, Perier found the ideal general agent, Mark Jennings. Jennings was an Englishman serving as director of Perier's Chaillot engineering workshops in Paris, which supplied Anzin with its steam engines.[71] Unlike Renard and Paulze, he was an experienced technician as well as an able administrator who was knowledgeable enough to direct mining operations, forceful enough to impose discipline on his subordinates, and diplomatic enough to keep on good terms with Perier. After six years of juggling men and positions, Perier had won his struggle to impose his will on Anzin's administration.

Even with a general agent like Jennings, who could be entrusted to take a certain amount of initiative, Perier's personal style of operation called for a constant flow of information between Paris and the Nord. As under Desandrouin, the board continued to hear reports and recommendations from its top managers at the three regular meetings of the year and to enter decisions in its Registres based on those reports. Furthermore, Perier called an increasing number of special board meetings at Paris on matters of urgent concern. Nevertheless, the board's decisions, like the managers' reports, must have been largely pro forma, given Perier's ascendency over the other directors and his complete command of the company's situation at every moment as a result of the intimate and almost daily correspondence between him or his brother Joseph and their general agent or other trusted employees in the Nord.

Indeed, the general agent's most critical responsibility was to gather information on every aspect of company operations and personnel and relay the data to Paris. Reports were of two kinds: formal statements of the inflows, outflows, and transfers of supplies, coal, labor, and cash drawn up by the chief office clerk under the direction of the general agent; and private letters written by the general agent himself, explaining the successes and failures reported in the formal statement, isolating critical problems, and discussing the effectiveness of various managers. Although Joseph Perier wrote most of the answers, he assured Paulze that Casimir took note of all his communications and that Joseph's replies were a product of their joint reflections.[72] In addition to the general agent's confidential correspondence with Paris, at least two other employees, Dournoy and Rolland, were under orders to pass on their impressions of the company's situation and especially their fellow workers, including the general agent. The fact that Dournoy, Rolland, and to some extent Renard and Paulze acted as spies for Casimir Perier helps explain the hostility and recalcitrance of the local managers.[73]

Dournoy and Rolland, however, had more to do than spying. They had come from the Perier bank in Paris to reform Anzin's system of record keeping and accounting. Unfortunately, none of Anzin's books survive, but it is clear that managers on all levels kept detailed accounts of the operations they supervised. The company made some attempt to correlate the various records to prevent theft and graft and assure the continuity of operations. Yet the general effect was of a growing number of records and record keepers that produced a "large number of wheels" without any "unity of results."[74] Except for the numerous duplicates, most of the records were kept in such a way that the company could not compare them or check them against each other. For example, it could not combine the separate records of inputs of various kinds of labor, supplies, and equipment and outputs of various kinds of coal at any pit to produce a (partial) figure for the cost of production at that pit. Nor could it determine for each pit the proportionate share of the cost of the surface workshops and other supporting units assigned to it, and the books lumped operating expenses with capital expenditures in both underground and surface operations. When the company managers could not tell whether or not a given pit made or lost money on current operations and the board could not discover whether or not the company made a profit except after checking its cash accounts at the end of each four-month period, the formulation of policy occurred in a cloud of obscurity.[75]

The board appointed Dournoy general inspector of all operations for the purpose of keeping "double-entry accounts so as to know in a more precise manner what each pit, chantier, piece of equipment, and tool costs us."[76] Rolland's task was to bring order into the bookkeeping system. Eventually he abolished, modified, or reduced eighty-one "registres, carnets, états, et tableaux" and replaced them with a single unified system of records and accounts.[77]

By the time Perier presented an extended report to the board in 1826 on the results of his stewardship over the previous five years, he was able to make quantitative comparisons of costs among the various supporting functional units (transport, warehouses, workshops, etc.) over periods of several years. It was still impossible to gain a true picture of the marginal profit produced by capital expenditures, since capital expenditures were not clearly separated from operating expenses.[78] Still, the partial cost-accounting system introduced by Perier the administrator had become an indispensable tool of Perier the entrepreneur.[79]

But was Casimir Perier an entrepreneur in the Schumpeterian sense of a dynamic, risk-taking innovator?[80] Some scholars, especially in the United States, have suggested that France has never produced an adequate supply of entrepreneurs. The French businessman, so the argument goes, has tended to place more emphasis on family than business per se, on security and saving rather than risk-taking and capital investment, on the rate of profit rather than its absolute size, and on social status rather than pecuniary success. He has tended to see and use business as a means to the "good life" rather than as an end in itself. His creative satisfactions have been outside the business and have often been achieved at the expense of expanded business, in contrast to the American, it is held, whose creative energy more commonly finds expression in building a business empire and who measures success in life by success in business. The French businessman tends to channel profit into banks, foreign investment, gilt-edged securities, business "reserve" funds, hoards, and consumption goods and services rather than invest it in new machines or increased plant. Both productivity and production remain low in consequence, the size of the firm remains small, ownership and management tend to be guarded jealously and kept securely in family hands. In such an environment, the ability to compete degenerates and the competitive process itself gives place to a spirit of "live and let live"; a petty bourgeois spirit animates even the largest firms. In addition, low production and high prices lead to a demand for high tariffs and governmental protection of the weak as well as the strong, protection which in turn reinforces the fundamental weaknesses of firms by shielding them from the necessity of meeting foreign competition.[81]

Although the stereotype fits Desandrouin better than Perier, it is true that the latter restricted dividends and built up a large reserve fund for the sake of security, opted for high prices and a high rate of profit, fought to maintain a tariff that sheltered Anzin prices and profits from Mons competition, and used business success as a political stepping-stone. As a chef de famille, he was virtually forced by Anzin's structure to regard himself as the steward of a family patrimony. At the same time, Perier markedly lacked Desandrouin's sense of tradition, his sentimental loyalty to old ways and faithful company servants, his conception of Anzin as a family in which he played the role of father. A calculating master of detail, driving, ruthless, and impersonal, sharing some of the visionary qualities of the Saint-Simonian businessman, Perier had the penetration to see Anzin's weaknesses, the imagination to conceive of solutions, and the personal will and forcefulness to impose his views.

Certainly one might question his unwillingness to delegate authority. "Clearly wherever entrepreneurs act like managers, wherever they concentrate on short-term activities to the . . . detriment of long-range planning, appraisal, and coordination, they have failed to carry out effectively their role."[82] Perier immersed himself in managerial detail, in day-to-day administration, because he had no alternative given the severe shortage above the lowest levels of able, willing, trained managers. In the early nineteenth century both entrepreneurs and managers were in short supply. Casimir Perier was that very common phenomenon of his time, a man who by temperament and need performed the functions of both and performed them well.

The Government and the Miners

The period from 1800 to 1833 was hardly one of political stability, yet the Anzin Company, after being severely buffeted by the storms of the 1790s, sailed serenely through three major changes of regime in 1799, 1815, and 1830. In fact, the governments of Napoleon, the Restoration, and the July Monarchy continually demonstrated their friendship toward a company whose private enemies were legion.

This favorable political climate prevailed partly because all three regimes were generally prepared to encourage big business in order to secure at least a minimal loyalty from its leaders, to insure social peace through full employment, and to stimulate economic growth. By modern standards, these regimes did very little for business, but each in its own way favored the propertied over labor and the French products over foreign. Institutional and personal continuity at the higher levels of administration lessened the impact of changes in constitutions and rulers.[83] As members of one of the most powerful of the new "bourgeois dynasties," the Periers inhabited the small world of Paris constituted by the salons of the great hostesses, the offices of the top bureaucrats, and the inner chambers of the major bankers and businessmen.[84] In these circles, political ideology was secondary to personal favor and the interests of property.

The Perier family was from the beginning among the most ardent supporters of the Napoleonic regime. Claude Perier, together with his fellow bankers and associates in Anzin, Lecouteulx and Sabatier, may have participated in the conspiracy that preceded the coup d'état of Brumaire.[85] In any case, in 1800 these three men were among the founders and original regents of the Bank of France, sponsored by Napoleon. In other ways, too, Napoleon's bureaucracy, the Bank of France, and Anzin's ownership were closely intertwined.[86] Scipion and Casimir Perier were on familiar terms with such stalwarts of the Imperial Regime as Montalivet and Gerando, and three of their brothers and a brother-in-law launched moderately successful careers in Napoleon's administration.[87] Not surprisingly, when Cambacérès, arch-chancellor of the Empire, asked for permission to increase his holdings in Anzin, the board "seized with eagerness the occasion to be agreeable."[88]

After prudently standing aloof from the political scene between 1814 and 1816, the Periers showed themselves prepared to cooperate with the new regime in the first years of the Restoration, and it was only after 1820 that Casimir became one of the leaders of the liberal opposition in the Chamber of Deputies. Elected in 1817, he fulminated against the weak foreign policy, inept financial administration, and open hostility to political and civil liberty of the governments of the twenties.[89] On the other hand, since 1816 he and Scipion had become members of many of the major public or quasi-public commissions, councils, and societies through which businessmen of the Restoration met and cooperated with important government officials on the crucial economic issues of the day.[90] As a result, the Periers and other business leaders "came as close as businessmen in France have ever been to the formulation of national economic policies."[91]

Like the other liberal leaders in the chambers, Casimir Perier was no revolutionary. He rallied to the Orleanist cause in 1830 in order to seize the revolution from the Paris "mob." Again reluctantly, he allowed himself to be named first minister in March 1831 as the only man capable of restoring political stability and social order. In ill health and weakened by the burdens of office, he died of cholera in May 1832.[92]

At no time in his career—whether in favor under Napoleon, in opposition during the Restoration, or as head of state in the July Monarchy—did Casimir Perier (or any other owner of Anzin) leave any trace of having overtly used his political power or influence to further Anzin's interests.[93] Nevertheless, on several occasions when Anzin's security was threatened the French government adopted policies consistent with Anzin's interests.[94] It is clear that the company's chief owners, as leaders in the social, business, and political worlds of the Empire and Restoration, were able to make their influence felt in the corridors of power.

The most obvious example of the influence of business on government economic policies was the protective tariff. "At the beginning of the Restoration, no industry, no important group was favorable to free trade."[95] The tariff law of 1816, approved overwhelmingly by a legislature dominated by reactionaries and conservatives, provided stiff import duties on most agricultural products, raw materials, and manufactured goods. The duty on coal varied according to the point of entry, with the highest rate along the Nord's land frontier set at 0.30 francs per hectoliter or over one-fifth the wholesale price of the majority of Anzin coal. Although the system of zones was often referred to as the "Anzin tariff," it was strongly supported by the mine owners of the Loire field, who felt threatened in their Paris markets by Belgian and English coal. "These duties . . . show clearly that the legislators wished to limit the introduction of English coal [and] maintain the situation acquired by Belgian coal [during the union with France] while protecting the Nord mines."[96]

Although the system of coal duties established in 1816 remained intact until 1835, it was under almost constant attack from coal consumers and Frenchmen who held interests in Belgian mines.[97] The administration held firm. As Louis Becquey, general director of bridges, roads, and mines from 1818 to 1830, wrote to the general director of customs, "You know . . . how much the large and useful mining establishments of the Nord department merit the protection of the government."[98] The Bureau of Mines' engineer in the Nord constantly supported Anzin as deserving "encouragement for its services to society and for its enormous risks," and the engineer supervising navigation in the Nord claimed that "Anzin merits in a special way the protection of the government."[99] Both during the Restoration, when the Periers were among the leaders of the political opposition, and under the Empire, which they staunchly supported, Anzin remained worthy of the favor bestowed by "un gouvernement paternel et d'une administration protectrice."[100]

While some Frenchmen were beginning to castigate what they called "the new industrial feudalism," most industrialists prided themselves on their own paternal and benevolent attitude toward their labor force and claimed that their workers' lot was superior to that of their English counterparts.[101] The Anzin owners provided free medical care, schools, heating, housing, pensions, and other fringe benefits worth an estimated one-sixth of the miners' basic wages.[102] Wage rates differed greatly from job to job, but on the whole, while lower than those of most skilled artisans and miners in some other coalfields, they were higher than those of textile workers or agricultural laborers.

On the other hand, the work was hard, the discipline harsh, and the miner's productive years few. "To train a miner, he must be put to work at the age of ten. At twenty he combines all the requisite skills, [but] at thirty he can no longer work except at tasks that require less suppleness."[103] Fringe benefits were regarded as generous gifts, theoretically revocable in times of declining profit, when the company's first duty was to its owners.[104] The workers' "loyalty" was assured by fines and the threat of dismissal, a threat made possible by an abundant supply of labor arising chiefly from the philoprogenitive instincts of the miners themselves.[105]

Except for a brief and partial strike in 1824, the Anzin mining population was quiescent from the time of the Revolution until 1833.[106] In May of that year a strike erupted spontaneously and spread until almost all the pits were shut down by extinguishing the boiler fires of the drainage engines. For eight days, some four thousand miners succeeded in maintaining a solid front against the company, the local authorities, and five thousand troops called in to "restore order." Then, lacking leaders, organization, funds, or a careful program, the movement collapsed as rapidly and spontaneously as it had begun. A handful of "leaders" were tried, convicted, and let off with what the company and authorities regarded as scandalously light sentences. Because of Anzin's immense unpopularity in the Nord, public opinion generally held that Anzin rather than the miners had stood trial.[107]

Despite the extensive testimony in court about Anzin's labor practices and the miners' motives, the real origins of the strike remained obscure. The initial outburst challenged the harsh fines levied by an unpopular manager, but restoration of wage rates, supposedly cut in 1824, soon became the main rallying cry.[108] Below the surface there lay a growing sense of discontent among workers who had believed Desandrouin when he claimed to be a true "père de famille," but who mistrusted those same words in the mouths of Paris bankers. At bottom, the great strike of 1833 was merely a symptom of the rationalization of procedures, the relentless search for economy, the unsentimental imposition of discipline that were at the heart of Casimir Perier's quest for control.[109]

2 The Mine and the Miners at Work

In a general way, most contemporary opinion attributed Anzin's success and bigness to its technical efficiency. From the late eighteenth century to the middle of the nineteenth, the near unanimity of opinion was impressive. In 1784 a government engineer reported that the Anzin mines were worked with the greatest intelligence,[1] and Mirabeau's praise in the National Assembly has already been mentioned.[2] At the turn of the century, the Nord prefect Dieudonné compared the present management unfavorably to that of the past and called it "vicieux et plus brilliant que solide."[3] While there was some truth in this statement, most mining experts did not agree with Dieudonné. The government engineer, Daubuisson, who had made a detailed visit to the mines in 1805, obviously considered them a technical marvel and regarded the management as extremely enlightened by contrast with those he had seen elsewhere in France and Belgium.[4] Clère, the engineer stationed permanently in the Nord, claimed in 1818 that the "art" of mining at Anzin was "pushed to a high degree of perfection" and cited it as a model for France and one of the best in Europe; he reiterated this statement in 1832.[5] More persuasive than the comments of these rather obscure men and especially of Clère, who often showed a strong bias in favor of Anzin, are the testimonies of Héron de Villefosse and Michel Chevalier, two of the most famous members of the Bureau of Mines in all its long history of worthy service to the French mining industry. Héron de Villefosse, author of *De la richesse minérale* (1812–19), the best survey of European mining techniques of its time, disliked Anzin as a "monopoly" and criticized the undiscriminating praise others lavished on it, yet he cited Anzin perforce time and again for its exemplary use of advanced technique.[6] Chevalier, a Saint-Simonian and the future promoter of the famous Cobden-Chevalier treaty of 1860, was a fledgling government engineer in 1832 and already a free-trader scandalized by the tariff shielding Anzin from Belgian competition. Although he compared the Anzin pits unfavorably (and on the whole unfairly) with those of Mons in the matter of size, he too had to admit that the impressive order and unity in the complex organization of its extensive surface and underground workings was "the principal cause of the high prosperity" of Anzin.[7] Finally, a local mining expert and historian could still claim at mid-century that Anzin and its newly organized competitors in the Nord, who patterned themselves after Anzin, left all other mines in France far behind them "from the point of view of good management."[8]

Both expansion in the area mined and improvements in technique were slowed by the harsh geology of the coal seams under the Anzin concessions. Geological conditions at Anzin were two-sided in their effect. The coal seams were difficult to work given the techniques of the early nineteenth century. They were mostly thin and highly contorted and lay deep under a thick, waterlogged covering of younger rocks. At the same time, they yielded coal of relatively good quality in all grades from semianthracite through the most volatile bituminous. Equally advantageous, in the long run, was the effect of harsh working conditions on mining technique. The very difficulty of working thin, deep, watery, and gassy seams required sophisticated techniques and machines and hence a large capital-intensive organization. Furthermore, the relative poverty of the seams encouraged Anzin to set out in a determined search for better conditions beyond its existing workings.

So far as mining technique is concerned, however, one must be careful not to attribute too revolutionary a character to the progress made between 1800 and 1833 or indeed at any time during the nineteenth century. A modern authority has written:

> The difference between the Anzin Company of 1790 and the Anzin Company of 1914 is more one of intensity than of kind. . . . As opposed to the eighteenth-century weaver or printer, let a miner of the same epoch go down today with a contemporary team of workers

and he will no doubt find that the surface workshops are equipped with machines of great power, . . . that the steam pump which drained the water then has received an infinite number of improvements, that the galleries have a new lighting system [electricity], that his own lamp affords him a security formerly unknown, but, finally, lifting his pick, he will not be uncomfortable among his co-workers.[9]

In the early years of the nineteenth century some coalfields and some firms went through a real revolution by finally adopting methods of coal production long standard in more progressive areas and firms. There was a general leveling up of technique throughout Europe as hillside drift gave way to vertical shaft, and man power gave way to horse power and then to steam for pulling, pumping, and hoisting. With deepened pits, miles of underground workings, which required scores of hewers and haulers, were developed off a single shaft. While many of these techniques were new elsewhere, Anzin had introduced them in the eighteenth century and carried them into the nineteenth with little change. Techniques of exploitation and the organization of work there were essentially the same in 1833 as in 1789. There was no technological revolution at Anzin between 1800 and 1833 in the sense of a significantly higher output per man-hour, much lower costs, or fundamental changes in a well-established and complex routine of exploitation.

Nor was Anzin really a pioneer in the changes that were taking place. Although Anzin was among the very first on the continent to introduce power winding of the coal from the pit bottom to the pit head, some large English mines had preceded it. Both Belgian and German mines, to say nothing of many in England, adopted advanced haulage techniques before Anzin, which was somewhat behind in this respect until the 1840s. And although Anzin adopted the safety lamp relatively early, it was adopted by government fiat and under the threat of physical extinction. Even when Anzin toyed with new techniques, it often postponed their full-scale use until its survival was threatened or competitive pressure forced its hand. There was a genuine reluctance to change until forced to do so, because change involved high costs either in cash or in the wrench it gave to ingrained habits of thought and work on the part of both managers and workers. Although Anzin was still a pacesetter in technique in 1833, as attested by the remarks of men like Chevalier, its lead had shrunk since 1800, partly because of its own conservatism but largely because so many mines that had been backward at the earlier date had now found rapid improvement to be the price of existence in a world where better transport was beginning to bring efficient giants like Anzin within reach of their local markets.[10]

Anzin's essentially conservative attitude toward technological change has been stressed here because, once in the midst of a detailed exposition of innovations made in drainage, winding, haulage, and other areas, it is easy to overlook the vast realm where the old ways sufficed. Nevertheless there were changes, important changes, made in the first third of the nineteenth century. The movement of coal from the cutting face to the surface was mechanized by the laying of rails in the underground ways and by the installation of steam engines capable of hoisting the rail-sped coal from pit bottom to the surface. The safety lamp permitted the working of deeper, gassier seams. Underground rails, steam hoisting, and the safety lamp widened and deepened the possible working area of a given pit and thereby lengthened its life as the seams close to the original shallow shaft were worked out. In addition, these innovations involved changes in various details of the overall organization of work, such as the widening of galleries and the elimination of the costly system of working by *bures*.[11]

The cumulative effect of such innovations on output per man-hour or cost per unit mined was not great. Their significance lay mainly in the fact that they permitted Anzin to continue working the old seams from the old pits under less advantageous conditions without a drastic decrease in productivity or an increase in cost per unit. For it is in the nature of coal mining, as in other extractive industries, to be subject to increasing costs per unit produced (given a constant technology and constant unit costs of such factors as labor), because the best quality coal and that most easily and cheaply extracted is normally mined first. As a miner attacks seams poorer in quality or farther from the pit head, his returns for a given input of labor diminish. Thus better drainage, haulage, winding, and lighting methods kept the cost of deep mining at the end of the 1800–1833 period as low as it had been at the shallower seams worked earlier. Although Anzin's costs did not fall significantly between 1800 and 1833, the fact that technological advances kept them from rising was something of a triumph, a triumph that enabled the older Anzin pits to continue to produce and at competitive prices.

The new techniques also helped Anzin expand its operations into new areas after 1800. Anzin explored the four old concessions, looked for new ones, and even considered purchasing already operating mines. Before 1820, there were a few minor successes, but most of these efforts failed. The reasons for the failure then are not quite clear. They seem to have been due in part to a lack of geological knowledge and insufficiently powerful drainage engines and in part to a management "plus brillante que solide."

What is abundantly clear, however, is that when success came in the 1820s it was due not only to better knowledge of local geology and improved exploration and drainage techniques, but also to the determination and ability of a new set of managers headed by the Perier brothers and to the incentive supplied by a growing market for high-grade bituminous coal, which Anzin could not supply from its old pits. It was the sinking of pits through the Torrent, the great subterranean lake overlaying the coal measures between Anzin and Denain, and the discovery of rich bituminous seams near Denain on the southern boundary of the Anzin concession several miles east of its old workings near Valenciennes that enabled Anzin to meet the growing demands of the market after about 1825 and especially after 1830. Improved mining technique seems to have played a role at the Torrent, but it had little to do with the success at Denain. In both cases, technical skills would have been of little use without the investment of large amounts of capital and a well-organized plan of attack. It is doubtful if any other mining company in France was big enough and wealthy enough and had a management sufficiently adroit, determined, and efficient to carry through an expansionary program as broad and successful as Anzin's. The geological and technological difficulties were great, so great that they proved insuperable in the first decade of the century, but it was other such difficulties, already solved under easier conditions at the older workings, that had created by 1800 an organization potentially capable of successful expansion with only slight improvements in knowledge and technique. Anzin's great size and relatively good management in 1800 were the necessary foundations for later expansion.

Geology

Coal is an organic mineral consisting of the remains of plants laid down in prehistoric swamps and more or less metamorphosed by subsequent pressure and heat. Normally it occurs in beds or seams separated by layers of shale and sandstone. The number of separate seams, their individual thickness, and the total thickness of seams in a single basin vary depending on the original conditions of deposition, but they are ordinarily great. Subsequent folding, faulting, and thrusting of the coal deposits and rock formations above and below them may seriously disturb, contort, and even destroy large portions of the original flat series of deposits. Such shifts in the earth's crust may also expose the coal measures to surface erosion. The geological history and structure of coal basins is normally complex and for the field of which Anzin is a part it is notoriously so.[12]

The deposits near Valenciennes form part of the western wing of a long, narrow coal basin stretching in a shallow arc westward from the Ruhr to Kent, in southern England, by way of Aachen, Liège, Namur, Charleroi, Mons, Valenciennes, Douai, Lens, Béthune, Hardinghen, and the English Channel. The field is far and away the largest of continental Europe in both total reserves and surface area.[13] In France, the basin extends almost one hundred kilometers westward from the Belgian border in a band varying from five to sixteen kilometers in width.[14] It enters France at Valenciennes, and its direction is directly westward to Douai, where it swings northwest as it enters the Pas-de-Calais. Beyond Béthune a gap separates the main deposits from the small outlying basin at Hardinghen near the Channel coast above Boulogne.

The beds of this field were laid down in their long, narrow, symmetrical basin during the Carboniferous period of the late Paleozoic era some 150 million years ago when most of the world's important coal deposits, including those of Europe and the United States, originated. After pressure and heat transformed the remains of a succession of semitropical swamp forests, movements of the earth's crust pushed the older rocks lying below the coal upward in great fault blocks both north and south of the basin. Subsequent erosion of these emerging *massifs* carried away most of the younger coal measures, and an overthrust of the older rocks hid much of the remaining coal.[15]

These disturbances also severely folded the remaining coal beds so that they now formed a series of five zigzag folds, resembling overturned Zs (see figure 1). The folds ran through the whole sequence of beds from top to bottom and formed a parallel series of *droits* ("rights") and *plats* ("flats). The steeply inclined parts of each seam, the droits, having revolved more than ninety degrees from their original position, inclined about seventy-five to eighty degrees from the horizontal. The plats connecting the droits were nearly level, being inclined about ten to twenty degrees from the horizontal, and were ordinarily about three times as long (ca. 800 meters). The conformation of droits and plats was such that a single vertical shaft might pass through a given seam two, three, four, or more times. Northward along the beds the system of droits and plats tended to flatten out. At Raismes, the beds inclined some forty-five degrees, while at Fresnes and Vieux-Condé, on the northernmost edge of the basin, and least exposed to the folding pressure from the south, the seams were almost flat. Unfortunately, the coal in greatest commercial demand was that from the more contorted droits and plats of the southern area.

A number of vertical faults, or fractures, extending through several seams further complicated their structure. One major fault (the Grand Retour) along the east-west axis of the basin raised the whole northern portion of the beds some hundreds of meters above their southern continuations. Many secondary faults and local irregularities contributed to the confused geology of the field.[16] Obviously a flat and uncomplicated system of seams would have simplified the miner's task. At Anzin, the highly contorted geology of the coal beds and especially the difficulty of working the droits necessitated a complex system of exploitation involving a high degree of technical sophistication and cooperation within the company as a whole, among different pits, and among the several cutting teams working the same seam within each pit.

While the broken-off southern ends of the seams lay deeply buried against the side of the fault block, the northern edges cut the surface in a series of bands that generally ran parallel, east to west, with the oldest (originally the lowest, now the "outside") beds outcropping furthest north. The quality of the coal changed as one surveyed the full sequence from oldest to youngest seams in passing from

Figure 1

Vertical Profile of Two Pits at Anzin (Anzin Company, 1795, AN, F14 7820). AA′ = line of surface (ca. 800 m.). BB′ = unconformity separating folded terrain houiller from flat-bedded terrains morts. x, y, z = galeries de traverse at 56, 120, and 216 m. 1 = première veine de six paumes. 2 = veine de quatre à cinq paumes. 3 = veine de trois à quatre paumes failleuse. 4 = grande veine. 5 = deuxième veine de six paumes.

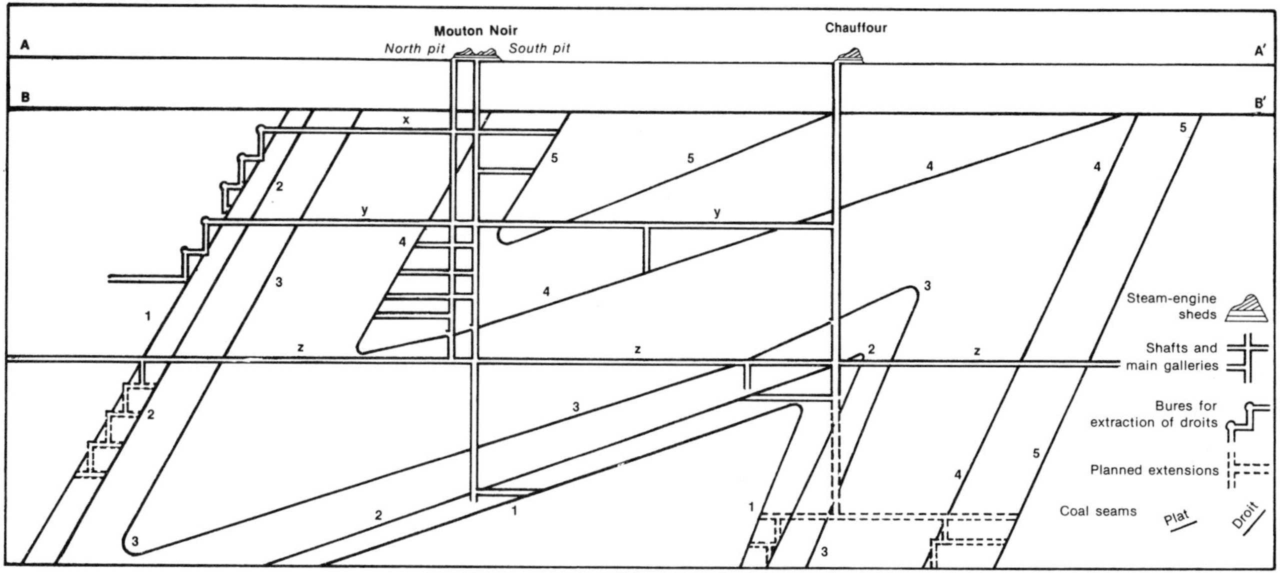

north to south across the basin at any given depth.[17] Happily, the Anzin Company concessions extended across the full range of the remaining coal measures, which lay in three major groupings (*faisceaux*). The most northerly, in the Vieux-Condé and Fresnes concessions, where coal was first discovered in 1720, consisted of a hard, low-volatile semianthracite (*charbon maigre*) occurring in seams much flatter and less faulted than those to the south. The middle group, which consisted of semibituminous coal (*charbon demi-gras*), lay two thousand meters further south and extended east northeast by west southwest on a line from Thivencelles passing just north of Anzin to Haveluy. The southernmost band contained relatively soft and volatile bituminous coal (*charbon gras*) and lay along the great axial fault, the Grand Retour. It began in the east between Anzin and Valenciennes and slanted to the southeast, passing by Rouvignies on the south and Hérin on the north and stretching to Rouelx and Abscon, which lay just inside its southern and northern edges, respectively, at its farthest extension west in the Anzin Company concession.[18] Each variety of coal has its own special qualities as a fuel, and although some types were interchangeable for some uses, each claimed its particular market. Much of the history of exploitation at Anzin is the story of attempts to adapt existing facilities or create new ones to produce the quantity of each variety in current demand.

The thinness of the individual seams also had an important effect on the system of exploitation. The surviving coal measures were very thick, over twenty-five hundred meters at Valenciennes, but the total thickness of the coal seams themselves was only fifty meters.[19] Of the seventy-seven seams now known, Anzin worked no more than a dozen or so at any one time in the early nineteenth century, although over forty were recognized as early as 1805.[20] The average thickness even of those worked was slight by comparison with, for example, the seams of western Pennsylvania, Yorkshire, Silesia, or even the Ruhr.[21] Of the eleven seams under extraction in 1805, only one measured over one meter, and three were only three-tenths of a meter.[22] That last figure represents the minimum thickness exploited in 1832; the average thickness was only six-tenths of a meter.[23] Such seams were advantageous because the miner could easily prop or fill the extracted areas to prevent subsidence of the mine "roof," and hence a vein could be completely extracted. Furthermore, the layout created by this system of extraction and the natural narrowness of the ways cut through the thin seams also helped control the flow of air through the mine and thus solve the vexing problem of deep-mining ventilation. On the other hand, the actual hewing of the coal was slower, more difficult, and therefore more costly in seams narrower than one meter. Equally serious, the expense of transporting the coal from the face to the foot of the hoisting shaft increased with the necessity of blasting a higher passage through the harder rock of the roof or floor of the seam.[24]

After faulting and erosion had carried away much of the original coal measures, a new subsidence below the sea had covered their stubs with younger rocks. These horizontal beds, mostly limestone chalks, had in turn eroded away in places, opening gaps through which the coal measures appeared in outcrops at the present surface. From the Ruhr to Mons the coal measures lay exposed where they abutted the massif of the great fault block. By far the greater part of the reserves in the Ruhr dipped under the chalk in the northern part of the field and remained unexploited during the early nineteenth century, but the massif to the north of the narrowly squeezed basin exposed the main Belgian deposits throughout their width. Between Mons and Valenciennes near the French border a limestone neck broke the coal measures. Where they resumed in France, they had dipped below the recently deposited calcareous sediments, and so the entire French portion of the northern field lay concealed beneath a thin shell of lime. This cover thickened regularly at a rate of one meter every four hundred, running from east to west; at Valenciennes it measured sixty to eighty meters.[25]

The concealed portions were at either end of the great coalfield; they were its last subfields to be discovered; and they were generally favored with thicker, flatter seams than the exposed subfields. In all three respects the Nord obviously held an intermediate position between the Pas-de-Calais and northern Ruhr on the one hand and Belgium on the other. The coal measures, though not so deeply buried as to prevent discovery with the mechanical and scientific means available in the early eighteenth century, hid beneath a cover that was not only fairly thick but also had a watery consistency.

Of the fifteen beds honored by a local name in the regular sequence of *terrains morts* (the Anzin term for the rocks above the coal measures, or *terrain houiller*), almost all contained faults or cracks that permitted surface water to percolate down to a thick, impermeable bed of shale, the *dièves*, which formed the bottom of a sort of subterranean lake known locally as the *niveau*. The niveau extended upward from the dièves to the surface at the low points of the valleys, where it formed the extensive marshes and peat bogs of the area. Particularly where associated with shifting sands, the niveau was the enemy of the miner trying to sink a shaft through it. The dièves, however, formed a sort of waterproof roof that prevented any surface water from leaking into the coal measures. Underground, the niveau was a costly nuisance to the coal miner of 1800; above ground, in some places it hampered movement of coal by road.[26]

A much more serious problem, and one special to Anzin, was the Torrent. The lowest bed of the terrains morts, resting on the surface of the coal measures, it consisted of water-filled shifting sands and shales. Since it was in immediate contact with the terrain houiller, it "poured its waters into the very interior of the underground workings."[27] The Torrent extended over an oval-shaped area of some six thousand acres stretching between Aubry and Denain.[28] Since it was apparently not in touch with the surface above the dièves, early mining engineers speculated that the Torrent would prove to be a subterranean lake that continuous pumping would eventually drain dry. This hope proved unfounded, for pumping was still going on in the late nineteenth century, though Anzin had attained sufficient mastery to extend workings below the Torrent during the years before 1833.

The Mine Shaft

At Anzin the first step in opening a new pit was the sinking of a vertical shaft an initial thirty-five to one hundred meters through the rocks covering the coal measures and beyond into the seams themselves. As the only entrance to the vast underground labyrinth of a mine, the shaft performed several functions. Most shafts were fitted with the steam engines used for extracting the coal and with the ladders used by the men in entering and leaving the mine. Some also had larger steam engines and pumps for drainage, and each played a part as either up shaft or down in the complicated system of ventilation.

Before digging a new shaft in unexplored ground, drillers sank a test bore or *sondage* with a special tool, the *sonde*, in order to determine the depth and arrangement of the coal measures (if any) and the presence of subterranean lakes or other obstacles.[29] The "sonde du nord de la France" consisted of one-meter sections of iron rod screwed into each other and tipped by a cutting point. It drilled a bore six centimeters in diameter and up to one hundred meters deep at a cost of four francs per meter in ordinary calcareous rocks.[30] Perhaps because of the delicacy of the operation, the company always used man power rather than horse power or a steam engine before 1860 to turn the crank that rotated the bit.[31] Anzin's sonde, with its specially trained crew of *sondeurs*, was apparently the only modern one in the area before 1830.[32]

Sinking the pit was a task for a highly paid crew of specialists, who used hand tools and black powder to loosen the rock. Above the water table on the plateau around Anzin, the digging went easily at first. At ground level in the Escaut valley,[33] however, and below about thirty meters on the plateau above the river, problems arose from cave-ins of waterlogged sands and from flooding by the niveaux that were scattered through the permeable layers of rock.[34] It took three years to reach the coal measures at the pit Bleuseborne in the 1780s, even with continuous pumping by three huge Newcomen engines whose combined efforts caused wells to dry up two miles away and aroused the ire of the commander of the citadel of Valenciennes, whose moat ran dry.[35] After 1822 the improved techniques developed to pass the Torrent enabled the crews at pits dug in favorable circumstances to reach coal in as little as eight months.[36] Ordinarily, one of the old Newcomen engines with its enormous sixty-inch cylinder still operated the pumps, but the water issuing from the fissures collected in a mobile reservoir which moved up or down so as to hang protectingly just over the heads of the excavating crew.[37] A mobile shaft lining now allowed the crew to pass the shifting sands safely, and on the Torrent itself even more heroic measures involved arrangements linking together the drainage systems of all the pits being dug in the area.[38]

Anzin was famous in technical circles for the complex perfection of its system of permanent *cuvelage* ("shaft lining"). Its installation involved a process called *picotage* invented in 1720 by the company's first technical director, Jacques Mathieu.[39] The shaft lining consisted of a massive wooden polygon of oak timbers chinked with tar that reached from the surface to the impermeable dièves that separated the coal measures from the niveaux above. Picotage was the operation in which *picots* ("wedges") of wood secured the cuvelage to the more solid of the rock beds along the sides of the pit. Thus the lining was a sort of tube within the tube of the rock walls, divided into a series of watertight compartments separated from each other by rings of picotage attaching each section securely to the walls at top and bottom. A liquid cement filled the spaces between the cuvelage and rock wall. Because the lining was under enormous horizontal pressures, it required constant repairs to control seepage and bulging timbers; only once between 1800 and 1833 did a serious accident occur of the sort that might have flooded out a pit and all its men and workings.[40]

Although Anzin had perfected its system of cuvelage and picotage in the eighteenth century, the company made several minor improvements in its pit shafts after 1800. It adapted octagonal shafts for economy and solidity after 1815,[41] and in 1822 it increased the diameter of new shafts from about six to "at least ten feet" in order to provide for multiple functions in a single shaft.[42] Iron shafting, although successful at a Newcastle mine as early as 1819, did not even begin to supplant the wooden cuvelage at Anzin until 1884.[43] But below the dièves, round brick walls slowly replaced the square wooden ones as the proper kinds of wood became scarce and expensive.[44]

The length of shafts also gradually increased. The bottom of the deepest pit in 1805 lay at 346 meters, but the average was probably closer to 150 meters.[45] By 1830 there were three pits near Anzin more than 400 meters deep and five others over 350 meters. Even the newer pits on the Torrent were already about 200 meters deep. These substantial increases required no special technical adjustments in shaft construction, however, since all the increase in depth occurred in the brick-walled sections below the cuvelage.[46]

The Interior Plan of the Mine

Once the shaft reached the coal measures, special crews began hewing and blasting horizontal galleries (see figure 1). These *galeries de traverse* intersected and passed through all the droits and plats of the coal seams in a given direction from the shaft. As a crew of miners worked out the seams at one level, another crew deepened the shaft and pierced new galleries that intersected approximately the same seams at a lower level.[47] These galleries, as well as the lesser passages also used for haulage, were theoretically at least 1.5 meters wide and 1.8 meters high. In practice, the special teams of cutters, paid at so much a meter of advance, often scamped the job, while coal dust piled up on the floor and the immense pressure of the overlying rock slowly squeezed down even the best-propped roof. The workers who propped, framed, and maintained the galleries and lesser passages with wood were generally poorly skilled men too old or too feeble to work at the face. Their crude workmanship contrasted oddly with the beautiful craftmanship of the cuvelage specialists. In 1805 all the galleries at Anzin were too low to allow a man to walk wholly upright, and in several one had to crawl on hands and knees.[48] The company kept the passages in better repair after 1800 when brick walls and vaulted roofs began to replace the wooden props and especially after 1822 when the introduction of railed haulage in the galleries required roofs at least 1.5 meters high and clean, straight walls.

Once the galerie de traverse intersected a workable seam of coal, three major systems of extraction were theoretically possible, but two of them, *grande tailles* ("longwall") and *massifs courts* or *longs* ("bord and pillar" or "panels"), demanded seams more regular, flatter, or thicker than those commonly mined at Anzin. As in the Mons area, where geological conditions were similar, the company employed the technique known as *gradins renversés* ("reverse steps"), so called because the workings resembled an upside-down staircase when seen from below (figure 1).[49]

From the galerie de traverse, cutters drove a new gallery at a right angle to the first and within the seam itself. This passage (*galerie d'alongement*, *chasse du fond*, or *voie du fond*) followed along the seam and formed the lowest of a series of such "ways." Above each way lay the *taille* ("cut") where a team of two to six miners extracted the coal along a face six to twenty meters long at right angles to the ways. The system produced the effect of a reversed staircase since the first way and its taille were driven some twenty meters ahead of the taille above, the second taille the same distance ahead of the third, and so on upward with three to ten tailles stacked one above the other.[50] Staggering the tailles prevented the extracted coal from rolling down onto the miners working in the tailles below, but the lead taille also served an exploratory function in discovering, for example, an unexpected interruption in the seam. The tailles, then, were separated vertically by the intermediary ways above and below each one and horizontally by about a twenty-meter advance or retreat from neighboring tailles on either side.

The miners worked the tailles from a *voie montante* ("ascending way") cut up into the seam at right angles to the voie du fond on a plat, and diagonally to lessen its slope on a droit. New voies montantes across the horizontal ways at intervals of less than fifty meters improved the circulation of air and shortened haulage distances to the voie du fond. On a perfectly regular plat the image left by the horizontal and ascending ways in an exhausted working area was a mosaic of rectangles of perhaps ten by thirty meters making up a much larger rectangle off a given main gallery, with the large rectangle constituting the chantier ("working area") of perhaps eighty by three hundred meters.

Extraction of droits by bures, an older method still commonly used in 1800, had almost disappeared by 1833, but its long survival testifies to considerable technical conservatism at Anzin in the earlier years, for even then its drawbacks were well recognized. Workers dug a short vertical pit (*bure*, *burc*, or *burcq*) just beyond the spot where a gallery traversed a seam. The bure, about eighteen meters long, passed through the seam at its mid-point. From the bure two tailles extended, one above and one below the point of intersection. At the bottom of the bure, a small gallery about ten meters long passed horizontally through the seam again and ended in another bure. A series of bures allowed miners to work droits without incurring the cost of running more than one gallery from the shaft to the seam, and, at the lowest level of extraction, to avoid the cost of deepening the shaft in order to cut a new gallery.

Even in 1800 these advantages were probably outweighed by a number of problems. The bures supposedly measured one meter square, but they were even less well cut and well maintained than the ordinary galleries. Air could circulate down through the bures only with the greatest difficulty. Evacuation of coal and water from the work face by a series of two-man hand winches located at the top of each bure and limited in lifting capacity raised labor costs. On very gassy or watery droits, bures could not be used at all. Nevertheless, the decisive move from extraction by bures to extraction by galleries on droits did not take place until the 1820s, when the coming of railed haulage so reduced the cost of transport by the horizontal galleries that bures disappeared except for the purpose of cleaning out odd corners.[51]

It was the cost of hauling the coal from the face down to the voie du fond, along it to the galerie de traverse, and hence back to the foot of the shaft that determined the distance from face to shaft. Until the 1820s the maximum distance was only three hundred to four hundred meters; it was cheaper to reach seams beyond this distance by sinking a new pit. After 1822 railed haulage increased the practicable distance to six hundred to eight hundred meters. Thus the "diameter" of an average pit and its workings, as well as the ideal distance between pit shafts, grew to about fifteen hundred meters or nearly one mile.[52] This increase together with the continuous deepening of the shafts and the elimination of bures constituted a significant improvement in mining efficiency with consequent reduction in costs. The essential layout of the mine, however, with its complex system of exploitation had already evolved in the eighteenth century in response to harsh geological conditions and survived basically unchanged until the mechanization of the mines in the twentieth century.

Work at the Coal Face

From the Middle Ages until the beginning of the twentieth century the methods and tools used in the actual wresting of the coal from its bed remained essentially the same. The miner at the face was a semiskilled workman employing pick and shovel at a job requiring a maximum of teamwork, experience, strength, and endurance. At Anzin, where the normal height of the taille was ten meters, three or four men composed the usual mining team stationed along the face at regular intervals. The Anzin miner's tools were few, simple, and typical for their time: a pick (*marteau pointu* or *rivelaine*) with a double head of iron and a wooden handle only one foot long, several iron wedges, and a sort of crowbar with a curved pointed tip (*havée*). In an average taille, the miners who had chosen to work together (often on the basis of family relationships) spaced themselves out along the face of the seam. Low plank barriers prevented dislodged coal and rock from rolling down on those working below and offered a solid foothold on a steeply pitched droit. Each miner worked bent double in seams up to one meter thick or lying twisted on his side in thinner ones. In this contorted position he used his pick to undercut the coal between the bedrock and the bottom of the seam along his part of the face. When the cut was one meter deep, he drove wedges into the joint along the roof of the seam and used his crowbar to break down the coal. If the coal was already fissured and of the proper consistency, the seam regular, its cleavage parallel to the face, and the undercutting well done, the entire thickness of the seam tended to give way at once under its own weight and fell to the floor in large, clean chunks; the miner was as happy as the mine owner, since his pay depended on his cutting speed. In less favorable circumstances, undercutting was only half the job, and the miner had to use his pick along with the levering tools to break the coal away from the seam.[53]

Each team had only partly completed its task with the felling of the coal, for it was also responsibile for *boisage,* the placing of wooden props at about one-meter intervals between the floor and roof of the seam in order to prevent subsidence of the roof. Where the roof was not solid, cross pieces spanned the gaps between the props to prevent sudden cave-ins. Each day the miner requested delivery to his taille of the correct number of props, cut to specified lengths to fit the seam. With his pick, the only cutting tool he knew how to use, he trimmed and fitted the props into notches gouged in the rock. The ineptitude of the miner as a carpenter and the fact that he was not paid for this required task insured that he did his share of the propping grudgingly and poorly, despite the fact that he himself had most to lose from a sudden collapse of the roof. The company fought against shoddy propping with close supervision of the tailles by the foremen and a system of steep fines. It even considered paying the miner for the time spent in boisage but rejected the idea because it considered itself generous enough in providing the props at its own expense.

The company was worried with good reason about the cost of the enormous numbers of props it purchased each year. Their cost was a dead loss, for they remained behind in the excavated seam. Each day the same gang of workers responsible for advancing the ways along the bottom of each taille filled the space around the props with loose rock up to the last two rows of props standing in the two meters of working space in front of the face. This process of *remblayage* used up the waste rock obtained from cutting the ways and galleries, from thin bands of schists running through the coal seam itself, and from the undercutting of the coal in the rock floor of the seam. In thick seams the supply of fill might have to be supplemented at considerable expense with rock transported from other parts of the mine. Nevertheless, remblayage was essential both to prevent the subsidence of all the rock formations above the excavated seam and to direct the circulation of air by limiting it to a prescribed system of passages.[54]

Haulage from Face to Shaft

While the techniques for sinking pit shafts, cutting passages, and hewing coal remained relatively immune to improvement in the early nineteenth century, the height and length of galleries, the size of chantiers, and the method of arranging the tailles changed significantly. These changes were largely consequences of major alterations in ways of moving the coal underground—in hauling it horizontally from the face to the shaft and in winding it vertically up the shaft to the surface. Alterations in hauling and winding required substantial investments of capital and at least with haulage involved serious problems of adjustment for workers and management alike, but the changes led to considerable savings in time and money.

In 1800 the system of haulage at Anzin was identical with that used in the early eighteenth century. From the face, the coal rolled down the slope of the taille to the way at its foot. On a droit, workers then hand-winched the coal up a series of bures or, in an ordinary chantier, shoveled it down to the voie du fond through a narrow "chimney" left open for this purpose during remblayage. The defects of extraction by bures have already been noted; in making the nearly vertical drop down a chimney, the coal often broke up and lost part of its commercial value.[55] Both practices tended to disappear as haulage methods improved.

Whether at the top of a bure or bottom of a chimney on a droit or in the way at the bottom of the taille on a plat, the coal was loaded by hand into an iron-hooped wicker basket with a capacity of about one hundred twenty kilograms (ca. one-eighth of a ton). The hauler (*hercheur*) lifted the basket onto a small wooden sledge fitted with iron-shod runners, hooked a sling that passed over his shoulders to a bar at the front of the sledge, and, leaning into his harness, drew his load about twenty meters over either the slippery uneven floor of the way, or, in the main passages, along two wooden planks "paving" the floor. The sledges of coal passed along from hauler to hauler in twenty-meter relays to the foot of the shaft. Ordinarily, the number of haulers necessary to evacuate the daily "cut" of a pit was about equal to the number of miners at the face, but where the distance from face to shaft was especially long, the ratio could be as high as five haulers for every miner.

Haulage did not require skill, but it was an arduous task usually performed by boys ten to seventeen years old. Up to the age of twelve or thirteen the boys worked in pairs, one pulling and one pushing. A very few girls served as haulers, but only to the age of fourteen, when the company removed them from the interior of the mine, apparently for the sake of their morals. A day's work, lasting eight to ten hours, consisted of 260 round trips of 40 meters each with a load of 120 kilograms, a total distance of 10,400 meters and a total haul of over 31 tons.[56]

In 1806 the company replaced the sledge runners with small iron wheels. Apparently the experiment won converts slowly, for it was only in 1811 that the board ordered the change introduced in all the pits within four months' time.[57] The next year Tournelle, the company "engineer," proposed another experiment, the substitution of iron rails for the wooden planks on which the wheeled carriers ran. Again, the company delayed an important innovation in haulage for several years; it was not until a decade later, in May 1822, that the board ordered rails for one pit in each of the four concessions. Six months later, because the new system had proved economical by reducing greatly the number of haulers as well as the amount of wood used to floor the galleries, the board required iron railways introduced immediately whenever conditions in the galleries would permit.[58] At the same time, since the rails reduced the effort necessary to push a given load, sided wooden carts carrying larger loads replaced flat sledges. The changeover was nearly complete by 1827 when the primitive, flat iron rails gave way to flanged ones and the capacity of the carts increased again to about five hundred kilograms.

The company made further unspecified improvements in the railways in 1833.[59] Many more years would pass, however, before it attempted to replace man power with animal power for drawing the wheeled carts. In 1811 the board proposed an experiment using donkeys for haulage in the main galleries, but the trial was probably never carried out.[60] Company records are full of information on the number, care, feeding, and stabling of horses used at the surface,[61] but there is not a single reference, beyond that in 1811, to the use of horses for interior haulage until 1847.[62]

The economic advantages of railed haulage are hard to calculate. The company directors stated in 1833 that without the changes in haulage made in the preceding decade, Anzin "could not have withstood the competition of foreign mines."[63] The statement was no doubt an exaggeration, made in the context of a great strike over wages, to prove to the workers that the company had had to undertake great financial sacrifices to insure its survival. Furthermore, Anzin was by no means in as dire peril from foreign competition as it claimed in its public statements. Nevertheless, railed haulage cut costs, although it is difficult for the modern historian (and perhaps for Anzin as well with its rather primitive system of cost accounting) to measure the savings accurately. It not only reduced the amounts of wood and labor used in haulage but also produced less direct benefits by enabling miners to work at a greater distance from the shaft without an increase in haulage costs. This in turn lengthened the life of older pits and thus reduced the capital expended on the sinking of new pits.

Why then were these improvements introduced so slowly and hesitantly at Anzin? Twenty-one years elapsed between the first trial of wheeled haulage in 1806 and the substantial completion of a network of interior railways in all the pits by 1827. And why were the changes made so late at Anzin in comparison with other large and progressive mines elsewhere? As the next chapter will demonstrate, there was no such hesitation, however, in the introduction of steam winding. The wheeled cart had long been in use in some English fields by 1750, at Liège and Mons in Belgium at least as early as 1802, and even earlier in some German metal mines. In all these mines rails quickly followed wheels.[64]

At first glance, it seems that the initial cost of installing the rails may have been an important factor, but Anzin was a rich company with large cash reserves and high yearly profits, and it was ready to spend far greater amounts for other technical changes, such as improved steam engines, which may not have been proved as clearly profitable as rails. It is not known where Anzin obtained the iron used in 1822. Iron objects in general, and especially new types of industrial equipment, were more expensive and less well made in France throughout the first half of the nineteenth century than in Belgium or especially England, and tariffs and prohibitions limited imports from those countries after 1815. Thus Casimir Perier and the other board members may well have considered the cost and quality of the iron rails then available in 1822, but the trial made in that year proved conclusively that the savings clearly outweighed initial costs. The cost of installing and maintaining the new iron rails for the five years from 1822 through 1826 amounted to 188,000 francs or 37,600 francs per year.[65] It is impossible to compare these costs with those for the older system. Wood prices were also high and rising in the early 1820s, however, and wooden planking required much more frequent maintenance and replacement than iron rails.

Perhaps the introduction of railed haulage in 1822 owed something to the hope that the company could cut the costs of installation and maintenance by manufacturing the rails in its own workshops. It is interesting to note that 1822 marked the substantial completion of the company's metal-working shops, which included a foundry, tilt-hammer, lathe, and rolling mill.[66] All the many references in the Registres to the workshops, however, appear in connection with steam-engine building and repair. In the same year that saw the shops fully established and at the meeting of the board of directors where it ordered the experiment with rails, the company changed its mind about the drive for self-sufficiency that had led to setting up the shops and reverted to a policy of outside purchase. Among the various types of wooden and metal equipment mentioned in this order, rails are conspicuously absent.[67] Three years later Anzin definitely was considering rail manufacture. The board authorized Renard, the general agent, either to negotiate the purchase of "des chemins de fer tout faits" or to buy fifty thousand kilograms of cast iron beyond ordinary needs.[68] The outcome of this order is not known, but if Anzin was manufacturing its own rails in 1826, it is certainly surprising that Casimir Perier's report to the board in that year, setting forth in great detail and with great pride the technical innovations of the previous five years, should omit any mention of such an important step.[69] The only other pieces of evidence that bear on the problem are the cost figures for supplies in Anzin's annual reports to the government for tax purposes. The figures for iron purchases rose considerably after 1822, but there appear no entries for rails. It was normal company procedure in these statements, however, to reduce manufactured commodities (such as rails) to their raw material constituents (such as iron).[70] At most then, there is evidence that Anzin considered, but probably rejected, a policy of company rail manufacture after 1822.

In any event, the speed with which the company introduced rails in the next few years makes it clear that the cost of iron or iron rails was not unfavorable by that time, although Anzin was probably still at a comparative cost disadvantage vis-à-vis its foreign competitors. It seems certain too that the disadvantage had been far greater before 1815. Until then Anzin was totally cut off from English iron, the French and Belgian iron industries were still relatively backward technologically, and their output was severely rationed to meet war needs. It may well be that the high cost, low quality, or limited availability of iron and iron rails was an important inhibiting factor until after 1815.

There were also strong technical reasons for Anzin's backwardness in haulage improvements. Anzin's old system was much superior to that used in a great many other mines, large and small, where coal moved all the way to the surface in sacks slung over the backs of women and children. Since Anzin had never used this method, it had less room for improvement.[71] More relevant were the indirect costs of installing rails. A gallery large enough to permit a boy to push a sledge was not high enough to permit the passage of large wagons or wide enough to permit the laying of rails. The company was obliged to spend money, not only for the purchase of new equipment, but also to flatten, straighten, refloor, enlarge, and better maintain the passages through which first wheeled and then railed carriers passed. The choice of donkeys rather than horses for the proposed experiment in 1811 was probably based on their smaller size, for there is no evidence that the cost or scarcity of horses was a limiting factor then or subsequently.

It is certain too that the haulers (or their miner fathers) opposed the introduction of railed haulage. The conservatism of the miners was as proverbial as that of the peasants.[72] Any change in work practices probably aroused their antagonism, even though it might apparently lighten their work load. In fact, although rails reduced the length and number of trips each hauler made, he now pulled a much greater weight of coal, and so his daily expenditure of muscular energy remained the same. The company claimed that the hauler's pay, based on a standard weight load hauled per day, also remained the same, although the rate per given weight was reduced. The haulers were not convinced that the company had converted the weight norms and pay rates equitably. Upon protesting, the haulers received in 1824 a diminution of the weight norm, but the company tied it to a corresponding diminution in pay.[73] That grievance was still alive in 1833 at the time of the great strike.[74] The haulers evidently believed that the company had introduced railed haulage as a means of forcing them to take a hidden cut in pay.

Even more significant, although incalculable quantitatively, was the haulers' grievance over the diminution of their numbers. Railed haulage did not greatly increase the output of individual pits, yet the number of haulers fell off sharply. The increased length of the galleries that rails made possible only partially compensated for this reduction in the haulage force. Unfortunately, there is no way of knowing how great a reduction occurred, but it must have been large. The effect went beyond the immediate loss of work for a group of boys, for haulage served as a sort of apprenticeship for miners' sons before they graduated to work at the face. Furthermore, their wages made up an important part of a mining family's income and might well make the difference between relative comfort and a standard of living at the subsistence level or between subsistence and outright destitution. It was one of Anzin's boasts that it furnished a living for its families almost literally from the cradle to the grave. Railed haulage struck a blow at its training program and forced it to find new occupations for a large group of boys with neither the skill nor the strength for many of the new jobs that opened when the company expanded with new pits.[75]

The board forced through the changes in work patterns, weight norms, wage rates, and number of haulers in 1824 despite the opposition of the haulers. It must have realized that such opposition was inevitable; it may have felt before that time that the risk and effort involved in overcoming the workers' opposition were not worth the possible savings. This hypothesis of managerial reluctance to push a measure harmful to a large proportion of its work force is strengthened by comparing the attitudes of those in control of company policy before and after about 1817. In the earlier period, control was largely in the hands of Stanislaus Desandrouin, the son of the original founder of the company, its savior in the Revolution, and a man whose roots were deep in Anzin and the Nord. A strong paternal attachment to the welfare of his workers may well have led him to hesitate when calculating the unrest and discontent that railed haulage would create. Casimir Perier, who controlled company policy after about 1821, had no connection with the Nord except his financial interest in Anzin. As Perier was to prove while prime minister of France after the July Revolution, he could be ruthless indeed in his determination to accomplish his will. He was not the man to be balked by humanitarian or paternal considerations nor by the foibles and shortcomings of the human material necessary to the accomplishment of his plans.

The change in Anzin's managerial policies under the Periers also helps explain the lag in introducing railed haulage from a more general point of view. The key role of the Periers will appear again and again in connection with various technical problems and the expansion of production. Innovations begun soon after 1800 but later abandoned or left uncompleted were pushed to success after 1820. To a large extent, the invention of new techniques and market conditions accounted for the rapidity of change after that date, but Perier's determination and aggressiveness played an extremely important role. The difference in the quality and goals of the leadership provided by Desandrouin and Perier probably holds first place among the factors that account for failure to introduce railed haulage before 1820.

3 Winding, Drainage, and Ventilation

Winding: Extraction of the Coal from the Mine

In contrast to a somewhat dilatory approach to the new techniques in haulage, the Anzin Company was the first important coal mining concern on the continent to employ steam engines to drive the winding machines that raised the coal up the shaft to the surface. Under Desandrouin, it successfully made the transition from horse-powered to steam-powered winding during the first decade of the nineteenth century and kept the lead when it adopted improved steam winding engines under the Periers between 1815 and 1826, even though the second changeover required a complete replacement of very costly equipment.

Through both experiments with improved sources of power, the operation of winding itself remained essentially what it had been in the eighteenth century. The galerie de traverse met the shaft at a good-sized room called the *accrochage*. Haulers dumped the coal in a pile, and a worker shoveled it into *tonnes*, iron-hooped wooden containers less than one meter square with a capacity of about one-half ton. A loaded tonne was hooked onto a thick hemp rope which extended up the shaft to the surface, where it passed over a pulley to a horizontal revolving drum, back off the drum over another pulley, and down the shaft again with an empty tonne, which, as it descended, partially balanced the weight of the ascending container. The method was basically no different from that of the small hand-winches in the bures, and the principle of winding remains unchanged today, although gearing and counterweighting systems have improved and the containers increased in size.[1]

The important changes in winding came in the source of power turning the drum. Until 1800 mines everywhere on the continent used horses to supply the power.[2] In the pits at Anzin, the revolving drum was geared to a strong horizontal wooden beam about ten meters long. A team of four to six horses turned the beam in a circle about the drum and raised six to seven tonnes an hour from a depth of two hundred to three hundred meters. Working day and night in four-hour relays, three teams could just raise the maximum daily production of sixty-five to seventy-five tons for a single pit.[3]

When a loaded tonne rested at the bottom of a three-hundred meter shaft, the weight of the rope accounted for over one-half the total load of almost one and one-half tons. The weight of the rope decreased as the loaded tonne rose and the rope descended with the empty tonne on the other side of the drum. The problem was to discover some means of counterbalance so that the weight raised remained equal to about sixteen hundred pounds throughout. The Anzin engineer Léonard Mathieu had patented a workable system involving a conical drum and extra chains in 1785. His invention was important even then, for it had enabled four-horse teams to do the work of teams double the size.[4] Even more important, Mathieu's system of counterweights was to prove the key to the application of the steam engine to winding.

Newcomen's atmospheric engine was one of the great inventions of the early Industrial Revolution in England, but its inefficiency practically restricted its use to pumping water in the coal fields, where fuel costs were at a minimum. Newcomen engines spread rapidly in England, and over eight hundred were in use by 1800,[5] but on the continent they never numbered more than one hundred in the eighteenth century, almost all for pumping in the deeper mines of Germany, Belgium, and northern France. One of the first was probably erected about 1725 at Lodelinsart near Charleroi, although not by its seigneur, Jacques Desandrouin, founder of Anzin.[6] To what extent the original idea, knowledge, and technicians for installing a Newcomen engine at Fresnes came from Belgian contacts, from England, or from within the Desandrouin Company itself, is in dispute.[7] What is certain is that Desandrouin's 1732 engine at Fresnes was the second in France and the first used in that country for mine pumping. More important, Anzin remained almost the sole exponent of mechanical pumping in eighteenth-century France.[8] Watt's improved steam engine (patents 1781 and 1782), equipped with a separate condenser, higher pressures, double-acting cylinders, and rotary motion, was potentially far more versatile and efficient than the Newcomen.[9] In theory it was an ideal source of power for winding coal as well as for many other industrial tasks, but its adoption for winding was very slow even in England.[10] The steam engine, unlike the horse, could not adjust its output of power to the changing weight of a load during a lifting operation. Mathieu's counterweighting system, invented in 1785 for use on horse winches, would solve the problem if it could be linked to the improved Watt engine.

It was the Perier brothers at Chaillot near Paris who brought these two inventions together.[11] Jacques-Constantin and Auguste-Charles Perier, engineers and machine makers, had erected the first Watt steam engine in France to pump water from the Seine for the Paris water supply and began to manufacture Watt engines themselves at their Chaillot workshops about 1780.[12] The Periers saw the possibilities of using the improved Watt engine for hoisting, and after several years of work they announced to the revolutionary government in July 1794 that they were prepared to sell engines fitted with Mathieu's system at 25,000 to 30,000 francs for the first and much more cheaply thereafter. The Committee of Public Safety issued a report recommending that the government insist on installation of Perier engines at the mine owner's expense wherever the depth of French mines and the wartime shortage of men and horses justified their use.[13]

Because of the stress and confusion of war and revolution, the Periers never carried out this project, and their new engine did not reach the market until 1798. The Littry mines bought the first one for both winding and drainage; it went into operation for winding in May 1800. The Periers took advantage of the fact that the sale had been made sight unseen to hold an impressive demonstration of the virtues of the machine at Paris for "members of the Institute, the Council of Mines, professors from several of the Ecoles, and representatives of several mining companies," including Anzin.[14]

The directors of Anzin were just then very much concerned with finding a way to reduce the expense of winding coal, and the demonstration of the Perier engine convinced them of its advantages from the standpoint of cost.[15] They contracted to buy twenty of the Periers' winding engines in February 1802. Jacques-Constantin Perier was personally to supervise the installation of the new engines and the rehabilitation of the existing Newcomen drainage engines. In the future he was to oversee the maintenance and improvement of all the company's machines for both winding and pumping. He was to visit Anzin once every three months for this purpose and to receive a yearly salary of 3,000 francs plus travel expenses, together with a bonus of 1,000 francs when ten of the new machines were satisfactorily in place and at work. One of the company's employees, Tournelle, was to correspond with Perier and to implement at Anzin his directions concerning the "machines méchaniques."[16] The cost of each engine is not known but was probably less than the 23,500 francs paid for the Littry engine, whose price included 5,200 francs installation and 300 francs for transportation to Littry.[17] The initial transition to steam winding was rapid, smooth, and successful. The company had already installed five of the new engines by the end of June 1803. Three or four more were probably operating by the end of that year, and the Periers had delivered all the machines by June 1806.[18]

Savings at Anzin were not as great as those at Littry, but they were still considerable. At Littry, each Perier engine replaced nine horses (two relays plus a spare), which, depreciation and replacement apart, cost 8,266 francs a year to keep. The steam engine burned in a year 1,290 francs' worth of coal figured at extraction cost; with the cost of maintenance and a tender added, the total yearly cost was still only 2,000 francs, or less than one-quarter the cost of the horses replaced.[19] At Anzin, where the cost of extraction was higher and the pits were deeper, it was said that the new engines proved economical only on the deeper pits with a large daily production, where it replaced twelve to twenty horses.[20] In fact, the lower limit must have been thirteen or fourteen horses, for in 1811, the only year for which the number of horses used for winding is known, the six remaining horse winches required eighty-two horses. As the pits grew deeper, even these last survivals disappeared. By 1815 only three horse winches were in use, and the last one vanished in 1819.[21]

Steam winding was evidently a response to other technical changes. In particular, as pits deepened, horse winding rapidly became more expensive and lowered the possible daily output of the mine. Although the initial costs were high, steam winding paid for itself not only by reducing costs immediately at existing pits but also, more important in the long run, by enabling Anzin to push shafts deeper and thus lengthen the life of old pits as railed haulage had done by extending the pit workings laterally. Unlike railed haulage, however, which had important repercussions on the system of galleries and chantiers and on the organization of the work force, steam winding caused few changes in other aspects of mine operation. The deepening of the pits was the cause and not the result of the introduction of steam winding and in itself required no special reorganization of work patterns or new technology. Undoubtedly, the winding engine, a costly piece of capital equipment combining working speed and power, was most efficient when used at capacity. Its introduction was therefore a further incentive to either expand output at the least productive pits or close them down, but such decisions seem ordinarily to have been made in response to other considerations such as the difficult geology of the seams, haulage costs, or the demand for certain types of coal. The means of expanding output at a single pit, like deepening the shaft, was more a matter of simple multiplication of existing units of men and technology than it was the introduction of a new technical organization of work. Nor did steam winding have serious direct effects on the work force. Unlike haulage, winding was a job requiring little labor; in the change from horses to steam engines, one form of nonhuman power merely replaced a less efficient one, and three or four men changed over from horse to engine tenders. Until the 1820s, Anzin made no attempt to train special crews. As before 1800, those charged with winding continued to be low-paid, elderly workers no longer fit for work at the face.

The Perier winding engine spread fairly slowly elsewhere after Aniche followed Anzin in adopting it in 1803. The first in Belgium began operating in 1814. The Loire had four in 1812, but it had not converted fully even in the deeper pits until 1830. Silesia, the Ruhr, and the Saar adopted the engine in the 1820s and Le Creusot in 1832.[22] As in other technical matters, however, winding changes begun in these fields took decades to work themselves out. Except for the Nord, no important field had the unity of ownership necessary to make a complete and rapid changeover.

The immediate success of the Perier engines at Anzin seems to have depended on a close and cordial business friendship between the leaders of the two firms. About 1806 the relationship began to deteriorate. In that year there was an acrimonious scuffle over payment for what Anzin claimed was a defective cylinder.[23] The Chaillot firm was languishing for lack of orders and found itself in serious financial difficulty.[24] In 1810, "considering that for several years the personal business of M. Constantin Perier, as well as his health, has not allowed him to give any attention to these machines," the board decided to relieve him of his position as engine overseer at 3,000 francs a year.[25] The entire burden of maintenance fell on Tournelle, one of the self-taught mechanic-engineers who were typical of the early stages of the Industrial Revolution everywhere.

As the local "master mechanic," Tournelle had helped Perier install his machines, and he later carried out successfully the change to wheeled haulage. During these years he seems to have thoroughly satisfied the company owners,[26] but after 1816 he came under increasing attack as the person most responsible for the run-down condition of the company's steam engines. Their excessive consumption of fuel, due to ill-maintained and badly arranged boiler systems, compounded the careless performance of the elderly and untrained engine tenders.[27] Significantly, Tournelle's effectiveness was first questioned when Scipion Perier, a professionally trained chemist and all-round practical scientist, took over from Desandrouin the leading role on the board of directors. Casimir Perier, Scipion's successor after the latter's death in 1821, was even more appalled than his brother by the amount of fuel used by the Perier engines and the dilapidated state of their foundations and housing sheds. He embarked on a rigid economy and rehabilitation drive that included the dismissal of Tournelle in 1823 for general ineffectiveness and petty peculation.[28] Thus the archaic and ill-supervised state of Anzin's mechanical service was one of the causes leading to an investigation of the means of improving the efficiency and lowering the cost of steam winding.

A related factor was the difficult problem of repairing the company steam engines and supplying them with the necessary quantities of spare parts. Previous attempts to set up complete company machine shops had been largely unsuccessful, but the company remained determined to become self-sufficient here as in other areas. Since the eighteenth century, Anzin itself had manufactured most of the materiel necessary to its complex operations. There were no equipment makers or dealers capable of supplying large quantities of such tools as picks at quantities and prices consonant with mass production and purchase, and the usual dependence of the small mine on local craftsmen was obviously not a satisfactory practice for Anzin in terms of either convenience or cost. Anzin had long used its own special equipment, whose traditional form derived from local conditions and experience, and which therefore differed in detail from that used at other mines. Thus the company had very early developed extensive workshops where clay, hemp, bar iron, and lumber were made into bricks, ropes, picks, and other ironware and a great variety of wooden articles and structures. In addition, company employees performed all the maintenance and repair work for existing facilities on company grounds. Anzin employed about one-third its work force on the surface (*au jour*) rather than in the pits (*au fond*),[29] and although by no means all of this surface population labored in the workshops, a large proportion were wood sawers, carpenters, joiners, brickmakers, wheelwrights, toolmakers, and metal workers. In 1811, out of a total work force of 2,571 workers, about 80 were "carpenters" and about 110 were "blacksmiths," the general designations for a variety of specialized workers in wood and metal.[30]

By 1800 a long tradition of self-sufficiency had grown up at Anzin. The practice rested partly on company pride and the desire for freedom from dependence of any kind on the outsider. To this extent, self-sufficiency was irrational. But fundamentally it was a consequence of the inability of small-scale and nonspecialized French industry and commerce to supply large quantities of specialized tools and materials when they were ordered and at costs competitive with self-production. Before 1792 and after 1815 foreign sources of supply were sometimes available in Belgium and England but at prices raised by protective tariffs and prohibitions. Between 1792 and 1815 regular supply by England was out of the question, while in the Belgian departments, as in the older areas of France, the government was a competitor with first option on whatever commodities were necessary for the prosecution of the war effort. Since all of these limits on the possibility of obtaining materiel through normal commercial channels held to an even greater degree for steam engines and engine parts, it is not surprising that Anzin should have early striven to make itself independent in this area too.

As noted earlier for Littry, purchasers of engines had to pay considerable amounts for installation and transport above the base price of the parts supplied by men such as the Periers of Chaillot.

The Newcomen engine, except for cylinder, piston, boiler, and the connecting tubes, rods, and chains, was largely a wooden machine. To a lesser extent, the same was true of the smaller and more compact hoisting engines. To buy an engine meant to buy these metal parts, have them shipped to the point where the engine was to be installed, and there, with the help of local labor and under the supervision of the parts supplier, to assemble the parts in the wooden and cast-iron framework constructed locally. Thus the Periers called themselves "engine builders" as well as "engine makers." It was to the advantage of the purchaser to do as much of the work of installation and to supply as many of the parts as he could.

In 1802, "in view of the mounting of a [large] number of hoisting engines . . . which require a daily supply of cast-iron pieces," Anzin decided to install its own casting furnaces. Jacques-Constantin Perier was to furnish the blueprints, equipment, and workshop as well as the information necessary for their proper operation. A few months later, the company lowered its sights and decided to settle for an "ordinary furnace." One of the company directors living in Paris was to search in the capital for an experienced iron founder and a worker in castings who would take charge of the completed foundry.[31] Meanwhile, the company drew on the resources of the cannon foundry of Douai for at least some of its cylinders, driving wheels, and other large castings.[32] The problem of parts supply became increasingly serious after 1806 when the Anzin-Chaillot relationship grew strained. Hence, between 1808 and 1811 Anzin tried to provide for the repair of some of the basic engine parts such as cylinders and boilers in its own shops. In 1808 the board decided to purchase a cylinder-boring machine through the Chaillot Periers.[33] Two years later it proposed to enlarge and reorganize the Anzin workshops in order to provide an easily supervised machine shop under one roof for the boring machine it would purchase, an enlarged foundry operation, and the boilermakers it would have to hire.[34] The project soon expanded again so that the company could free itself of "its present dependence on foreign [i.e., Belgian] workshops for certain steam-engine parts." The new plans ambitiously aimed at an integrated operation, including the long-sought boring machine, a tilt-hammer, and a rolling mill. Thus equipped, Anzin could cast and forge large iron parts, bore cylinders, and roll boilerplate. The result, it was hoped, would be cheaper parts available when needed.[35] Several months later, in early 1811, the company realized that years might pass before it could produce such parts in the necessary volume and therefore proceeded to buy two new spare cylinders on the regular market.[36] In June a bonus of 1,000 francs to Tournelle for the repair of a cylinder and the promise of more money if he should succeed in casting and boring a fresh one in the company's shops showed both Anzin's determination and a measure of despair over the solution of the problem of supplying parts. It is unclear how successful Tournelle's efforts were.[37] In 1812 and again in 1815 he had orders to "construct" new drainage engines with parts manufactured by the company,[38] presumably in its "reverberatory furnace" for "casting the pieces of iron and copper necessary for the maintenance of its steam engines."[39] Certainly the basic problem remained unsolved in 1816 when Anzin entered into a close relationship with Edwards and Aitken, Englishmen seeking a French base for the manufacture of their improved steam engine.

The engine built by the Periers of Chaillot was the Watt design operating at normal atmospheric pressure. Toward the end of the eighteenth century, several men began experimenting with engines working at two or more "atmospheres." Richard Trevithick, a Cornishman, and Oliver Evans, an American, produced their "high pressure" models soon after 1800. In 1803 another Cornishman, Arthur Woolf, patented an engine combining the high pressures of the Trevithick and Evans engines with the efficient two-cylinder ("compound") arrangement worked out for Watt engines by Richard Hornblower some twenty years earlier. The success of Woolf's design was partly in the idea of combining these two arrangements and partly in perfecting a new type of cast-iron boiler capable of withstanding high steam pressures without exploding.[40]

Much of the credit for the new boiler and the practical working out of the whole scheme should probably go to Humphrey Edwards, a millwright at Lambeth near London and Woolf's partner for a short time after 1806. When Edwards broke with Woolf in 1811, he continued to make machines at Lambeth according to Woolf's design. Finding his best market in France, he went to Paris in 1815 and obtained the French import license for the Woolf engine. In the next two years he and his new partner, Hugh Aitken, brought fifteen engines into France from the Lambeth workshops, mostly for sale to textile manufacturers in Paris and the northern departments.[41]

Woolf's engines were never very popular in England,[42] but in France, where coal was expensive and good mechanics scarce, the easily maintained and fuel-saving Woolfs were the first steam engines built in large numbers, and they remained the standard French model until about 1870.[43]

While Edwards was importing his engines from England, he looked for a base for the manufacture of the Woolf engine in France. How he struck up the initial acquaintance with Anzin is not known, but it was probably through Scipion Perier, a member of various learned, business, and governmental organizations. Already in March 1816 the Englishman was building one of his engines for the company in Anzin's own workshops and at the company's expense. Anzin was to pay 700 francs per horsepower for the use of the patent on condition that the trial made of the engine resulted in "a saving of at least half in the consumption of fuel and in all maintenance costs as compared to the company's other machines."[44] Moreover, Perier was to arrange with Edwards and Aitken for constructing and equipping at company expense, up to a maximum of 30,000 francs, a workshop on company property that the partners could use rent free. Edwards and Aitken agreed merely to pay 6 percent interest on the sums advanced by the company and the future taxes on the property leased. If, when the lease expired in ten years, either of the two parties refused to renew it, they were to reach an agreement about which of the parties would buy out the other's interest in the venture.[45]

The generous terms offered by Anzin are not difficult to understand in the circumstances. Anzin suddenly and desperately felt the need of replacing or renovating its existing steam engines. Edwards and Aitken were Englishmen at a time when France regarded England as a nation of mechanical geniuses, particularly with regard to the steam engine. These Englishmen promised much in the way of fuel savings. And there was no evident alternative French manufacturer of engines, since the French engine-building industry was practically limited to the moribund Perier works at Chaillot.

Although the solution of the parts problem now seemed assured, it had become secondary to a more grandiose vision. The terms of the contract actually signed on 14 June 1817 plunged Anzin even deeper into the business of constructing steam engines. Edwards ceded a share in his patent to Anzin in return for a payment of 400 francs for each horsepower built until the expiration of the patent in 1825. In fact, the first engine Anzin bought under the agreement was a large one with twenty-four horsepower, installed to operate the machines in the company workshop.[46] It proved so economical, "especially in fuel consumption," that the board decided to build as soon as possible a sixty-horsepower drainage engine to replace an old Newcomen engine.[47]

Nevertheless, the new workshop engine would pay only if the company supplied it with sufficient work, but Anzin again faced problems in its attempts to equip a modern machine shop. In 1816 and 1817 its plans included once more the purchase of a boring machine and tilt-hammer, as well as a triple-headed lathe. Tournelle was to install the new machines under Edwards's paid supervision, or Edwards was to contract to do the job himself for a lump sum.[48] But in 1818 the machines remained unbought, with the possible exception of the boring machine, of which no more is heard. For the hammer and lathe Anzin decided to turn to England. Renard, the general agent, was therefore to determine their price as of completed delivery at Dunkerque and to take the necessary steps to get plans for a machine to move the lathe heads. Apparently Anzin expected legal and financial difficulties in importing the machines because of England's long-standing ban on such exportations. The wording in the Registres anticipates the need to smuggle the machines in or bribe the customs officials in order to obtain an export license.[49]

In 1820 Scipion Perier negotiated a revision of the 1817 contract with Edwards. In the new version Anzin dropped the 400 francs per horsepower payment in return for a flat sum of 60,000 francs for its share of Edwards's patents and 15,000 francs more for the right to build and sell his machines.[50] Anzin then seemed more determined than ever to become self-sufficient for the supply of its most expensive capital equipment and, in addition, to make the venture pay by setting itself up as steam-engine builder to northern France. It had no rivals in the field as yet, and it must have hoped to get in at the beginning of what looked like a "growth industry" in the expansive and technologically advancing conditions of Nord manufacturing after about 1818.

This large-scale divergence from Anzin's essential function and field of expertise failed for a number of reasons. Among the most important was the continued difficulty in obtaining the necessary men and machines to set up a complete engine shop. In September 1821 Anzin was again considering the purchase of a casting furnace and rolling mill.[51] One year later the company paid Edwards for fulfilling the 1818 contract for the hammer and lathe.[52] It also claimed that in its workshops it produced the large structural parts for drainage and winding engines as well as copper and forged-iron boilers, big iron castings, and general hardware.[53] Yet, in 1823 and again in 1827, Anzin ordered the purchase of spare Newcomen cylinders from Chaillot.[54] The only reference to the machine shop after 1822 occurs ten years later when the existing furnace and its blower were to be replaced and its Edwards engine stepped up in power in order to "complete" the workshop and "render it capable of meeting the company's present needs."[55] By that time its "needs" were merely maintenance, for in 1822 Anzin had officially surrendered its policy of engine building and had reunited its fortunes with Chaillot, where the Anzin Periers and Edwards were now in charge.

It is doubtful that Anzin ever built any complete engines in its own shops. Even the famous workshop engine was most likely "constructed" there of parts brought from England. The three Edwards winding engines probably in operation at Anzin by the end of 1819 were also most likely essentially English.[56] Certainly Anzin never sold engines to other Nord manufacturers; the ten Edwards engines sold to Nord textile firms between 1815 and 1820 must all have come from England or possibly Chaillot.[57]

Anzin had broken all connections with the Perier brothers at Chaillot in 1810, but in 1818, on the death of Jacques, the last of the brothers there, Anzin's own Perier brothers, Scipion and Casimir, bought the Chaillot foundry and installed Humphrey Edwards as manager.[58] Chaillot began to build Woolf engines. It is not clear exactly how the Periers found themselves in a position to purchase Chaillot, although debts incurred to their Paris bank by the engineering firm in its failing days may have been the lever.[59] Nor is it clear whether the new owners already conceived of Chaillot in 1818 as the supplier of Edwards engines for Anzin. Until 1822 the Anzin Company was ostensibly working at setting up its own workshops, but the Periers and Edwards may have considered the Anzin workshop a branch of the Paris establishment, capable of supplying engines to manufacturers in the Nord and nearby departments. It seems certain, however, that soon after 1818 they regarded Chaillot, the largest and most reputable machine shop in France, as the center of their engine-building activities and may even have purposely allowed the Anzin workshop to remain inactive in hopes that it would quietly disappear.

In any event, the death in 1821 of Scipion Perier also spelled the death of his vision of a self-sufficient Anzin. Under the leadership of his brother, the cost-conscious Casimir, Anzin formally abandoned the manufacture of engines in May 1822. The board of directors cited "the immense advantage resulting from the use of steam engines of the Edwards system" and the pressing need to replace existing engines at all of its pits. At the same time, the board pointed out, the company's own workshops could not make the requisite number of engines at the speed or cost offered by Chaillot, which promised immediate delivery at bargain prices. Anzin ordered ten of Edwards's ten-horsepower winding engines at 18,000 francs each, a saving of 4,000 francs on the market price and undoubtedly far below the cost of construction in Anzin's own shops. At the same time, Chaillot received a special discount on coal purchased from Anzin, which was particularly interested just then in opening up the Paris market.[60]

Certainly there existed a potential conflict of interest between the Periers' holdings in Chaillot and Anzin but the relationship might as easily be said to have led to a profitable community of interest favorable to both parties. In any case, one may be sure the Periers did not lose by the arrangement, and it is probable that in the long run Anzin did not lose either by abandoning its projected engine business. Anzin's early attempt at vertical integration was probably a mistake given the meagre mechanical knowledge of owners, management, and workers. The attempt was perhaps even more foolish if the goal is interpreted as "diversification" in the sense used by modern corporations. The repeated large-scale financial failures in this direction by such famous contemporary giants as Cockerill's enterprises and Le Creusot demonstrate convincingly the weaknesses of big "integrated" or "diversified" industrial complexes at a time when organizational and sales techniques were embryonic.[61]

Chaillot lived up to its promises not only on the initial cost but also on speed of delivery and fuel savings. It delivered five engines within a year of the first order, and by the end of 1826 there were sixteen Edwards engines operating at Anzin, although the same number of Perier engines were still in use.[62] Their initial cost was not negligible. In 1826 Casimir Perier put the total cost of the steam engines bought from 1821 through 1825 at 457,186 francs. This figure included the giant eighty-horsepower drainage engine costing 101,000 francs and probably fourteen winding engines. Each of the latter, then, cost about 25,500 francs, a figure which included the basic engine price of 18,000 francs plus the cost of transport from Chaillot, installation, and a supply of spare parts.[63] This estimate accords with that of the engineer Clère, who claimed in 1830 that Anzin obtained its winding engines for 25,000 to 30,000 francs as compared with the usual installed cost of 40,000 to 50,000 francs.[64]

Some of the new engines required adjustments when at first they did not "produce the desired effect." The company workmen strengthened the parts connecting the engines to the winches by substituting cast iron for wood in gear wheels, gear teeth, rope drums, and engine beams. They raised the winch pulleys and placed foundation floors under the whole winding system. Castiau, who had replaced Tournelle as the company's master mechanic, suggested and carried out most of these changes. Two improvements to the engine itself, suggested by Edwards, were successful.[65]

By 1826 the Edwards engines were working smoothly, and Casimir Perier claimed that they burned only one-third as much coal as the Perier types and did the same amount of work in a shorter time. The figures he gave indicate a decline from 1822 to 1826 of only a little over 15 percent in fuel consumption per engine-hour worked, but those figures included coal burned by the eleven huge drainage engines and the sixteen Perier engines, and some of the Edwards engines were already operating in 1822. Most of the 15 percent saving, then, was due to the eight to twelve Edwards engines introduced between 1822 and 1826.[66] Anzin ordered more engines from Chaillot after 1826,[67] and by 1830 conversion from the Perier engines was substantially complete, although a few of the latter continued to operate on the shallower pits at Fresnes and Vieux-Condé.[68]

Mine Safety: Drainage, Ventilation, and Lighting

The great enemies of the mine and its miners were the four elements of earth, air, fire, and water in the form of falling rock, exploding gas, and flooding. The miners partly parried the danger of falling rock by their own somewhat negligent efforts at propping and filling the tailles behind them, but falls still took their toll in injuries and deaths. The second source of danger, the gas known as the *grisou*, was far more spectacular, for its explosions brought death to large numbers of workers at a time by burning, burying, asphyxiating, or blowing up anyone in its long reach. Anzin had its share of such tragedies. Flooding was of much less concern to the miners, for although the danger of sudden flooding by breaking into a large underground pocket of stagnant or running water or of the collapse of the shaft lining was always present, Anzin was knowledgeable enough and fortunate enough to escape such catastrophes. Nevertheless, part of the price paid for deep mining was continuous pumping to combat water seeping through the shaft lining from the niveaux and down to the workings, as well as the extraordinary efforts which Anzin made to permit exploration under the Torrent.

Water seeping through the cuvelage trickled down each shaft to a reservoir at the bottom, from which a sloping drainage gallery led to a main reservoir at the base of a shaft fitted with a pumping engine. A single drainage system might link several pits, as in the older workings near Anzin, where about 150 cubic meters of water collected in each twenty-four hour period at the reservoir of the pit Beaujardin from six or seven nearby pits. On the other hand, an equal quantity might accumulate in a single pit where water seeped from the coal measures themselves, as at the pit Bleuseborne, which required a pumping engine for its use alone. And when bures were still commonly in use, water collecting in them had to be laboriously hand-winched up in casks and then sledged to the shaft or a drainage gallery. In one case, the cost of drainage from a series of bures amounted to one-fourth the total cost of extracting the coal. The tendency after 1800 was to consolidate the drainage of each grouping of pits at a single shaft by eliminating bures, extending the system of drainage galleries, and increasing the size of the central reservoirs.[69]

From a central reservoir a series of suction pumps ranged one above the other raised the water to the surface. At the pit Beaujardin in 1805, then 240 meters deep, there were six pumps, two of 43 meters in length, three of 117, and one of 37. Each pump consisted of a cast-iron casing about six inches in diameter containing a copper piston attached by a connecting rod to the main pump rod.

The rod of bolted wooden beams hung from one end of the great balance beam of a Newcomen engine. The other end of the beam was attached to the engine piston. Seven and one-half times a minute, as the steam in its cylinder condensed, the huge piston, more than one meter in diameter, descended and tilted the balance beam, which in turn raised all the pistons of the series of pumps about one and one-half meters. The resulting vacuum sucked the water up from the reservoir at the lower end of the pump and eventually spilled it out into the reservoir above. Thus the water from the large reservoir at the bottom of the shaft finally found its way up through the series of pumps to tanks atop the engine, where it condensed the steam in the cylinder.[70]

This description of the system of drainage at Anzin would not have to be changed much to fit the situation as it existed twenty-five years later—or twenty-five years earlier. So far as the pumps themselves were concerned, the only important change made before 1825 was to decrease their diameter as the pits deepened in order to save on coal consumption.[71] From 1825 onward there seems to have been some experimentation aimed at applying "the processes of the Engineer Woolf," which apparently involved force pumping from below rather than suction pumping from above. After a demand by the directors in 1827 for a report on its success from Castiau, who had suggested the new system, the subject drops from the records.[72]

As in winding, although to a much lesser extent, the important improvements in drainage came in the engines furnishing the power. Despite considerable improvements in pumping efficiency connected with the same three men—the Periers of Chaillot and Humphrey Edwards—who had been responsible for two revolutionary changes in winding engines in a generation, the engines used for drainage in 1833 and long after were essentially the same as the one erected at Fresnes in 1731. The reason for this seeming conservatism was simple; the switch from organic to inorganic energy for drainage was the great technical achievement of the early eighteenth-century coal industry, an achievement in which Anzin was one of the leaders. The Periers had enabled Anzin to replace horses with steam for winding coal about 1800, and Edwards had provided a much more efficient compound, high-pressure steam engine to replace the early Watt engines of the Periers, but the old single-acting, to-and-fro, low pressure engines of the Newcomen and early Watt types were cheaper to build, maintain, and operate at the very high horsepower ratings necessary for drainage.[73] These engines were simple in construction and easy to repair. Most of their parts were of wood, which was cheaper to buy and work than iron; both wooden and metal parts could fit poorly and still operate reasonably well; and the cost of low-pressure boilers was naturally much less than of those heavily reinforced for high pressure.[74] Where technical skill was low, the cost of well-machined iron parts high, and the cost of coal relatively low, as on the French coalfields, the simpler old engines were preferable to more complicated new ones.[75]

All the pumping engines at Anzin until 1800 were of the Newcomen type. Since the departure of the trained engineer Mathieu during the Revolution, they had fallen into dilapidation through negligence and ignorance of their proper care and operation. When Anzin invited Jacques-Constantin Perier to install twenty of his new winding machines in 1802, the board also contracted with him to repair and "make the changes and improvements to which they are susceptible" in the pumping engines.[76]

Perier apparently advised the company that some of the Newcomens ought to be replaced and counseled them to do so with the original single-acting Watt engines rather than the double-acting, rotary Watt winding engines which he sold as "Perier engines." Chaillot wholly constructed the smaller Perier engines and shipped them to the buyer, who usually also hired Jacques-Constantin Perier to supervise their installation. For the older type pumping engines, however, still known in France and Belgium as "Watt engines," Chaillot (or perhaps a Douai foundry) seems merely to have furnished the chief metal parts, such as cylinder and piston, which company workmen then fitted into a locally built, heavy wooden framework under the supervision of Jacques-Constantin. Thus, a pumping engine was "constructed . . . according to the new principles" at Vieux-Condé in 1802. Anzin employed the same system on at least two other occasions, in 1815 and 1830.[77] Other Watt engines soon followed the original one of 1802, and by 1810 when Anzin dropped Jacques-Constantin Perier from its payroll, the company probably had four Watt and only three Newcomen engines.[78] The changeover was complete by 1817 when, because of a better arrangement of the drainage pattern of the pits, five Watt engines were able to perform all the pumping chores.[79]

That same year, the economy promised by Edwards's high-pressure engines prompted Anzin to decide to construct one of the same type at Anzin for pumping. This engine was probably never built, and the only pumping engine ever bought from Edwards was a very large low-pressure engine built at Chaillot in 1822 and purchased for 101,000 francs.[80] Yet the need for pumping engines grew steadily after 1822 because of increased production, the opening of several new pits, and, related to both of these efforts, the determined new attack on the seams below the water-logged sands of the Torrent. One solution was to bring the old Newcomen engines out of storage. Perier, in addition to introducing the Watt engine and repairing the old Newcomens, had devised a portable carriage for both types. The mobilized Newcomens helped drain the niveaux from new shafts under construction, and the company assigned three of them to aid the Watt engines in the struggle to drain the Torrent.[81]

Another means of stretching existing pumping capacity was to improve the performance of both Watt and Newcomen types, which had again deteriorated in effectiveness since Tournelle had taken over responsibility for the engines from Jacques-Constantin Perier. Under Casimir Perier's unrelenting pressure, Edwards's tutelage, and Castiau's direct supervision, changes in the makeup of chimneys, boilers, firepots, and engine crews led to greater working efficiency, which cut both the amount of time and fuel consumed in pumping. Engines could now pump in eight hours the same amount of water that had formerly required twenty to twenty-four hours. This improvement helped make it possible to pass the Torrent, where several engines had to pump continuously. In other parts of the field, the pump had operated only eight to ten hours each day because the main reservoirs could contain only a single day's flow of water. With the reservoirs enlarged and pump efficiency improved, two days' continuous pumping could handle a week's flow. By using a single crew for two or three engines and reducing the number of expensive engine firings, the company achieved considerable labor and fuel savings.[82]

A serious problem at some of the engines was the lack of an abundant supply of soft water for the boilers. The water pumped from the mines usually contained minerals leached from the rocks and left a scale on boiler tubing that lessened pumping efficiency, raised fuel costs, and required frequent shutdowns for cleaning.[83] It was, however, expensive to cart soft water from the river. After 1825 the company solved the problem by building wooden conduits from the Escaut River to the engine emplacements.[84]

By 1833 Anzin had carried out a series of individually small but cumulatively important changes in its drainage techniques. Its motives were the desire to lessen fuel costs, the need to maintain close contacts with leading French engine builders and, above all, the necessity of pursuing the coal seams down ever deepening and more watery pits and of opening new shafts on the rich seams below the Torrent. Although the changes were less decisive than those in winding, they enabled Anzin to keep the lead over its continental rivals, a lead achieved in 1731 with the installation of the first steam drainage pump in France.

As in pumping, so in ventilation and lighting Anzin had inherited advanced techniques from the eighteenth century, but now it found itself forced to adopt new methods as a result of the steady deepening of the pits. One ventilation problem exacerbated by deeper shafts was that underground temperatures increase at the rate of one degree centigrade for every twenty-five meters of depth.[85] Lighting, ventilation furnaces, and the miners' respiration added to the natural heat of the mine. Worse yet, the air in the mine contained fine coal particles and saturated air whose moisture condensed in drops on the walls of the workings and galleries. It was little wonder that the miner working in this hot, humid, dust-filled atmosphere was liable to chronic miner's diseases such as anemia and "weak chest," nor that he demanded and received a premium for working at the especially torrid depths of the deeper pits.[86]

Still, owners and men alike regarded uncomfortable working conditions and occupational ailments as the inevitable accompaniment of life in the mines. The primary purpose of ventilating systems was rather to combat mine gas, a more insidious and immediately dangerous foe. Most of the many mine gases, such as choke damp, killed by asphyxiation. They were sometimes troublesome and occasionally the cause of major disasters, but over the centuries miners had developed precautionary techniques for detecting and dealing with them. Increasingly important in the deepening shafts of the eighteenth and nineteenth centuries was a far more dangerous gas, firedamp, which the miners in the Nord and Mons fields called the *grisou*.

The grisou consists mostly of methane gas generated by the decomposition of organic materials in the coal itself. Although it is odorless and colorless, its occurrence was fairly predictable. Small quantities often seeped silently out of the seams, but sometimes, "when one listens attentively, one hears in the coal . . . a rustling noise like that of the wind passing across a slit, and at other times it resembles a slow bubbling sound."[87] Sometimes it "whistled" out of crevices suddenly opened by a blow of the miner's pick. The grisou was most abundant during storms when the barometric pressure in the mine fell suddenly. Being lighter than air, it also tended to collect near the roof of the more humid areas of the mine, especially in corners and in old, abandoned parts of the workings. Seldom found in anthracite or semibituminous seams, it was almost always present in the richer bituminous seams below a depth that varied from mine to mine.

Like other mine gases, the grisou in heavy concentrations can kill by simple asphyxiation, but its fearsome reputation derived from the fact that it is explosive when mixed in certain proportions with air in the presence of a spark. An explosion might be relatively local and harmless, even a small pop like a balloon, or it could set off a train of explosions along galleries and so through a wide area of a mine. Preceded by a fireball, such an explosion blew over everything in its path—props, men, and carts—and often brought down the roof and walls of the mine in great cave-ins.[88]

Although somewhat dangerous at Anzin in the middle of the eighteenth century, the grisou, by 1800, "no longer inconvenienced" the miners except at one pit.[89] In 1811 however, the grisou had become thick enough in some pits that Anzin experimented with a primitive safety lamp. In 1816, in the first fatal explosion at Anzin "in several years," a worker burned to death while touring the workings with a leather flapper and candle to clear out pockets of the grisou that had collected before his shift began. This task, although unpleasant, was a standard precaution and usually safe enough, but on this particular Monday an unusual amount of gas had built up over the Sunday holiday, and the worker had incautiously not worn his protective leather suiting.[90] Two years later, a gas explosion killed a boy of eleven by throwing him against a gallery wall.[91] Although the grisou caused no further fatalities until 1823, the danger of explosion was building up to a critical point in some pits. The anthracite seams of Fresnes and Vieux-Condé remained "clean," as did the semibituminous seams near Anzin and four of the newer pits on the southern faisceau. But some of the other seven pits in extraction on the southern faisceau shut down for long periods after about 1818 because of the increasing danger. For example, Saint Jean remained closed almost continuously after 1820 while the company awaited the outcome of the "efforts which are being undertaken to purify it, if it is possible."[92]

The pit Chauffour, closed for most of the winter of 1822–23 because sales were slack, reopened on 3 April 1823. The company took extraordinary precautions, since several miners had suffered burns there in the previous year. For the next six days it carefully inspected the workings for dangerous concentrations of gas before each shift. On 9 April three workers made the rounds successively and found nothing out of the ordinary. Then at 10:00 A.M. a miner's candle touched off an explosion. Of the twenty-two miners nearby, nineteen were dragged out of the pit dead, two died soon afterward, and the other was found buried under the rubble some months later.[93] This terrible accident, by far the worst in Anzin's history to that time, decided the government to close all of the pits subject to the grisou. Since these pits produced all but a fraction of the bituminous coal that accounted for most of the Anzin production and was in most demand, the future of the company hung on finding a solution to the problem of the grisou.

In the more primitive centuries of mining, miners had known no weapon against this mass killer of men and wrecker of mines: "This horrible and diabolical fire which we do not know how to vanquish has rendered our workings deserted and to fight it we have only our prayers."[94] By 1795 practical experience had provided a set of simple ground rules for fighting the foe. A government circular of that date written by a mining inspector was entitled "Method of Working the Veins of Coal Called *à grisou* and Means of Preventing the Terrible Accidents Caused by Flammable Gasses."[95] The number of similar publications which followed in the next few years indicated a growing seriousness about and greater awareness of the problem. By about 1815 the developing theory of gases provided scientific corroboration of simple empirical rules painfully learned.[96] It was not until 1839, however, that a full-scale discussion of all aspects of the problem, both theoretical and practical, appeared in France.[97]

And much later in the century, following a series of disastrous explosions, it was still possible to assert that "the rules to follow . . . have been established for a long time, but they are not always closely observed."[98] As in so many instances, Anzin followed more closely and earlier than almost any other mining company on the continent the pattern of perfection set down by the government experts from 1795 on.

Theoretically, there were two means of preventing an explosion of the grisou. One was to set up a powerful draft of fresh air through the mine workings to "wash" the gas from its hiding places and lower its concentration below the explosion point. The other was to prevent the grisou from coming in contact with a spark or flame that would set it off. The first solution gradually evolved during the eighteenth century with the development of artificial systems of ventilation, the second after 1815 with the introduction of the safety lamp. Neither was sufficient by itself in constantly deepening workings; used together and in conjunction with the steam engine for draining and hoisting, they made possible an enormous expansion of coal production in the nineteenth century.

To be effective, a system of ventilation had to solve two associated problems: finding a means of setting up a sufficient air current and directing the current most efficiently through the workings. Shallower and more primitive mines usually had natural ventilation. Air circulated spontaneously because of differences in temperature between the air outside and the air inside the mine.[99] These more or less complex systems of natural ventilation still served most French and German mines as late as 1840, including those of the Saint Etienne and Rive-de-Gier sections of the Loire field. After 1800 the deeper and gassier pits near Mons found natural ventilation inadequate and switched to the furnace system, the method practiced since the middle of the eighteenth century in the Newcastle and Nord fields.[100]

The furnace system worked on the same principles as the natural system but raised their effectiveness by increasing the temperature differential between surface and mine air. Anzin placed its furnaces in an opening hewn in the coal measures. At a point where a haulage gallery abutted on the shaft, a small gallery connected the furnace room to the "room" where the coal was loaded for hoisting. Two or three doors along the connecting gallery prevented the contaminated air of the galleries from coming into contact with the furnace. Clean air for combustion reached the furnace by a short slanting chimney leading down off the shaft from the bottom of the cuvelage. The surface air heated by the furnace set up a strong updraft through the chimney and upper shaft which drew the vitiated air up and out with it.[101] The system was a great improvement over natural circulation since it was much less dependent on the seasons and weather: "almost everywhere in ordinary weather, the air there [at Anzin] is rather fresh and the circulation rather active," because of the efficiency of its furnaces.[102]

Yet, "in spite of these precautions, it sometimes happens during heat waves, and especially when the weather is heavy and stormy, that the air does not circulate in the pits."[103] A government inspector attributed an explosion of grisou that killed ten persons at Anzin in 1832 to the interruption of circulation by stormy weather.[104] There was also the possibility that a grave explosion in the galleries would blow out the furnace, despite the intervening doors, and eliminate circulation at the very moment the mines needed it most.

Anzin was thus led, along with numerous inventors and government officials, to consider the possibility of using mechanical ventilators.[105] The French Council of Mines regarded the great water-powered ventilators used in some German metal mines as insufficiently powerful for larger coal mines, and similar English experiments found no imitators.[106] In 1808 Anzin's own Tournelle produced a model ventilator, whose blueprints the directors sent to the Periers in order that they might "get advice at Paris as to the efficacy of this means of renewing the air without using fire."[107] Apparently the advice was negative, for no more is heard of the project. In 1827, when the prefect of the Nord requested the opinion of the Anzin general agent on the air pump of Dutouquet, who hoped to exhibit it at the Exposition of 1827, the reply was:

> Several others have already advised the use of pumps for this purpose, but the large-scale application of this means appears impracticable because of the considerable expenses that the introduction of this system of ventilation would occasion; moreover [ventilation] is advantageously achieved by furnaces which, giving the effect of a siphon, set up at the bottom of the shafts a more considerable circulation of air than it would be possible to obtain with any pumps that might be employed.[108]

Mechanical ventilation had to wait another quarter century in the Nord because it was less effective, more costly, and more subject to breakdowns than the time-tested furnace, whose size could be increased, which used cheap fuel produced on the premises, and which, having no moveable parts, never broke down.[109]

As with the drainage pump, one furnace was usually enough by 1800 to ventilate the workings off several shafts, each of which functioned as a point of entry for the fresh air pulled down into the system. The fresh air followed the sinuous path blocked out for it through every gallery, way, and taille. By 1833 each group of pits usually had several routes leading away from the entrance shaft and only rejoining each other at the ventilation gallery leading to the exit shaft. Wherever galleries converged, sets of double doors produced the effect of air locks, so that the passage of a haulage team from one gallery to another did not allow the air current to deviate from its established path. Packing the tailles with rock waste helped prevent large pockets of the grisou from accumulating in the abandoned parts of the workings, and the solid masonry of the walls of the ways between tailles also kept the system air tight. There was a continuous elaboration of the circulation pattern in the period 1800 to 1833 with considerable route straightening and shortening by dividing off more and more paths in each pit as the workings went down and the galleries out in ever more complex patterns.[110]

The chief improvements, however, came after the disaster of 1823. The government investigators ordered the system rearranged so that the fresh air sank immediately to the bottom of the entry shaft and followed an ascending path throughout its route in order to "wash out" more thoroughly the naturally rising grisou. In addition, it ordered the furnaces at the exit shafts placed deeper in order to produce a stronger updraft. The local engineer Clère, however, always solicitous of the welfare of the company owners, pointed out that such a rearrangement would require a considerable investment. At the moment, he claimed, the company's financial situation was not bright; only its present policy of rigorous economy enabled it to compete with its Belgian rivals, whose seams were so much richer. Since it would hardly be wise for an enlightened and farseeing administration to endanger the very existence of an enterprise so valuable to France, he proposed that the government merely request that the new system be installed in all new pits immediately and in the older ones "as rapidly as possible."[111] The general director of mines accepted Clère's proposal and incorporated it in the prefect's order of 1825 entitled "Lighting and Ventilation in the Nord."[112] How well Anzin complied with the order and how quickly it went beyond it to rework the system in its older pits is not certain, but in 1833 the company claimed to have spent large sums in improving its furnaces.[113] Certainly by 1839 Anzin's ventilation practice was regarded as a model of its kind.[114]

However well planned and maintained a ventilation system might be, it was not sufficient in itself to meet the threat of the grisou, which persisted in collecting at unexpected times and in unsuspected places. It was also necessary to prevent a spark from reaching the gas, while providing some sort of light for the miner to work by. From earliest times, miners had used a variety of lighting devices including pine torches, coal braziers, and decaying fish, which gave off a dim phosphorescent glow. By 1880 the two most commonly used lights were the suet candle and the small, simple, earthen-pot lamp filled with oil in which floated a cotton wick. Both were equipped with a hook so that they could hang from a prop, the miner's vest button, or a crack in the rock wall. Where the gas was thick, the miner usually placed the lamp or candle some distance from the face in a way or gallery where the air circulated better than in the tailles. Often he assigned a helper to keep the air moving around the light and to watch the color and length of the flame as a telltale sign of the concentration of gas. The lamp was somewhat safer than the candle but also gave less light, a very unsatisfactory feature for the miner who, working in semidarkness, was usually paid by the amount of coal he cut. Miners thus developed a fatalistic attitude toward explosions and tended to ignore the most commonplace security measures in the attempt to shed as much light as possible on their work. Their supervisors had constantly to be on the alert to prevent them from smoking their pipes on the job and even carrying hot coals to the face to light their pipes when the lamp or candle burned a few time-consuming meters away.[115]

At Anzin, the miners had always used candles made of beef suet with hemp wicks. Each miner received four candles of four ounces each at the beginning of every eight-hour shift.[116] According to the mining inspector Brice, who investigated the tragedy at Chauffour in 1823, the use of such candles in a pit where grisou had long threatened, together with the workers' imprudent use of them, were the "sole causes" of the explosion.[117] Brice claimed he had warned the Anzin general agent, Renard, of the danger several times and had advised him to anticipate such an event by providing the miners in gassy pits with the new Davy safety lamps.[118]

As early as 1811, when the danger of the grisou in the deepening pits first seems to have become a matter of serious concern, Anzin had dabbled with the idea of the safety lamp. Tournelle ordered a primitive version known as the Argand lamp made to his specifications by a Brussels firm and put it to the test in a gassy gallery in competition with a candle. While the candle either touched off small explosions or flickered out, the lamp continued to burn safely with a steady flame. The lamp's only failing, a serious one, was a tendency to overheat and melt its tin cover. Requesting the replacement of the tin by iron or copper and a few other, minor adjustments, Renard ordered ten more lamps from the Brussels manufacturer. The eminent engineer Héron de Villefosse had witnessed the trial and reported it in the *Journal des Mines*. Yet, whether because the adjustments requested proved unsatisfactory, the lamp too delicate or expensive, the miners recalcitrant, or the threat of the grisou not serious enough, nothing further is heard of this interesting anticipation.[119] Like Tournelle's trials with railed haulage, his experiments with the safety lamp proved ephemeral. Tournelle perhaps deserved something better than the opprobrium heaped on him by the Anzin management of the 1820s. Responsibility for the lack of determination and patience in failing to put these two vital innovations to use at a much earlier date probably belongs to the company's leaders, especially to Desandrouin, and not to Tournelle.

The Englishman Sir Humphrey Davy invented the lamp to which he gave his name in response to demands from Northumberland where pits were deep and growing rapidly deeper and where gas explosions occurred with alarming frequency in the first years of the nineteenth century. By 1815 he had developed his principle, similar to that of the Argand lamp, of enclosing an ordinary oil lamp in a wire gauze cylinder across which the gas would not pass in explosive quantities. In the next year, his new lamp proved effective in practice. It was speedily adopted in England and apparently also in Belgium but was almost untried in France before the Anzin catastrophe of 1823.[120]

After the tragic explosion, Renard claimed that he had ordered a "case" of Davy lamps from England in the previous year but that the Calais customs had permitted only two to enter as samples, whereupon the rest of the case had been returned to England. He had then attempted to get more from Belgium "par contrabande," but, despite his "lack of any scruple," the risks involved in smuggling in an object of so little intrinsic value had created "difficulties."[121] It is easy to believe that, at a time when Casimir Perier was imposing the most rigorous economies in the company's purchasing policies, no one was willing to take the responsibility for the high cost of smuggling in lamps designed to prevent a merely potential danger. It is even easier to suppose on the basis of two curious entries in the Registres, the only ones between 1800 and 1833 dealing directly with the lighting system, that, whatever the truth of Renard's story, Anzin was not seriously concerned in 1822–23 with the safety features of lamps. On 22 September 1822 the board proposed a "serious and careful" experiment with oil lighting in order to determine whether or not it could be substituted for candles. On 13 March 1823, less than a month before the Chauffour tragedy, the experiments with oil lighting at Fresnes and Vieux-Condé were declared a success from the point of view of economy; the use of oil lamps was to be generalized in those two concessions. Neither entry contains any reference to safety, safety lamps, the grisou, or the gassy pits at Anzin and Raismes.[122]

The immediate response to the disaster of 9 April by the various interested government officials was a spate of demands for enforcement of the safety provisions of the 1813 imperial decree on the mines. In the report of 17 April to the Nord prefect, as required by the decree whenever fatal accidents occurred, the local mining inspector Brice, after reviewing the events and causes of the explosion, recommended immediate steps to prevent a recurrence. Anzin was to close all grisou pits until safety lamps could be obtained and was to use no other sort of lighting henceforth under any circumstances in such pits. The gauze cylinders shielding the flames from the open air were to be locked on by keys kept by the pit foremen. The foremen were to exercise a strict surveillance over the workers to prevent their bringing smoking equipment, tinder, live coals, or open lamps into the pit and were to inspect each miner's lamp every day before and after each shift, make repairs at the slightest necessity, and fill the lamp and lock it.[123] The strict provisions concerning supervision of the workers' use of the new lamp were necessary to prevent their unscrewing or wrenching off the cylinder in order to light their pipes or provide better illumination for their work.[124]

Anzin's difficulties in complying with the order to equip itself with safety lamps and thus reopen its pits make it clear how unprepared the company, the government, and French industry were to deal with a potential danger that had suddenly become a horrible reality. The Nord prefect formally decreed enforcement of Brice's recommendations on 22 April. Taffin de Sorel, the sole member of Anzin's board of directors in the Nord at the time, wrote to the prefect on 6 May to plead "in the name of humanity and a multitude of poor miners without work" for a softening of his order until the company had had time to make for itself the necessary lamps. The prefect naturally enough refused to rescind his order closing the pits and answered that the general director of mines had pointed out to him that M. Dumotiez of Paris made lamps as good as any that could be bought in England. Renard replied that, although the company had ordered one hundred safety lamps from another Paris firm the very day after the disaster, he had discovered that none of the "numerous" French firms manufacturing them could deliver more than tiny quantities in the near future.[125]

The company found itself forced to make the lamps in its own workshops in order to reopen most of its pits. By 16 July the workers at the four pits on the southern faisceau that had remained open because they were free of grisou were using the company's new lamps, as were those on three of the pits closed by the prefect's order in April. All together, 918 miners on these seven pits were equipped with 1,015 lamps. The number of workers there was rising toward normal, and the other four pits still closed would reopen as more lamps became available. By October most of the pits were evidently back in operation at full capacity, for 1,800 miners, all with safety lamps, were working in the company's grisou pits at that date.[126]

The introduction of the new lamps had led, however, to serious difficulties. The first group of workers equipped with them refused to make their descent. They claimed that the lamps shed too little light and especially objected to the strict enforcement of the rule against smoking in the pits. The company quelled this "mutiny" by substituting other miners from another pit who descended quietly and without objections. The next day the first group "begged" the company to allow them to take up their old jobs again.[127]

A year later the Bureau of Mines, taking cognizance of recent "terrible accidents" in gassy mines due to "routine and lack of watchfulness," despite the fact that the means of preventing them were well known, cheap, and easy to use,[128] issued instructions on the use of safety lamps. The "Instruction" gathered together for the first time the numerous notes on safety lamps that had appeared in the *Annales des Mines* since 1816 and combined them with the provisions of Title II of the imperial decree of 1813. The decree had detailed the responsibilities of owners and mining inspectors in preventing accidents, the criminal penalties for noncompliance, and the emergency powers of the prefects in urgent cases of negligence or accident.[129] In accordance with the detailed information contained in "Instruction," which in effect justified the closing of Anzin the previous year, the Nord prefect issued a revised order in September 1824 which permitted the use of candles or open lamps at Fresnes and Vieux-Condé but authorized the working of all the other company pits only on the condition that the miners use Davy lamps. He tacitly admitted the failure of French industry to provide the necessary lamps by ordering Anzin to make its own, and he also ordered Anzin to rearrange its ventilation system so as to direct the air current from the bottom to the top of the workings, with the results already noted.[130] The prefect issued in 1825 a third and final order, "Lighting and Ventilation in the Nord." It was intended to be a permanent guide to good practice along the lines indicated by "Instruction" and the various inspectors' reports since 1823 on the specific problems at Anzin.[131]

Although the order that safety lamps be obtained before the gassy pits could be reopened was put into effect under the now watchful eyes of the alerted mining inspectors, accidents due to the grisou continued to occur frequently during the next ten years. In May 1824, even before the first revision of the prefect's regulatory order, one worker was killed and five others nearly asphyxiated in Chauffour again. The worker, trying to extinguish his lamp in a heavy concentration of gas, had attempted to blow out the flame instead of turning down the wick and so had brought it into contact with the gas across the gauze.[132] Two years later, three workers died because, despite regulations, they had not been using Davy lamps. They had been working at shallow depth in an area near running water where the presence of the grisou was "unprecedented."[133] Apparently neither the company, the inspectors, nor the workers themselves were prepared to abide by the strict letter of the regulations set down by the prefect in 1824 and 1825. In this case, their combined experience and judgment proved fatal. In 1827 three more miners were seriously burned, although not fatally, despite their being equipped with unopened Davy lamps.[134] Finally, in 1832 the limitations of the safety lamp and the extent to which incalculable accident played a part in the prevention of explosions were pointed up by the death of ten miners, killed when a sudden shift in atmospheric conditions caused a rapid concentration of gas before the miners could evacuate the pit.[135]

Thus explosions could occur even when miners used safety lamps correctly, as they often did not. Anzin was not the only company that had fatalities after the introduction of the safety lamp and the renewed sense of mission on the part of the government. At Aniche, in 1827, eleven miners were killed by asphyxiation, and in the Gard and Loire other deaths were reported.[136] In fact, the thirty-eight men who died by the grisou at Anzin between 1816 and 1833 were part of an overall safety record that was superior to that of other mining companies of the period.[137] Moreover, as pits pushed deeper and deeper during the nineteenth century, explosions continued to occur with ever greater frequency and on an ever greater scale throughout the European mine fields. It can be argued that the new lamp, far from decreasing the accident rate in the mines, raised it in the long run by encouraging the mining of seams which companies would otherwise have considered too dangerous to work.[138] Like the Newcomen engine in the eighteenth century, the Davy lamp at the beginning of the nineteenth saved the mining industry. In the process, however, it proved to be not so much the miners' friend as the ally of the Industrial Revolution with its insatiable demand for coal.

4 Expansion of Workings and Concessions

In the early 1820s Anzin faced perhaps its most serious crisis of the early nineteenth century. Its reserves of good forge coal were running low just as the demand for coal broke sharply upward. Looking back to those critical days, Casimir Perier would later claim, "We were not able to extract all the forge coal we required. The pits of old Anzin . . . were becoming exhausted with frightening speed," and as the search for the remaining coal sent shafts ever deeper, extraction became ever more costly and difficult. "Most of our customers, put off by our frequent inability to supply them with either the kinds or quantities of coal they asked for, left us and supplied themselves at Mons; those still attached to us by old habits ceaselessly manifested their repugnance for the bad quality of our coals."[1]

Perier no doubt somewhat exaggerated the immediate seriousness of the problem out of a natural desire to magnify the eventual success of his solution. Anzin was in no immediate danger of extinction. Reserves in the already exploited parts of its concessions would probably have sufficed to maintain production levels for another decade. And despite the universal "repugnance" for Anzin's forge coals, they would continue to find a market so long as the company remained shielded from its competitors by high tariffs and transport costs; in a seller's market Anzin could continue to hold its own. Nonetheless, Perier was basically correct; the problem of reserves was critically urgent for Anzin's future expansion and prosperity. Moreover, time was pressing. Years would pass before explorations would pay off in coal extracted, no one could be certain that explorations would turn up fresh seams, and, where untapped reserves were known to exist, the technical obstacles to extracting them were formidable. Racing time in an attempt to parry the potential threat of cheaper transport or a change in tariff policy and to tap Anzin's share of a growing market for forge coal, Perier pursued a determined policy of rapid expansion.

Two decades earlier, in the opening years of the century, Stanislaus Desandrouin, following his own vision of Anzin's future greatness, had also attacked the Torrent, explored the company's vast territories, and sought new concessions. Although Desandrouin had failed, thwarted by technical difficulties and his own lack of ability and perseverance, his defeats marked the path Perier was to follow and laid the foundation for his successor's greatest triumphs, the passing of the Torrent and the discovery of the rich seams near Denain.

Stagnation on "Old Anzin": The Northern, Lower, and Anzin Cantons

The company's chief workings on the Anzin and Raismes concessions clustered in a shallow arc some three kilometers long and half as deep along the Escaut River on either side of the citadel of Valenciennes (see maps 1 and 2). Of the five more or less separate cantons, or groups of pits, in this area, the least important by 1833 were the two on Raismes, the northern and lower cantons, and the Anzin canton. For both commercial and geological reasons, the northern canton was never an important center of expansion, with the exception of a few years after 1800. The company discovered seams of semibituminous coal in 1783 at the pit Bleuseborne, but the niveaux were so abundant and so difficult to pass that even with "enormous expenses" the pit produced no coal until 1788.[2] Three attempts in the years after 1798 to sink a shaft near Bleuseborne failed before a successful effort in 1805 somewhat to the west at the new pit Moulin eased the drainage problem at both places.[3]

Once solidly established at Moulin and Bleuseborne, the company sought to extend its knowledge of the surrounding area by explorations in all directions over the next thirty years.[4] All these attempts proved futile; the northern seams were isolated. Only to the west was the continuation of the Bleuseborne-Moulin seams found at the pits

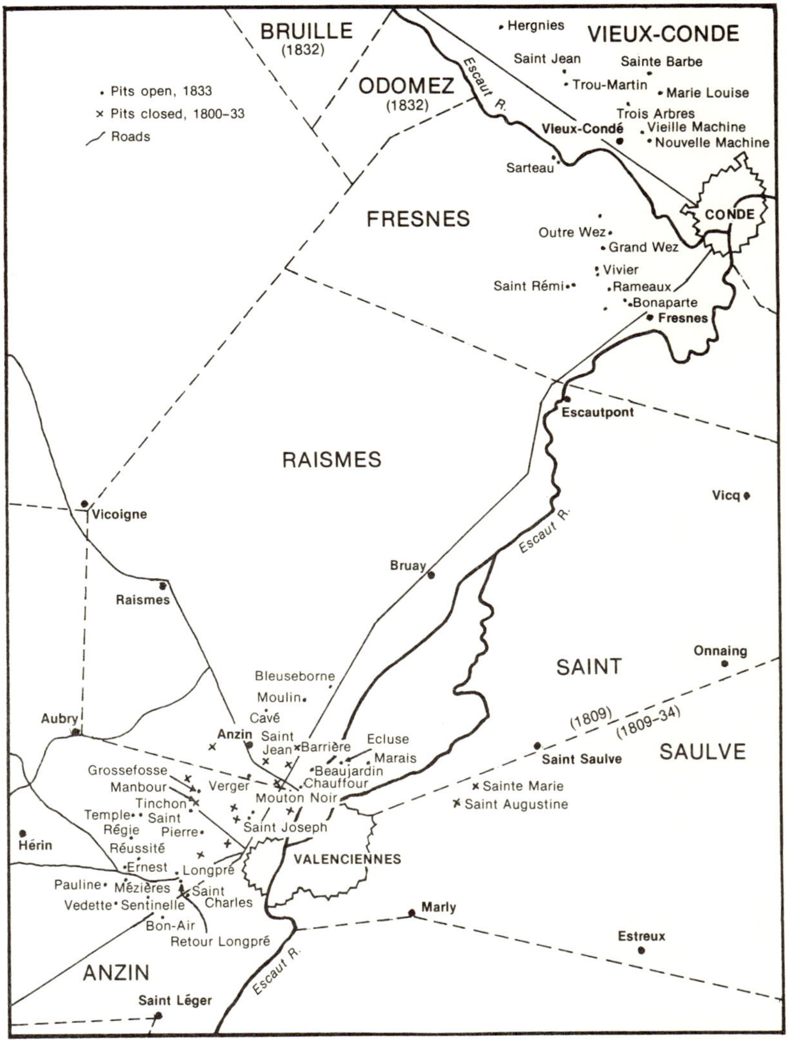

Map 2

Detail of Anzin Company, ca. 1833. Only the most important pits on Fresnes–Vieux-Condé are shown.

Cavé (1816) and Saint Louis (1821). Even there, however, the niveaux were dangerous, drainage costs high, the seams broken, and the coal stony.[5] In any case, this harder, less volatile semibituminous coal was much more difficult to dispose of on the market than the bituminous type found to the south across a band of sterile rocks. For these reasons, except for brief periods as in 1816, Anzin never pushed its activities in the northern canton with much energy after the initial impetus of the years 1798–1805 had died away. Cavé was only "about" to go into production in 1819, three years after the initial steps had been taken.[6] The board suspended work on Saint Louis in 1822 because there were "enough pits on the northern canton . . . to supply [its] market," and it was still incomplete in 1825.[7] Between 1825 and 1832, while the company expanded and improved its workings in the area and its four pits continued to produce a substantial amount of mediocre coal, sold mainly for domestic heating, the northern canton remained a backwater isolated from the more active areas to the south.[8]

The lower canton was more important than the northern one, mainly because the coal was of forge-type bituminous, although not always of the best quality. In addition to Poirier, Chauffour, and Beaujardin, three old pits along the north bank of the Escaut River, the company sank Marais (1782) and Ecluse (1806) in the marshy ground to the south between the two arms of the river. Exploratory galleries that were pushed out north, east, and south from Beaujardin and Marais (1799, 1812, 1831) failed to discover any new seams beyond the canton's frontiers. In 1831, the board of directors ordered a halt to all further exploration and closed Beaujardin in 1833 and Marais in the following year.[9] At Poirier a collapse of the cuvelage that, because of expert emergency repairs, just missed letting the Escaut into all the pits of the canton by "a couple of hours," led to the abandonment of Poirier in 1823.[10] Only Chauffour on the north bank and Ecluse in the marshes to the east remained active in the lower canton, which, like the northern one, had opened the century with great activity and high hopes.[11] By 1833 the company had resigned its expectations for future expansion in both cantons and had accepted the minor role these cantons were to play in producing a steadily declining share of the company's output.

Just to the north and west of the lower canton, in an angle formed by the junction of the roads from Valenciennes to Saint Amand and

Condé, Jean-Jacques Desandrouin had discovered the first forge-coal seams in the Nord in 1733, but the original pits were pretty well worked out by the early nineteenth century. La Croix was apparently inactive by 1812 and perhaps since the Revolution,[12] and Saint Jean closed down in 1821. Southeast of Verger and just west of the lower canton near the citadel of Valenciennes were the pits Mouton Noir and Barrière. Here too extraction was difficult and costly, and the coal contained much sulfurous matter and stone. The company found the coal unsaleable by itself, and fearing to hurt its commercial reputation by mixing the coal with that from other pits, it decided to close Mouton Noir in 1807 and Barrière in 1811.[13] Anticipating the need to replace them, the company had opened the double pit Saint Joseph to the west in 1803.[14] Although Saint Joseph produced some good forge coal, much of it was sulfurous as at the older two pits.[15] Saint Joseph and another new pit, Verger, which was opened in 1797 west of the Saint Amand road on the boundary between the Anzin and Raismes concessions, remained the only pits in this area where faults were common; much stone was mixed with the coal and the abandoned and unmapped workings from the previous century constituted a serious hazard.

The Attack on the Torrent: The Saint Vaast and Torrent Cantons

Only one pit, Grossefosse, was open at Anzin west of Saint Joseph and Verger in 1800, although the pits in this area had made the major contribution to the company's expansion between 1757 and 1791. Henri and Saint Christophe were worked out even before the Revolution; Pied and du Bois were not deemed worth reopening after the Revolution; and Manbour closed for good in 1808.[16] Still further west, on the edge of the Torrent, another group of pits (Tinchon, Saint Pierre, Longpré, Mézières, and Temple), dug in the 1760s and 1770s, had reached the waters of the Torrent. Only the first three had begun production before the Revolution and then at a very low level; their revival after 1800 proceeded slowly.[17] Even in 1813 there were still only four pits in extraction on the Anzin concession as compared to nine on Raismes.[18] From 1800 to 1820, Anzin pits (Saint Vaast and Anzin cantons) regularly produced only one-third as much as Raismes pits (northern and lower cantons).[19] While the company expended considerable time and money in attempts to push operations eastward from the northern and lower cantons, westward expansion toward the Torrent raised both greater hopes and, for many years, greater despair for the Anzin Company's future.

The company first discovered the Torrent in 1764 while trying to sink the pit Temple. This extraordinarily thick layer of water-logged shifting sands that overlay the coal measures also halted efforts at Mézières and Longpré in the same year. At first the company tried by continuous pumping to deal with the Torrent as it had dealt everywhere with the lesser niveaux of the upper layers of rock, but apparently the success of the other pits of the 1760s further to the east prevented any prolonged effort at the abortive Torrent shafts.[20] At the end of the Revolution, then, the company was not really certain of the extent of the Torrent or whether it could be mastered by traditional methods.

In the years after 1798 the company renewed its efforts there with vigor under the leadership of Desandrouin. Convinced that the veins of forge coal extended straight west from the known seams at Anzin, he hoped to leap the Torrent by establishing a new pit, Désirée, nearly one kilometer beyond Temple, and renewing drainage efforts at Temple to facilitate the sinking of the new pit. In 1800, while Désirée was being sunk, a blow of a worker's pick accidentally broke through into the Torrent. With a terrible stench, the water rushed forth with such force that in ten minutes it rose a hundred feet up the half-finished shaft. Similar trouble at Temple did not discourage Desandrouin; although this "prétendu Torrent" had "frightened the workers for a long time" and although the difficulties there had previously been judged "insuperable,"

> the concessionaires have decided that these operations will be resumed and manned with all the skill and courage our foremen are capable of . . . since all the seams certainly run off in this direction. . . . Citizen Desandrouin especially brings to this job all the interest with which explorations inspire him, [and] there is not an inch of the terrain he does not know or a movement of the waters he cannot calculate.[21]

These boastful words seemed borne out the next year when several pumps working together continuously appeared to lower the water level at Temple, but both there and at Désirée attempts to deepen the pits to the coal measures met only broken, shifting, and watery terrain. By 1802 the company's optimism was flagging, and by 1803 it admitted in its annual report to the Bureau of Mines that the water was proving "undrainable."[22] In 1805 the government engineer Daubuisson pointed out that the Torrent was not "prétendu" and that it was radically different from and in no way connected with the niveaux with which the company reports continued to equate the Torrent. He added in a footnote that "the obstinancy of one man who has a great deal of influence in the direction of these mines forces the continuation of an expense disapproved of by everyone."[23]

Perhaps this Desandrouin wished to repeat the triumphs of his father, whose success had often been attributed to stubbornness in the face of jeers and doubts. Both Desandrouins had owed much, however, to the expert knowledge and experience of their technical directors of the Mathieu family, whose interest in Anzin had lapsed through emigration during the Revolution. The Mathieus's traditional post of works director had disappeared with them, and its functions combined with those of commercial manager in the single post of general agent for all the company's affairs. Renard, the general agent, was primarily a businessman, and the several younger men now in charge of various areas of operations, while adequate for supervision of day-to-day work at the face or the sinking of the average new pit, were too inexperienced, too ill-trained, or too lacking in influence either to bring the Torrent project to a successful conclusion or to cut short the company's losses there. The project was finally given up altogether in 1807, about the same time efforts in the northern and lower cantons ceased for a time. For the next twelve years, Temple and Désirée lay abandoned and neglected while the company turned to surer projects to the east.

In 1791 there had been about fifteen pits in extraction on Anzin and Raismes.[24] In July 1805 there were only eleven, but five more were being sunk and were expected to be in extraction by the end of the year.[25] Two of these, Ecluse and Saint Joseph, were well to the east on proven ground, but the other three pits, located on the edge of the Torrent along the road from Valenciennes to Cambrai that extended southwest from the town citadel, quickly ran into trouble. The reopened Longpré and its double, Retour Longpré, together with the new shaft Saint Charles experienced strong influxes of water. Standing far out to the west on the continuation of Longpré's seams, Sentinelle, a southern outpost on the Torrent like Temple in the north, was officially opened in 1806, but even more than at Saint Charles and Longpré the waters of the Torrent continued to block extraction.[26] None of these three new pits was a significant producer, and all three were periodically put in suspension and probably did not begin to return initial costs until the 1820s.[27] Moreover an awareness was growing that the great fault Grand Retour, which cut off the southward extension of the seams of the lower canton at Marais, Ecluse, and Beaujardin, extended westward past Mouton Noir and Saint Joseph to Saint Charles. The Torrent could not be skirted by pushing south.

Evidently to compensate for failing hopes on and near the Torrent, two older pits further east, Tinchon and Saint Pierre, not reopened after the Revolution, were ordered into production again in 1812 at the same time work began once more at Saint Charles. Here too, however, production seems to have languished until the 1820s.[28]

The Anzin Company had proclaimed in 1803 that it was in the midst of a program of rejuvenation and expansion. The program included the Torrent project, the reopening of pits closed since the Revolution, and the deepening of all pits on all its concessions. Such an effort, Renard pointed out, was very expensive and could not be "l'affaire d'un jour."[29] Despite delays and outright failures, the company had accomplished a good deal by about 1811. To the east on Raismes, Bleuseborne and Moulin had solidly established production on the semibituminous faisceau for the first time, and Ecluse had temporarily rejuvenated production on the lower canton. Production on Raismes was almost one hundred thousand tons at ten pits employing over one thousand workers below ground.[30] On Anzin, the results had been less solid despite more extensive and more spectacular efforts. Production was only a little more than thirty thousand tons at nine pits with six hundred workers.[31] Results at both the old and new pits east of the Torrent had been variable but mediocre on the whole. On the Torrent itself, the company had exerted great efforts for slight rewards. It was still struggling for a foothold on the watery frontier along the great arc of outposts running from Saint Charles and Longpré to Sentinelle on the south and thence northward past Mézières to Temple. Yet only at the first of these pits was there any hope for extraction in the immediate future, and at the last two the situation appeared hopeless. The enterprising leap in the dark at Désirée had failed utterly.

Production had risen slightly after 1807 but was to return to earlier levels after 1813 until 1818. Although the company's chief hopes in 1800 had lain in pushing its operations westward as far and as fast as possible, it had gradually fallen back to the east as one effort after another failed. From Désirée and Temple in 1800 to Sentinelle and Saint Charles in 1806 to Saint Pierre and Tinchon in 1812—the stages of the retreat were plainly marked.

Immediately following the major reconstruction after the Revolution, the years of unrewarded toil against the Torrent appear to have sapped even Desandrouin's will to succeed. The company accepted the Torrent as in fact "undrainable," a permanent barrier to westward expansion. For some twelve years after 1807 it was content to tap gently at its edges in the hope that an inexpensive miracle would retrieve something of past losses.[32] Between 1813 and 1818 it attempted about ten exploratory efforts, apparently soundings, either in the triangle Saint Pierre-Longpré-Temple or to the south of Saint Charles across the Grand Retour fault.[33] The suspicion that these efforts were rather half-hearted is confirmed by the far-fetched decision taken in 1816 to attempt the drainage of Temple by a windmill for reasons of economy.[34] The windmill episode represented a last wish to try easy solutions before the company turned to negotiations for powerful new Edwards steam engines, and in somewhat the same way the tentative soundings at the edge of the Torrent were the first steps toward the revival of the Torrent project on a larger scale than ever before.

The renewed aggressive policy of westward expansion in the twenties was the result of two factors: the company's growing difficulty in meeting the rising demand for good forge coal and the replacement of Desandrouin by the Periers and their Parisian group as the real leaders of the company. Desandrouin was seventy-seven years old in 1815, and his personal financial situation was thoroughly snarled. Moreover he had become increasingly ineffective and unimaginative in his unofficial but well-defined role as the company's chief policy maker since the Revolution. Over the next six years, this role was gradually assumed by Scipion Perier and his brother Casimir.

The Periers seem to have come to the conclusion that the high profits the company continued to earn masked serious weaknesses and that Desandrouin's lax leadership was insufficient in the face of an impending crisis. Anzin production had continued to stagnate below prerevolutionary levels even after the political separation of the rival Mons field in 1815 and the moderately protective tariff of the next year. By 1818 the industrialists of northern France were buying more coal, but imports from Belgium, easily leaping the tariff barrier, filled much of the new demand. The opportunity provided by the tariff and by the increasing demand for coal found Anzin unable to expand rapidly enough its supply of the good forge coals required by the new consumers, and the company's tariff-protected prices put it in an unfavorable political light. Increasingly, local industrialists accused Anzin of operating a vicious and unpatriotic monopoly, and these men were not slow in putting pressure on the French Chambers to reduce or abolish "the Anzin tariff." Moreover, even its best coal was universally believed inferior in quality to Mons coal. Anzin's clientele of wholesale merchants sometimes felt forced to buy Mons coal because of their own customers' preferences.

It may briefly be pointed out that the company never ceased to make large profits and a high rate of profit during every year from 1800 to 1833 and that these profits were sufficiently large to permit the financing out of current income of several large-scale technical improvements, the rehabilitation and maintenance of old workings, and the development of major new ones, with enough left over to permit a regular policy of large dividends. It also maintained output levels, or nearly so, even through the most serious years of the supposed crisis in the early 1820s. Imports from Mons grew but only enough to satisfy most of the new demand created by industrial development after about 1818. The problems facing Anzin at the beginning of the 1820s were real enough, however, especially in their long-run implications and in comparison with the profits that might be made on an expanded output. Furthermore, by increasing production Anzin could avoid much of the opprobrium it suffered as a so-called monopolist while in fact strengthening its monopoly. The timid, conservative management of a Desandrouin ill suited the considerable dangers and even greater opportunities confronting Anzin in 1819.

Under the leadership of the Periers and especially of Casimir after 1821, Anzin embarked on a program of severe operational economies coupled with large capital investments in technical improvements, such as the Edwards engines and railed haulage, but the most extensive and significant effort to meet the new challenge was the opening to production of new workings of good forge coal. The Periers were not especially knowledgeable about mining technology, nor had they been personally very closely involved in the prolonged futilities on the Torrent or the failure of explorations to the east and south during the first years of the century. One must suspect, therefore, that much of the audacity with which they renewed the attack on the Torrent and the exploration of the company's broad concessions was born of ignorance coupled with a stubbornness of will worthy of either of the Desandrouins. Realizing that since before the Revolution all attempts to pass the Torrent or to find bituminous seams elsewhere in the Anzin concessions had failed ignominiously, the company's mining experts were unanimous in regarding any further efforts to reach new forge-coal seams as totally useless. Apparently most older members of the board of directors were no more sanguine about the success of the Periers' plans. Only Renard, the company's general agent, was later credited by Casimir with having supported his aggressive policy of expansion "au milieu de l'incrédulité," and Renard, like Casimir, was a businessman and no technical expert.[35] Yet expansion was basically an intelligent and foresighted policy designed to meet the host of commercial problems facing the company by attacking them at their root—the pressing need for more and better forge coal.

In the next decade the company achieved success in the search for forge coal in four areas of varying importance. Three of them were new, the most immediately important being the Torrent area itself. The other two were yet farther west in the Anzin concession, at Denain on the western edge of the Torrent and at Abscon not far from the Aniche Company pits. Even the new Torrent pits, however, did not significantly enter into production until some time after 1826. The Denain workings, although very rich, contributed little until 1830, and Abscon was still far from fulfilling its initial promise as late as 1860. All the more important, then, in meeting the immediate needs of the critical years was increased production in the fourth area, the pits of "old Anzin" already discussed, although in the documentary record this less spectacular achievement has left only passing references.

At the outset Anzin described the projected new attack on the Torrent in fateful tones:

To the south of the concession [of Anzin] the company is going ahead with what is perhaps the boldest attempt ever undertaken by this establishment. It consists in trying to master a second mass of water [i.e., other than the niveaux], improperly named "Torrent," which rests directly on the coal measures at a depth of forty meters below the usual surface in this area. . . . This vast project, which previous administrations have tried vainly to accomplish, will assure possession of the finest faisceau of seams in the whole region, if we are lucky enough to succeed, but it will also require a heavy investment of funds.[36]

It is a question of draining a subterranean lake covering all the coal seams to the west of the Anzin concession, an operation without which it would be necessary not only to renounce forever the working of these seams but also to witness the end of any production in this concession in a few years, because the northern [and eastern] part, where extraction has gone on for eighty-six years, has been pierced to a depth of nearly 1,200 feet, a limit at which men must stop. The drainage of this lake is perhaps at once the most hazardous and boldest enterprise ever conceived till now by any mining company. We estimate the extent of this mass of water at two square leagues on the basis of soundings and unsuccessful attempts [at sinking pits], which have cost us half a million [francs] in sheer loss over the last fifteen years.[37]

The plan to pass the Torrent was indeed massive and imaginative; it looked to the complete drainage of the Torrent *from below*. In this way, the company would bypass the difficulties encountered in sinking a pit in saturated shifting sands and in pumping vast amounts of water to the surface on the spot. Instead it would throw the burden of drainage on a pit safely established beyond the reach of the Torrent. The first step was to sink a new pit called Bon-Air (1819) south of the Grand Retour fault below Sentinelle and Saint Charles and just east of the edge of the Torrent, located beyond the coal measures in dry, easily pierced limestone. At a depth of ninety meters workmen drove a *traverse* ("horizontal gallery") northwest below the Torrent. One can only imagine the Periers' anxiety and doubt as the tunnel crept north toward Sentinelle.[38] As late as May

1822 the outcome was doubtful: "In the present uncertainty about the possibility of rejuvenating the workings of Anzin by pits toward the west . . . because of the Torrent," an attempt to establish a pit near Abscon was to be initiated as soon as possible.[39] At the same time, the tunnel having now reached the coal measures below Sentinelle, the directors gave the order for the key experiment of driving a bore upward from the tunnel through the Torrent to the bottom of Sentinelle, still stalled at a depth of forty meters after sixteen years. The bore was successful, for three months later they ordered Sentinelle deepened to the drainage gallery and the gallery itself extended toward Temple by way of another abandoned pit, Mézières.[40]

Work continued slowly into 1823. Again the directors ordered Sentinelle's shaft deepened, this time below the level of the drainage gallery into the coal measures, and preparations were made to begin extraction there "in the near future." The gallery reached the foot of Mézières.[41] Yet the drainage tunnel plan, while apparently assured of success in the long run, was proceeding too slowly to solve "the pressing problems of Belgian competition, lack of forge coal, and the growing demands of commerce."[42] The company therefore revived its efforts to pass the Torrent directly from above at the Temple, which was believed (wrongly) to be placed in the center of the richest extension of forge-coal seams. Casimir Perier by implication attributed to himself the decision to "attack with vigor" at the Temple, for, probably over the objections of the company's veteran technical employees and perhaps of the general agent, Renard, as well, he later stated that "the impulsion was given by the board of directors," which he himself dominated.[43] For a man of impatient character, avid of quick results once he had fixed his mind on a goal, the decision was a natural one. Early in 1824, a second large drainage engine was placed at Temple, and by September the workers there received bonuses for their efforts and courage in passing the Torrent.[44] The board credited this sudden sparkling success, seemingly against all the odds piled up by the failures of the previous sixty years, to "a firm will, a better maintained drainage system, and the use of a new and ingenious mobile cuvelage." The new cuvelage, it maintained, would henceforth permit the passage of sands and water of any thickness.[45]

By 1825 the line Temple–Mézières–Sentinelle–Bon-Air was solidly established on the coal measures along the drainage tunnel leading to the pumps of Temple and Bon-Air at both ends, but the "Torrent canton" was not yet helping solve the shortage of forge coal. While production on Raismes had held at just over one hundred thousand tons since the early years of the century, production on Anzin had actually fallen in the years 1820–22 from about one-third the Raismes level to less than one-fourth. The sharp upward movement on Anzin thereafter owed very little to the new pits until about 1829. There were eight pits most commonly in extraction on Anzin between 1823 and 1828. One was Pensée in the far west at Abscon, and two others were the veterans Grossefosse and Saint Joseph. The four most responsible for rising output after 1824 were Tinchon, Saint Pierre, Saint Charles, and the double pit Longpré–Retour Longpré, all pits on "old Anzin" put into full production for the first time during these years.[46] Despite the improvements of the 1820s in winding and haulage, these pits, like those on the lower and northern cantons, were working at maximum depth and at near maximum distance from the shaft by 1830.[47] One could hope for no further expansion from them, and exhaustion of their seams in the near future appeared likely.[48] The burden of forge-coal production was about to pass to the Torrent and Denain.

The only pit officially opened for extraction on the Torrent by 1825 was Sentinelle, and the quality and quantity of the seams there was so disappointing and production so small that there was serious discussion of abandoning it. Joseph Perier, Casimir's brother, and Count Hocquart, another of the Paris group, called a special board meeting at Paris on 5 August 1825 to consult with the new general agent, Paulze d'Ivoy, a Perier appointee, in order to find a solution for the company's inability in the near future to fill all its orders for forge coal. They drew up an exceedingly detailed program approving or revising existing plans or proposing new ones, all aimed at pushing the Torrent project to a speedy and productive conclusion.[49] Between 1825 and 1827, the company began five new pits, Vedette, Pauline, Ernest, Réussité, and Régie, all on the Torrent plateau. Each of their shafts easily passed the Torrent, especially at the last three, where another long drainage tunnel to the pump at Longpré helped for the first time actually to lower its level.[50]

By 1828 these new pits, with the exception of Pauline, were ready for production, but evidently increasing output on the older Anzin pits was sufficient to supply most of the company's commercial needs, since Mézières, Temple, Réussité, Ernest, Régie, and Vedette were all declared inactive at the end of that year. The last three named were in production by the next year, but even in 1831 Pauline had no winding engine because the existing pits on the Torrent were capable of furnishing all the coal needed.[51] The company was naturally prone to exaggerate its difficulties in the new canton for purposes of obtaining new concessions from the government and went so far as to claim as late as 1830 that "it is impossible that we can long maintain ourselves" on the Torrent and that "numerous difficulties will not allow us to give our project [there] as great a development as had been planned."[52] Yet it was in this same year that, aided by the new pits at Denain, production on Anzin almost doubled and soared past the output on Raismes for the first time.[53] It seems evident that after 1826, or 1828 at the latest, it was the market rather than production capacities that set the limits on Anzin's production of forge coal. The determination of Casimir Perier had at last carried the Torrent project through to a triumphant conclusion and broken through the production bottleneck.

Two Leaps in the Dark: The Abscon and Denain Cantons

While the fate of the Torrent project hung in the balance, the company sought to find alternative deposits of forge coal by exploring the empty spaces of the Anzin concession. Various attempts beginning from the northern or lower cantons have already been mentioned. Farther west, exploratory tunnels that were pushed northwest from Temple as far as the boundary of the Raismes concession found nothing. Nor were similar attempts south of the Grand Retour from Bon-Air and Saint Charles any more successful. Much farther south, two soundings near Saint Léger in 1819 and 1824 reached the Torrent but did not find the coal measures beneath it.[54] There remained the immense but largely unexplored area to the west of the Torrent pits, especially the part of it more or less on a line between Valenciennes and the Aniche workings twenty kilometers away. Hopefully soundings could pick up the trace of the known seams beyond the Torrent.

Two earlier attempts at exploration west of the Torrent deserve mention, although neither was successful. In 1776 the company sank at least two soundings about twelve hundred meters northeast of the village of Escaudain in the center of the Anzin concession. Then in 1795 and 1796, in the very midst of recovery from the effects of the Revolution, five or more shallow soundings failed farther north and east near Hornaing. These failures, together with the apparent opportunities for expansion from the existing workings around Valenciennes and the increasingly conservative attitude of Desandrouin after his failure on the Torrent, discouraged any further leaps in the dark until 1822.[55]

In that year, again on the initiative of Casimir Perier, seconded by Renard, a sounding reached the coal measures near the village of Abscon on the far western edge of the Anzin concession not far from the pits near Aniche. Immediately, the company ordered the sinking of a double pit there as insurance against the possible failure of the Torrent project. Although something like the Torrent was found at 137 meters, workmen passed it without much difficulty, and the two pits, together called Pensée, began extraction soon after 1824.[56] The coal was of forge quality but excessively broken and filled with stone, and the seams were very irregular. Together with high drainage costs, these difficulties raised the cost of extraction well above that in the eastern Anzin pits. In fact the geology, coal, and working costs at Abscon resembled those at troubled Aniche much more closely than those near Valenciennes. For this reason, Pensée served mainly as a base for further exploration. It remained the sole pit in the area long after 1833, despite plans (apparently never carried out) to place a new pit there in 1830.[57]

The company made its final attempt to locate coal in the northern part of the Anzin concession near Wallers in 1827 in connection with explorations beyond Abscon. The soundings failed, but in failing helped confirm what the earlier explorations at Hornaing, Helesmes, and Aubry had already indicated—that the whole northern area of the concession lay outside the coal measures.[58]

By far the most important result of the western explorations during the 1820s was the discovery of seams of excellent forge coal in the area around Denain on the southern edge of the Anzin concession midway between the Torrent pits and Abscon. Encouraged by the discovery of the southern faisceau at Abscon, the company ordered two soundings in this area in 1824, one between Rouvignies and Hérin, the other just inside Anzin territory at the great bend of the Escaut River near Denain. Both were abandoned when they struck the Torrent, which was much thicker here than farther east. The company must have wondered then whether it would ever succeed in jumping the Torrent east of Abscon, but in 1826 another sounding somewhat to the west of the first two reached the coal measures while missing the Torrent altogether. As soon as possible the pit Villars was sunk there. In the next two years, all nine soundings made over a large fan-shaped area north of Villars reached coal, and two of them became the pits Turenne (1828) and Bayard (1829). A gallery emptying into the Escaut linked all three pits in a common drainage system. Between 1830 and 1833, the company began four more pits (Saint Marc, Jeanbart, Mathilde, and Napoléon) west of the original three, and by 1836 all seven were in full operation.[59]

Villars rested on the southern boundary of the Anzin concession; beyond it lay unconceded territory. In 1824, while making the first sounding on Anzin near Denain, the company had also sounded across the boundary to the southeast near Rouvignies. It struck the Torrent there as in its two more northerly soundings of that year. With the sinking of Villars between 1826 and 1828, however, a series of soundings and exploratory tunnels from Villars to the southwest established the existence of the continuation of the coal seams beyond the Torrent on the unconceded territory. In competition with the rival Dumas Company, formed expressly to explore the southern edges of the Aniche and Anzin concessions after Anzin's discovery at Villars, the Anzin Company obtained in 1831 a thin sliver of a concession adjoining the Anzin concession to the east and west of Denain in the area of its explorations. It immediately began sinking a shaft there.[60]

Even the pessimistic engineer Clère, who in 1829 was still doubtful about the success of the new Torrent pits and who foresaw the imminent exhaustion of all the rest of the company's workings, was forced to admit that the pits near Denain gave "positive hopes." The coal was of a better quality and could be mined more cheaply than that of the Torrent because the seams were flat and dry.[61] After 1830 and even more after 1833, the bulk of the company's output would come from the Torrent pits and those of Denain both above and below the old concession boundary.[62]

Second Priority: The Fresnes and Vieux-Condé Cantons

The workings on the concessions of Fresnes and Vieux-Condé had a character and history different from those to the south. Physically, nearly four kilometers of terrain barren of coal separated them from the pits around Valenciennes. Commercially, the semianthracite of Fresnes and Vieux-Condé had its own special, limited markets. Administratively, the two concessions constituted a separate unit. Geologically, their seams were relatively flat, thick, dry, and shallow; their coal was easily and cheaply extracted. Hence, pits there tended to be easily sunk and as easily suspended, and the capital investment involved was small. Since it was cheaper to sink a new pit than to run galleries out beyond about two hundred meters from the bottom of the shaft, the pits were closely placed and rapidly exhausted, and the number of pits was large relative to total production.[63] Expert observers of the company's operations usually ignored Fresnes and Vieux-Condé as being of minor importance from the point of view of either mining technique or significance for France's industrial future. Nevertheless, these two concessions were important enough to the Anzin Company fortunes, especially in the earlier years, to warrant an attempt to trace some general developments there, particularly the almost constant attempt to "renovate," "resuscitate," and generally rationalize the rather old-fashioned pattern of workings and to extend operations beyond the nearly exhausted known seams.

In the early years of the century Fresnes–Vieux-Condé produced about one-half as much coal as Anzin and Raismes from the same number of pits. The pits were old and scattered helter-skelter with little regard to the most economical placement of shafts. During the period 1800–1807, when the company was vigorously trying to expand and renew its operations on all its concessions, three new pits were sunk on Fresnes (Bonaparte, Saint Jacques, and Saint Rémi) and two on Vieux-Condé (Trou-Martin and Sainte Barbe). In 1810 the company determined to push its workings on Fresnes northward on the continuation of the seams nearing exhaustion at the pit Vivier (finally closed in 1826). As a result, Grand Wez and Outre Wez were opened in 1812 and 1817. Although production had remained stable, by 1820 the number of pits had decreased from ten to seven and from nine to six on Fresnes and Vieux-Condé, respectively, and about half the pits in operation were new ones.[64]

The condition of the workings, while somewhat improved, was not yet equal to that of the pits near Valenciennes. Under the critical eyes of the new Perier management, the old accusation of decrepitude was again flung at the operations on Fresnes and Vieux-Condé. Furthermore, their commercial situation was about to take a turn for the worse. The Belgian coal tariff of 1821, imposed in retaliation for France's failure to reduce its tariff of 1816, cut production by over 20 percent between 1820 and 1824; the very tariff that had helped protect Anzin's French forge-coal market from Mons was partly responsible for the closing of a large part of its hard-coal market in Belgium.[65]

Casimir Perier launched another project of renovation and expansion in 1822. Only the "necessary" pits were to be kept open, preferably the newer ones opened since 1800. In fact, only six pits, three on each concession, were in operation by 1824.[66] The company united the administration of the two concessions for reasons of economy and more rational management and reduced the number of drainage engines by combining the drainage systems of several pits. Despite these measures of economy, the existing Fresnes seams were rapidly becoming exhausted, and the pits of both concessions lay a long and expensive wagon haul from their loading docks on the Escaut. Hoping that the Fresnes seams bent northward from their known location so as to join the seams of Vieux-Condé across the Escaut, the company began explorations at a point on the river called the "Sarteau." Several times the board of directors, influenced by discouraging reports from the men in the field, was on the point of giving up the project. Probably on the insistence of Perier, efforts persisted to a successful conclusion, and a new pit was in full operation at the Sarteau by 1825.[67] Yet once again, the company was calling for a sustained effort directed at rejuvenation of the two concessions. Some steps were taken in the next five years, but in 1830 the same call went forth once more.[68] Evidently modernization of operations amid the immensely complex tangle of old abandoned workings was easier said than done. Probably the much brighter technical and commercial prospects at the Torrent and Denain in these years also led to lower priorities and second-best efforts being given to improvements on the hard-coal field.

Some success in expanding operations northward followed the establishment of the Sarteau pit. The direction of the seams there had indicated that the hidden seams connecting those known at Fresnes and Vieux-Condé might lie in the unconceded corner between the two concessions on the communes of Bruille and Odomez. Soundings on Odomez in 1829 reached coal, and a battle ensued between Anzin and two other companies for concession of the territory. As in the similar contest for the area south of Denain, the government issued two new concessions, and Anzin obtained Odomez. Anzin ordered a pit sunk on its new concession during 1833, but neither Odomez or Bruille showed brilliant prospects.[69] Other explorations in the tongue of Fresnes across the Escaut were totally discouraging.[70]

Despite a record of some partial successes and several outright failures in its quest for renovation and expansion of Fresnes and Vieux-Condé, the Anzin Company could congratulate itself on output figures that recovered their old level after 1824 and began to rise beyond earlier peaks after 1831 when the Belgian tariff was moderated. Production continued to rise throughout the 1830s, although at a slower rate than on the southern faisceau. By 1836 the proportion of hard coal produced had as yet barely fallen below 30 percent of the company's overall total, and output had almost doubled in absolute figures over the nearly static level of the first thirty years of the century.[71] Although the company's efforts at renovation and expansion were partial failures, the growing market for hard coals succeeded in preserving most of the importance of the venerable workings on Fresnes and Vieux-Condé.

The Struggle for New Concessions

Before the Revolution, there were only four mining companies operating in all of northern France: Littry in western Normandy, Hardinghen near Calais, Aniche, and Anzin. Only Anzin had a substantial output or served a more than purely local market. Until 1831, these four continued to hold their monopoly in the area, despite extensive explorations by a large number of hopeful new companies.[72] The French government tried to stimulate the search for new coal deposits through the provisions of the mining laws of

1791 and 1810 and through the counsel and encouragement of the Bureau of Mines. Anzin's resounding financial success encouraged new ventures,[73] and groups of local businessmen and other notables hoped to free their town, department, or local industry from dependence on high-priced foreign coal.[74]

The reasons for the almost universal failure of these efforts are clear. Possessed by a rather naive notion of the mineral riches to be found everywhere below French soil by whomever would take the trouble to dig a hole, most new companies failed simply because there was no coal under the soil of most of northern France.[75] In addition, they often lacked adequate technical skills, sufficient financial backing, or enough determination to find the coal that was there but that would be discovered only after 1830.[76] Some of the companies from Paris, Rouen, Lille, Douai, Valenciennes, Cambrai, and Mulhouse never began operations;[77] others spent tens of thousands of francs on one futile sounding or exploratory pit after another.[78]

On the whole, the four active companies were less forward in exploration than one might have expected. The smaller three were too burdened with their own organizational, technical, or financial problems.[79] Anzin was far more richly situated, but before the Revolution it had all it could manage in expanding its operations around Valenciennes,[80] while during the Revolution and its immediate aftermath it was totally occupied in saving itself from its enemies and repairing the ravages of war and governmental mismanagement. Then, after 1800 Anzin tended to seek expansion chiefly by exploring the geology of the vast unknown areas on its four existing concessions rather than by adding to the territory under its control.

Nonetheless, there were two periods between 1800 and 1833 when the company bent serious efforts toward rounding out its territory or anticipating inconvenient new competitors who proposed to set up shop as its neighbors. Between 1805 and 1810, at a time when its strenuous efforts to expand its old workings around Valenciennes were coming to an unsuccessful halt and it had achieved a degree of renovation on Fresnes and Vieux-Condé, Anzin bought up the rights to the Saint Saulve concession, obtained exploration permits for the areas around Thivencelles and Vicoigne-Bruille–Saint Amand, and pushed exploration in each of these. None of these efforts bore fruit, and by 1814 hopes in all these areas had virtually ceased.

Between 1825 and 1832, when the success of the new workings on the Torrent and Denain appeared assured, the company once more took up its explorations north of the old concessions near Bruille and also began probing the area south of Denain. Discoveries of coal in both places were a logical consequence of following seams found on the old concessions. Both areas were contested by rival companies, but the government shared out the disputed territories between Anzin and its rivals. The granting of concessions for Odomez and Bruille to the north and of Denain and Douchy to the south marked the beginning of a new rush of exploration and speculation. Both the earlier and later periods of exploration also found Anzin involved in attempts to buy up mines outside the Nord, while, as a token of its lack of interest in external expansion between the two periods, it refused the opportunity to buy out its old rival, Aniche, in 1819–20.

In the eighteenth century a veritable mining fever had followed the discovery of coal in the Nord. Ephemeral companies sieved the subsoil near Solesmes, Le Quesnoy, Bavay, Landrecies, Maubeuge, Marchiennes, Bouchain, and elsewhere. The only real success, at Aniche in 1778, seems to have marked the end of the fever in the Nord by focusing attention on the Pas-de-Calais as the presumed location of the continuation of the Nord field. Explorations on territories contiguous to the Anzin concessions near Bruille and Saint Saulve continued under the several reorganized Mortagne and Martho companies, respectively, but here too, although traces of coal were found in both places, efforts were languishing in the decade before the Revolution broke out.[81]

Nor was the period from 1800 to 1825 one of much active exploration in the unconceded parts of the Nord department. The only important new attempts were undertaken by companies in the southernmost district, Avesnes, which, unable to supply itself with Anzin coal because of the lack of water transport, continued to use charcoal for its metal-working establishments and was "dependent on foreign sources [of coal] for its heating."[82] Government mining engineers were never under any illusions about the possibility of coal being found there, however, and as early as 1808 they advised the local men involved to direct their efforts elsewhere.[83] An even less prepossessing venture at Lambersart near Lille led to the entirely false rumor that coal had been found in 1822.[84] Anzin paid no attention to these rather amateurish explorations far south or north of the coal measures. The lack of serious challenges by outsiders helps explain Anzin's concentration on internal expansion until the late 1820s when new rivals appeared, partly in response to new discoveries made by Anzin itself on the borders of its old concessions.

Perhaps the most explored, contested, and disappointing area of the Nord was the large thumb of territory enclosed by the Belgian frontier and the Escaut below Condé. The Desandrouin Company had sounded it in the 1720s and 1730s, and in 1767 Laurent, an engineer and owner of a small share in the Anzin Company, obtained a concession there. In 1770 the government turned the concession over to the Martho Company for thirty years on the popular grounds that Laurent's concession would further Anzin's monopoly of a commodity "of absolute necessity for domestic heating." By 1780 the Martho Company was preparing to abandon active operations. A claimed expenditure of 1,600,000 livres had resulted in sales of only 50,000 livres' worth of coal extracted from two watery pits, Sainte Marie and Saint Augustine, located along the road from Valenciennes to Mons. After a company reorganization in 1789, four more pits were opened, all of which proved unproductive. In the next few years, the Revolution ruined several of the owners financially, guillotined two others, and forced many of the rest to emigrate. When its concession expired in 1800, the company seemed dead.[85]

Then, in 1804, Thieffries de Beauvoir, one of the new members of the reorganized company of 1789 and a member of the board of directors of Anzin, set about reviving the defunct company under a new name. The Saint Saulve Company obtained a fifty-year renewal of its old concessions, but Thieffries's hopes of obtaining badly needed new financial backing on the strength of the new concession failed, as did his efforts to reactivate the pits during the next three years. By 1807 total expenditures of the Martho and Saint Saulve companies since 1770 had risen to 2,000,000 francs against only 100,00 francs in sales.[86] Thieffries, under pressure to relieve his associates of responsibility for any new debts the company might incur, bought up most of their interests for derisory sums in 1805 and 1807.[87]

It is probable that Thieffries had already sounded out the possibility of selling the company to Anzin, the only party which could possibly be interested, for reasons of its own, in taking over a losing proposition like Saint Saulve. Who actually opened the negotiations is not clear, but either party could easily have done so since Thieffries was in the inner councils of Anzin, a member of the dominant faction of owners headed by Desandrouin. It seems likely that Anzin had encouraged or at least approved Thieffries's part in the reorganizations of the Martho and Saint Saulve companies in 1789 and 1804. His leadership was a guarantee that this strategic territory on Anzin's eastern border would be in friendly hands. Now that Thieffries could not find fresh capital or continue to bear the astronomical losses himself, he naturally looked to Anzin to relieve him of his burden. Anzin in turn was prepared to shoulder the necessary losses itself when its more or less conscious puppet refused to assume them for it any longer.

At any rate, in October 1807 Anzin's board of directors authorized its general agent, Renard, to negotiate with the owners of Saint Saulve for its purchase at not more than 50,000 francs.[88] Thieffries drove a good bargain, and the two necessary purchases, from Thieffries and a relative by marriage, the widow Duquesne, cost Anzin 72,000 francs, including 13,000 for Saint Saulve's Newcomen pumping engine. The previous owners assumed responsibility for all of the old company's outstanding debts.[89]

Although Anzin now owned the Saint Saulve Company, the purchase in itself did not give it the right to mine coal on the Saint Saulve concession, a right belonging only to the original concessionaire until transferred to another by the government. Anzin bought up the Saint Saulve Company interests quietly by private contracts (31 October 1807 and 8 February 1808) in order not to alert the Bureau of Mines until it could present the purchase as a fait accompli. On 27 April 1808 Renard informed the government of Anzin's purchase and requested transfer of the Saint Saulve concession to the Anzin Company.[90] Some minor qualms on the part of various government officials over the danger of increasing Anzin's power to set coal prices in northern France were effectively quashed by the engineer Miché's remark that the Mons companies could supply better and cheaper coal than Anzin's to anyone enterprising enough to go and get it. As Miché put it, only Anzin had the "means and experience" to turn Saint Saulve to account without too much endangering its own prosperity.[91]

Anzin's interest in the Saint Saulve concession was natural. It touched on all four of the older concessions and completed the company's occupation of all the possibly productive parts of the whole Belgian frontier area. The active workings on the four older concessions were all near the Saint Saulve territory, and the company hoped to find on it the extension of their seams. Saint Saulve thus constituted a reserve for the future, a check to rivals along Anzin's eastern border, and the natural direction toward which to push the several explorations which Anzin had rather unsuccessfully tried since 1800.

In any event, the decision to buy up Saint Saulve proved unwise so far as hopes of finding coal there were concerned. For six years after the purchase Anzin put considerable effort into an attempt to explore Saint Saulve's chaotic geology. It deepened the six old pits, began two new ones in 1810 and 1813, and made new soundings at various points in the extensive territory (Crespin, Vicq, Quarouble, Rombies, Onnaing, and Marly). In 1812 new explorations were begun by means of underground galleries pushed out into the new concession from the pits Marais and Bleuseborne.[92] None of these approaches met with any success. By 1820 the company's efforts on Saint Saulve had become perfunctory, and any thoughts of further attempts there were ended by the costly new attack on the Torrent.[93]

It seems certain that the concession was ignored throughout the 1820s. Anzin's repeated statements to the government that underground explorations from Raismes were continuing can be explained by the necessity in mining law of justifying its continuing to hold the Saint Saulve territory locked away.[94] The engineer Clère declared Saint Saulve hopeless in 1829, just before Anzin finally found a use for its white elephant.[95] It abandoned the southeastern two-thirds of the concession in 1831 as bait for its claim to the disputed territory south of Denain. The government was exceedingly size conscious, especially about Anzin, which it had already permitted to exceed the legal maximum. Therefore the abandonment of most of Saint Saulve's acreage may have helped Anzin's case, although the Saint Saulve territory was known to contain only traces of broken seams and those mostly in the area Anzin retained, while explorations had already proved that the new Denain concession held rich deposits of easily mined, high-quality forge coal.[96] The remaining one-third of Saint Saulve, lying northwest of a line from Valenciennes to Crespin, remained in Anzin's possession until the government suspended the rump concession in 1845 because Anzin had ceased to attempt to put it to use.[97]

Anzin also showed considerable interest in extending its workings and maintaining its monopoly along its northern borders, especially in the pocket formed where the four older concessions approached each other. About the time that Anzin was beginning to maneuver for possession of Saint Saulve, it was starting explorations on the commune of Bruille. In 1806 it asked the government for exploration permits for the Bruille area and for Thivencelles, the unconceded "thumbnail" above the Saint Saulve concession bounded by the bend of the Escaut, the Mons-Condé Canal, and the Belgian border. Even before it obtained the permits in 1809, Anzin had abandoned a sounding between Vicoigne and Saint Amand, and four equally unsuccessful soundings on Thivencelles had resulted only in the loss in each case of the lower part of the drilling rod, snapped off in the bore by shifting sands.[98] Renard hinted that the company would employ "all its resources and efforts" in these two areas if the government would anticipate its expenditures by granting it concessions there, but the Council of Mines dryly commented that any concessions would be made in the usual way and for the usual reasons.[99] The company never made any formal demands for concessions, apparently because it never obtained "the usual reasons," that is, proof of the existence of workable seams. Something of Anzin's motives in these maneuvers was revealed by the decision taken in 1810 to renew its exploration permit in these two areas "in order to continue the soundings so that . . . the board of directors can judge if it is in the interest of the company, *in order to avoid having a dangerous neighbor*, to ask for a greater extension of territory in concession."[100] By 1814 both explorations and research permits had lapsed in both areas.[101]

In addition to the expansionist efforts on company borders, in 1807 the board authorized Renard to negotiate with one of the groups attempting to obtain a concession of "the territory called le Flenu." Le Flenu was located on the richest seams of forge coal near Mons, then in the Jemmappes department and a part of Napoleon's greater France. This attempt to arrange an association with a group of "Belgian" capitalists represented a venture outside Anzin's home grounds into what was in many respects foreign territory and into the very stronghold of its bitterest commercial rivals. Anzin apparently never followed up the proposal, but it is a further example of the company's expansionary verve in the first decade of the nineteenth century and an interesting illustration of the lengths to which it felt itself forced by the failure of the Torrent project.[102]

A somewhat similar proposal in 1809 led to a split in opinion between the older and newer factions of owners. The Fins and Noyant mines near Moulins in the Allier department of central France had been established in 1770 by Christophe Mathieu, son of the original Mathieu of Anzin and brother of Léonard Mathieu, who had served as works director at Anzin until his emigration during the Revolution. In 1809 Christophe's heirs invited Anzin to buy a two-thirds interest in the two mines for 100,000 francs. The purchase price was to be reinvested in the mines and Anzin was to have complete control over management. The board appointed Scipion Perier and Thieffries de Beauvoir to represent Anzin in the negotiations. Perier was the leader of the Paris faction of owners eager for the purchase, and Thieffries was perhaps the most "liberal" of the Desandrouin-Taffin group. They reported that the coal was of sufficiently good quality and easy extraction and had small but well-established markets at Orléans and Nantes by way of the Loire. The real incentives to purchase were a desire to acquire a foothold in the Paris market, which Anzin could not reach until the long-delayed completion of the Saint Quentin Canal, and a hope of finding enough iron ore intermixed with the coal seams to establish the kind of vertically integrated, low-cost iron-smelting operation that Frenchmen believed to be a chief cause of England's industrial might. Despite the favorable tenor of the report, the more cautious and provincial of the Anzin owners were unconvinced. After what seems to have been a lively argument among the members of the board, Desandrouin, backed by the Taffin interests, gave a "formal refusal" to participate in the purchase himself or to permit the Anzin Company to take part as such, although he gave his blessing to those members of Anzin who might wish to purchase Fins and Noyant as individuals. Accordingly, a Perier group bought the two mines in 1810.[103]

On the whole, interest in external expansion languished from about 1810 to about 1825. In 1819 explorations on northern Fresnes at the Sarteau led to a brief revival of interest in the Bruille area. Anzin contemplated asking the government to permit it to annex Bruille to the Fresnes concession, but its interest lapsed before any new soundings were attempted, probably because of the extent and greater importance of the reopened drive on the Torrent.[104]

The cost and attentions required by the crucial Torrent project may also help explain why Anzin failed to carry through successfully an opportunity to purchase the Aniche Company in 1819–20. If completed, the purchase would have been the most extensive and eventually the most profitable addition of territory since the original union in 1757. As with Fins and Noyant, and probably Saint Saulve, the advances came from a weak company whose owners were not prepared to put up the new capital necessary to find out whether or not known coal seams could be made to pay. Never prosperous even before the Revolution, Aniche had struggled vainly since 1800 to compete with Anzin and Belgian coal. Its extraction costs were high, its production low, its seams watery, and its equipment outmoded. Many of its owners were discouraged not only by its failure to pay dividends but also by the constant calls for more capital to meet unexpected disasters and to permit technical renovation. Neither its short-run nor its long-run prospects were bright in 1819 when several of its owners made "evasive overtures" to some of Anzin's owners concerning an association between the only two active mining companies in the Nord. Renard was instructed to keep the negotiations open and to obtain as much information as possible on Aniche's technical and financial condition. Anzin then proposed either an outright purchase for about 2,000,000 francs or an association of the two companies by which it would take over half ownership and exclusive management powers in return for an immediate outlay of 300,000 francs to step up production. It soon withdrew the offer of association, probably because it had no real interest in involving itself in such unpropitious operations at a time when it was launching its own expensive campaign on the Torrent. By outright purchase, Anzin would have rid itself of a minor competitor, but in the past it had found a weak Aniche a useful rebuttal to charges of monopoly. Purchase would also have protected Anzin's eastern flank against the threat of a revitalized Aniche bought by a group of Anzin's numerous enemies, but, like Saint Saulve, Aniche had apparently exhausted all alternative sources of capital before turning to Anzin. Therefore, the price offered by Anzin represented the value of eliminating a possible future danger rather than a calculation of present commercial profit. It was not sufficient to tempt Aniche's owners, who struggled along unsuccessfully for another twenty-five years before a transfusion of new men and money from Cambrai finally put the company on its feet in the 1840s.[105]

After 1825, with the Torrent project nearing completion, Anzin felt free to renew its interest in external expansion when it was stimulated to do so by the results of explorations near Denain and on northern Fresnes. Beyond the Sarteau pit on Fresnes, new and deeper soundings in the Bruille area reached an apparently workable vein of hard coal in 1828–29. This discovery led to a long and acrimonious dispute with two rival companies who had jumped into the area when Anzin's soundings showed promise.[106] The struggle was settled by division of the contested territory into two concessions on 6 October 1832. The southern concession, Odomez, went to Anzin, and the northern one, Bruille, was given to one of the rival companies. Neither of the new concessions proved profitable in the near future. The only productive pit opened on either territory for some years was located near Anzin's preliminary sounding on its Odomez concession.[107]

Far more important were the explorations to the south of the Anzin concession near Denain. There a group of merchants of Lille, organized as the Dumas Company, took the initiative. A successful sounding in 1829 led to a bitter contest between the old champion and the challenger. Again, the government ordered division of the disputed area. In 1831 Anzin received the part where coal had been reached as the Denain concession. The other and larger part of the area, without proven coal deposits, was conceded to the Dumas Company as the Douchy concession in 1832.[108] Since both new concessions occurred during the period when Casimir Perier was prime minister, it is not surprising that "the interests of the Anzin Company were not sacrified, but the government had to take into account the desires expressed by the [government] mining engineers and the Nord Chambers of Commerce that competition be stimulated by the establishment of new companies" (see map 3).[109]

Both of these new concessions developed rapidly. Anzin had placed a pit on Denain by the end of 1833, and Douchy discovered rich forge coal veins in 1834 and opened extraction the next year.

Map 3 Nord Coalfield, 1862

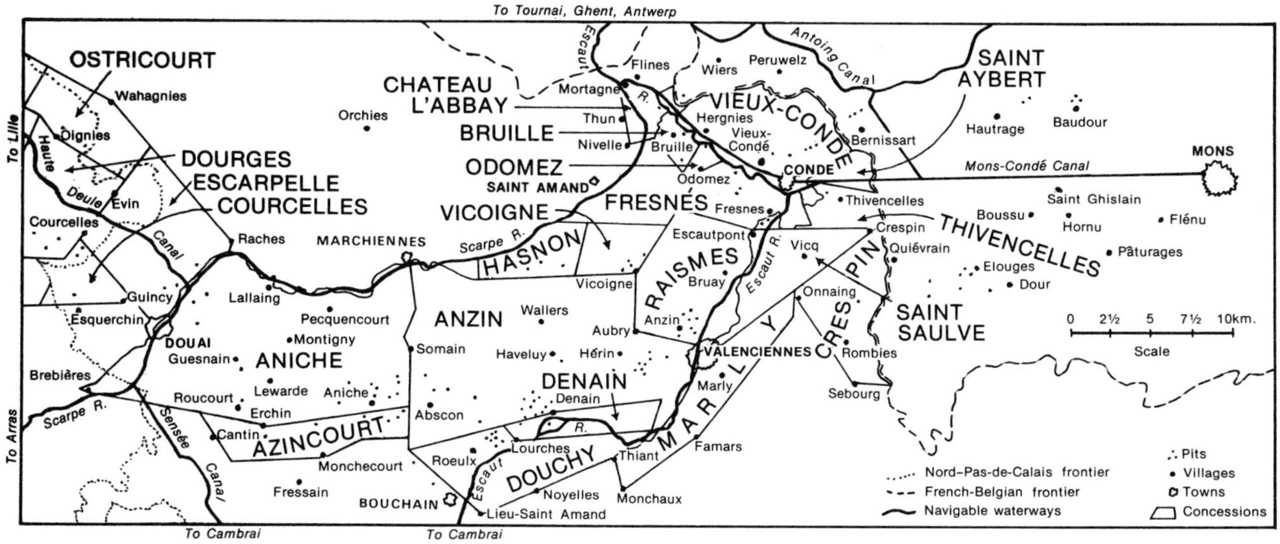

Douchy's success touched off a prodigious fever of mining speculation in the Nord. Anzin's sols almost never entered the market and Aniche's were not worth buying. None of the other dozens of companies which had sprung up and then disappeared in the past fifty years had ever given solid reasons for "mine fever," and by 1830 general public interest in mining companies and mining exploration in northern France was at a low ebb. Now, on the announcement of Douchy's discovery in 1834, the public's real appetite for mining ventures was quickly evinced. An almost hysterical wave of speculation lapped over the normally staid inhabitants of the Nord and the Pas-de-Calais. Douchy's sols rose to one hundred times their face value in a matter of days, and "Cambrai and Valenciennes gave witness to scenes recalling those of the rue Quincampaix."[110]

In the next five years, as Douchy's early success proved well-founded, dozens of new companies sprang up and began indiscriminately sinking soundings throughout the two departments. Most of these ventures were as unsuccessful as those of earlier years, but a few succeeded in establishing concessions and a few more by 1850 were prospering. Among the more prosperous were a refinanced Aniche and Azincourt. Azincourt's position and seams were related to Aniche as Denain's and Douchy's were to Anzin. Most of the other new concessions lay north and east of Anzin on territory already explored several times before 1833. Given somewhat better mining techniques, the concessions of Marly, Crespin, Thivencelles, Saint Aybert, Hasnon, and Château l'Abbay succeeded in finding coal there but not in extracting it profitably. Only Vicoigne, in which Anzin acquired a one-fourth interest in 1841, prospered on its good deposit of easily mined hard coal.[111] The speculative fever of the 1830s had barely died down and a disappointed reaction set in when the magnificent deposits of the Pas-de-Calais northwest of the Aniche concession were discovered in the 1840s and opened a new rage of company formation and exploration in the 1850s.[112]

In all this speculation and expansion after 1833 Anzin took little part. Its last important attempt to expand for many years was the purchase of a Belgian mine near Mons in September 1831. The reasons for the purchase of the "Nord du Bois de Boussu" are not clear. Perhaps the company wanted to obtain a foothold in the Belgian market at about the same time that it was establishing itself in Paris. Perhaps it was providing for a further reserve of good forge coal in the event the Torrent and Denain seams suddenly proved less satisfactory than they appeared to be. Perhaps, as its enemies claimed, it was seeking good Belgian *flenu* coals to mix with its own inferior qualities.[113] Or it may have been hedging against a reduction of the French coal tariff. Certainly, French investment in Belgian mines, especially by men from the Valenciennes and Lille areas, was far from rare.[114] At least one of the Anzin owners had bought a mine in the Liège area during the Revolution, and for "outsiders" such an investment represented one means of outflanking Anzin's hated "monopoly."[115]

In any event, the Bois de Boussu mine proved a disappointment. The seams were gassy and wet, and the system of extraction was primitive. Anzin was forced to introduce its own system of ventilation, supply Watt drainage engines, and send supervisory personnel to train the Belgian miners in modern extraction techniques. This attempt at thorough renovation failed. Within a year of the purchase, most of the ten shallow pits were shut down. Exploration of the seams by means of underground galleries was begun, but production must have been slight by 1833.[116] After Anzin had spent large sums "in pure loss" in the next few years, it resold the mine to Belgian interests in 1837.[117]

Despite false starts, setbacks, and occasional displays of bad judgment and apathy, Anzin demonstrated considerable expansionary verve in the three decades after 1800. Several notable breakthroughs during this period laid the basis for the company's prosperity throughout the nineteenth century. By 1833 Anzin had soared far beyond its prerevolutionary performance records. Production stood at 542,000 tons, nearly double the 1791 figure. The number of workers employed had increased by 25 percent to about five thousand and there were some thirty-six pits in extraction. More important perhaps than the statistical levels reached was the fact that Anzin was now able to expand rapidly to meet an expanding market. Its reserves of good forge coal gave some assurance of increasing prosperity in the years after 1833.

5 Output, Costs, and Productivity

Technological innovation and expansion were justified only in so far as they enhanced Anzin's financial position by increasing the total output of coal or lowering the average cost of each ton produced. To meet the opportunity posed by growing markets and to parry the growing threat of a shortage of the company's best coal, Casimir Perier launched a drive to expand output. After a long period of stagnation, production grew at a rapid rate from 1818 to 1833 and beyond. Perier's efforts to lower costs and raise productivity were superficially less successful. Anzin's costs fell perhaps 10 percent and its productivity rose by somewhat less during his period of leadership. Nevertheless, by chopping away ruthlessly at the fat in Anzin's operating expenses while investing heavily in expanded production facilities and up-to-date equipment, Perier forced his company to remain competitive with its rivals. Thus Anzin increased its profits by meeting the expanding demand for coal without greatly improving the marginal efficiency of capital or labor.

Output

The figures for output in table 1 and the graphs in figure 2 reveal a clear-cut pattern; two periods emerge between 1800 and 1833 separated by a sharp breakthrough in 1817–18.[1] Production had reached a peak of some 300,000 tons in the last five years before the Revolution, fell precipitously from this level in 1791 to a low of 65,000 tons in 1795, but recovered fairly quickly by 1797 to 185,000 tons. There then ensued the long period of stagnation that

Table 1 Output of the Anzin Company, 1789–1850

Year	Output in Metric Tons (× 1000)	Output Index (1789 to 1791 = 100)	Annual % Gain or Loss	Five-Year Moving Average
1789	290	97	...	...
1790	310	103	+ 7	...
1791	291	97	− 6	249
1792	275	92	− 5	204
1793	80	27	−71	167
1794	65	22	−19	137
1795	124	41	+91	119
1796	139	46	+12	145
1797	185	62	+33	182
1798	214	71	+16	200
1799	248	83	+16	213
1800	214	71	−14	219
1801	205	68	− 4	222
1802	216	72	+ 5	214
1803	229	76	+ 6	217
1804	208	69	− 9	222
1805	226	75	+ 9	218
1806	231	77	+ 2	222
1807	194	65	−16	233
1808	252	84	+30	243
1809	260	87	+ 3	251
1810	280	93	+ 8	261
1811	269	90	− 4	269
1812	245	82	− 9	263
1813	290	97	+18	257
1814	233	78	−20	253
1815	247	82	+ 6	249
1816	250	83	+ 1	258
1817	227	76	− 9	277
1818	335	112	+48	293
1819	324	108	− 3	314
1820	330	110	+ 2	337
1821	354	118	+ 7	333
1822	340	113	− 4	343
1823	319	106	− 6	349
1824	372	124	+17	353
1825	359	120	− 6	366
1826	377	126	+ 5	385
1827	401	134	+ 6	393
1828	417	139	+ 4	423

Table 1—Continued Output of the Anzin Company, 1789–1850

Year	Output in Metric Tons (× 1000)	Output Index (1789 to 1791 = 100)	Annual % Gain or Loss	Five-Year Moving Average
1829	411	137	− 1	440
1830	509	170	+24	454
1831	460	153	−10	479
1832	473	158	+ 3	511
1833	542	181	+15	528
1834	573	191	+ 6	561
1835	592	197	+ 3	597
1836	624	208	+ 5	624
1837	652	217	+ 4	651
1838	680	227	+ 4	657
1839	708	236	+ 4	661
1840	623	208	−12	675
1841	644	215	+ 3	668
1842	721	240	+12	646
1843	642	214	−11	664
1844	598	199	− 7	696
1845	715	238	+20	707
1846	804	268	+12	702
1847	775	258	− 4	706
1848	619	206	−20	699
1849	615	205	− 1	. . .
1850	670	223	+ 9	. . .

SOURCE: Michel and Renouard, p. 111.

lasted throughout the Napoleonic era and into the first years of the Restoration, during which production never reached the level or achieved the expansive élan characterizing the last years of the Old Regime. The yearly average was only 234,000 tons for the period 1797 to 1817.

After 1817 output suddenly rose above the high watermark of the early 1790s and began a substantial although seldom spectacular rise that extended throughout the 1820s and 1830s. Average output per year in this period of expansion was 97 percent higher than for the earlier period. Although output rose 153 percent from 1798 to 1833, it rose only 6 percent for 1798–1817 as compared to 62 percent for 1818–33. Output per year grew only 0.35 percent from 1798 to 1817 as compared to 5.10 percent from 1818 to 1830.

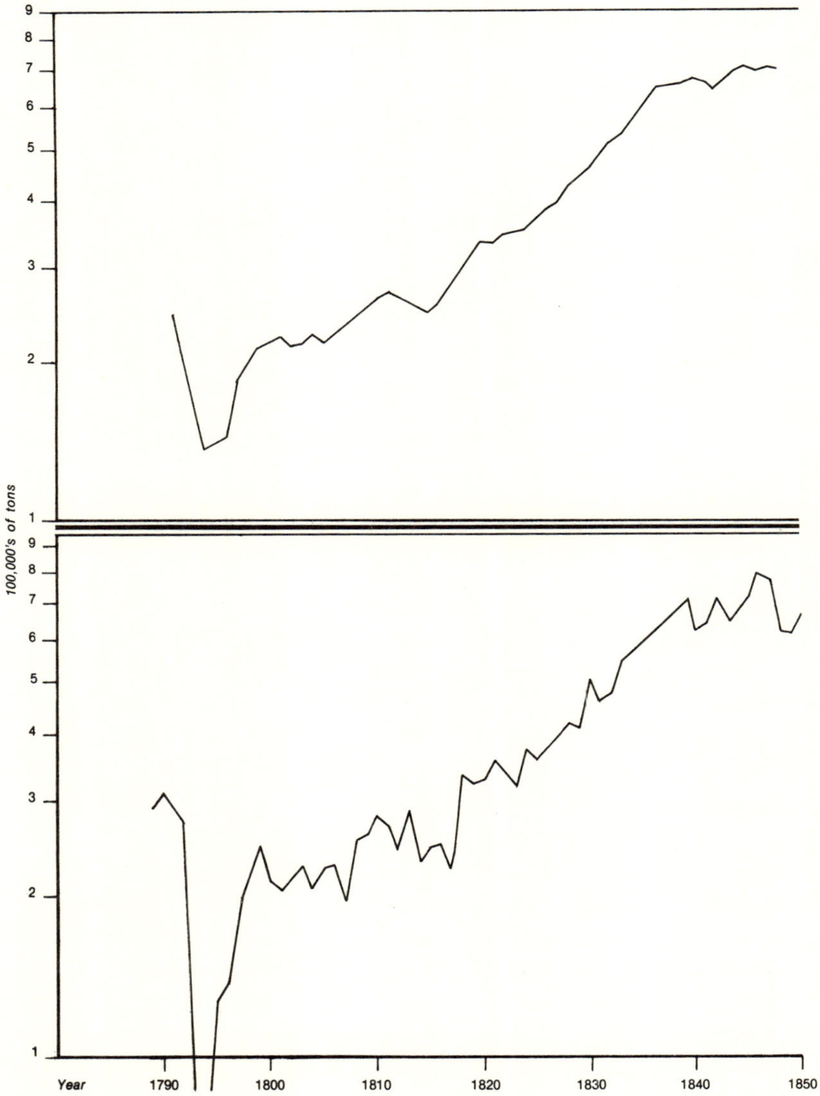

Figure 2

Production Levels for Anzin Company, 1789 to 1850. Top: five-year moving averages. Bottom: yearly output.

The output pattern can be summarized as follows: stagnation somewhere below prerevolutionary levels from the turn of the century to 1818; then a series of modest upward jumps about every five years until 1833, followed by steady growth to about 1840; and a new period of stagnation stretching into the 1850s, followed by a strong new period of expansion under the Second Empire. Recent calculations have shown that the average annual geometric growth rate for the French coal industry during the first two decades of the nineteenth century was only 2.2 percent, while the rates for the next two decades were 5.9 and 5.6 percent, respectively.[2] Clearly, it was Anzin's poor level of performance that held down the overall progress of the industry until 1818. Thereafter, Anzin advanced at a rate only slightly less than that for the industry as a whole.

It is not surprising that a turning point in the company's fortunes should have occurred sometime between 1815 and 1820. The partial isolation of the Mons competitors behind new political and tariff barriers, the speeding up of the process of industrialization in France, and the growth of the market for coal were all products of the early years of the Restoration that stimulated the demand for Anzin's output. Another factor, however, seems to have helped determine the timing of the company's take-off into a long period of rapid, sustained growth. The first years of the Restoration coincide with the assumption of leadership by the aggressive innovators Scipion and Casimir Perier. The general revision and tightening of the organization and control of production that they initiated were among the changes on the supply side that made it possible to slightly improve unit costs of production and output per man-hour while rapidly increasing the total output of coal.

Costs

The two main sources of information on costs at Anzin are the Perier Report and the Etats de Redevance. The former, a summary of the company's activities in the early 1820s by the leading member of the board of directors, gives considerable insight into Anzin's financial situation during those years. Since the report is a confidential one to the board for the purpose of elucidating Anzin's true situation, it can be accepted as accurate, and since it spans years of crucial expansion and technological change, it yields information on a highly important period. To an extent, the report is useful beyond the few years it deals with, for Anzin's situation in 1821–22 was in a sense an epitome of the "old way" of doing things, and its situation in 1824–25 forecast a new state of affairs that would continue through the early 1830s. Obviously, however, such trend projection into the past and future becomes more and more a matter of guesswork as one departs from the base years of the report. In addition, the form in which Perier set out his cost figures is not always as useful as it might be. For some purposes, then, the Etats de Redevance are a necessary supplement.

The Etats were statements furnished by the company to the government that contained a variety of statistical information on the company's operation.[3] For each year after 1812, they supply a figure for total annual expenditures. In addition, detailed lists breaking down global costs into various categories such as wages have survived for 1813 and for the years from 1820 onward (see table 2). As has already been pointed out, the Etats are not at all satisfactory as a statistical source for output. The same is true in the matter of costs. A comparison between the total cost figures given by Perier and the Etats for 1822 shows that in the public format of the Etats, the company undervalued its costs and output alike by about one-third. In 1825 it similarly undervalued output, but the difference in costs was much less.[4] On the other hand, the proportions between detailed cost items in the Etat lists are probably substantially correct. Since the categories used by Perier and the Etats in breaking down total costs were often different, the Etats remain very useful in estimating the proportions between some cost items (see table 3).

By far the largest single expenditure at Anzin was the wage bill. Until the 1820s wages probably accounted for about 50 to 55 percent of total costs. By the end of that decade, laborsaving techniques such as railed haulage, a slight decrease in piece-work rates for most underground workers, a tendency to contract out some surface jobs formerly done by company men, and rising expenditures on new pits and machines had lowered this item to about 45 percent. The rising cost of pensions, medical services, and a variety of other worker benefits only partly offset any decline in direct wage payments. The cost of salaried employees also gradually rose throughout the twenty years from 1813 to 1833. Together, however, these two items, salaries and benefits, never grew beyond about 10 percent of total costs or about 25 percent of wages. Expenditures for all employees' serv-

Table 2 Costs at Anzin (In Francs)

Items	1829	1831	1833
Employees			
Salaried staff	173,154	190,932	215,000
Wage labor	1,403,933	1,520,254	1,570,000
Employee benefits			
Pensions	89,042	112,993	160,000
Medical services	8,601	. . .	. . .
Total	1,674,730	1,824,179	1,945,000
Supplies			
Wood			
Propping	327,975	350,000	340,000
Carpentry	60,497	81,499	96,000
Fuel	281,326	331,845	324,000
Horses	74,953	77,615	77,000
Wrought iron	97,655	90,385	90,000
Cast iron	44,613	59,990	60,000
Copper	19,513	20,468	37,000
Lead, zinc, and tin	3,451	5,400	4,800
Steel	4,116	3,034	3,200
Leather	6,706	15,903	12,000
Rope	86,461	80,817	80,000
Blasting powder	25,701	20,000	2,000
Lighting (oil)	68,100	92,646	95,000
Miscellaneous	44,323	40,000	3,200
Transport of supplies[a]	19,294	22,448	25,000
Total	1,164,684	1,292,050	1,249,200
Capital expenditures			
Roads and buildings	77,386	60,000	50,000
Machines	. . .	80,000	70,000
A gare	. . .	70,000	. . .
Total	77,386	210,000	120,000
Nonproductive expenditures			
Commercial discounts	77,417	100,500	. . .
Land indemnities	14,485	42,567	38,000
Taxes	44,525	48,396	64,000
Total	136,427	191,463	92,000
Miscellaneous	162,778	24,891	20,000
GRAND TOTAL	3,216,005	3,542,583	3,426,200

SOURCE: AN, F14 4030, Etats de Redevance.
a. This item not strictly a supply cost; evidently represents payments made to haulage contractors.

ices together seem to have averaged about 55 percent of total costs by the early 1830s (see table 3).

These figures are, if anything, surprisingly low. While large-scale mining is a "capital intensive" industry with respect to the cost of

Table 3 Costs at Anzin (By Percentages)

Items	1813	1820	1829	1833
Employees				
Salaried staff	3	4	5	6
Wage labor	49	54	44	45
Pensions and medical services	2	5	3	5
Total	54	63	52	56
Supplies				
Wood	15	8	12	13
Fuel	12	7	9	9
Horses	6	6	2	2
Metals	3	7	5	6
Lighting (oil)	3	4	2	3
Other supplies	2	6	6	4
Total	42	37	37	37
Capital expenditures				
Roads and buildings	1.0	...	2	1
Machines	0.5	...	...	2
Total	1.5	...	2	3
Nonproductive expenditures				
Commercial discounts	...	...	2.0	...
Land indemnities	...	...	0.5	1
Taxes	0.1	...	1.0	2
Total	0.1	...	3.5	3
Miscellaneous	3	...	5	1

SOURCE: Based on AN, F14 4029, Etats de Redevance.
NOTE: Cost figures for 1813 do not include Raismes.

establishing or expanding production, it is extremely "labor intensive" in its day-to-day operations. In spite of a wide range of technical, geological, and wage-rate conditions, large-scale mining operations in various times and places have generally had high labor costs: Aniche, 1855—60 percent; the Ruhr, 1913—60 percent; the United States, 1940—60 to 65 percent; the Nord, Pas-de-Calais basin, 1958—68 percent.[5] By way of comparison, labor costs in the late eighteenth century in some other industries amounted to only 38 percent for papermaking, 20 percent for glovemaking, 10 percent for silk spinning, 8 percent for tanning, 7 percent for blast furnace operations, and 3 percent for iron founding.[6]

One basic reason why the mining industry is heavily burdened with labor costs is that the industry's raw material, coal in the ground, is the free gift of nature rather than a commodity to be bought for manufacture. Nevertheless, the cost of a wide range of material necessary to the operation of the mines accounted for most of the rest of Anzin's expenditures. Throughout the 1820s and early 1830s such costs represented about 37 percent of total expenditures (see table 3). Supply costs at Anzin were relatively high compared to wages because the system of exploitation required larger amounts of wood for propping, metals for machinery, and coal for ventilating furnaces and steam engines than were necessary where mines operated on a less advanced system or on more easily exploitable seams. Furthermore, Anzin paid more for most of its supplies than did its rivals across the frontier near Mons.[7] French metals, metal products, and machinery were of poorer quality or higher price than those available in Belgium, and tariffs prevented Anzin from finding much advantage in importing Belgian or English goods. The same was true for wood for propping or carpentry, and Anzin coal was a more expensive fuel for its engines and furnaces than that which the Mons companies had available from their own pits. Supply costs, then, were high both in relation to Mons and in proportion to total costs. Hence, Anzin was steadily concerned with finding cheaper sources of supply.

Wood was easily the most important of the supplies purchased by Anzin; it comprised from 10 to 15 percent of all costs.[8] About 80 percent of the wood by value went for the temporary uprights and crosspieces (*combles* and *perches*) used to prop the mine roof at the face. Since all this wood was eventually lost beneath the debris filling the excavated seam, the company substituted cheaper varieties, especially unbarked birch saplings, trimmed and cut to proper length. Anzin bought nearly one-third of a million combles in 1825 as well as better-grade hardwood and pine that was sawed into conventional lumber and used for a variety of building purposes.[9]

From at least the beginning of the nineteenth century, one of the most serious problems on the northern French and Belgian coalfields was the rising cost and diminishing supply of the proper types of wood. The seacoast areas of northeastern France, Belgium, and Holland had never been heavily timbered, and even the forested areas inland had early given way to dense agricultural and industrial populations. The coalfields had to compete with town and country for the remaining local supplies or pay the high cost of transporting this bulky commodity from distant regions where abundant forests still flourished. In either case, the result was high-cost wood, an intensely competitive wood market, and a scramble for cheaper sources of supply.

The Revolution had not improved the situation. Over half the remaining Nord forests had become nationalized by 1800 as a result of confiscations from the church and émigrés, but the revolutionary governments had permitted the Old Regime's system of forest management to break down under the impact of war needs and administrative anarchy. Reckless, premature, and excessive cutting, both by various governmental bodies and by the local population, in state and private forests alike, had further reduced the already depleted reserves of good timber.[10]

The Valenciennes area was especially hard hit by the destruction of forest lands during the Revolution. In 1789 very little usable timber had remained in the Nord department's four northern districts (Dunkerque, Lille, Hazebrouck, Cambrai). Over 50 percent of the department's reserves were at its southern tip in the Avesnes district on the flanks of the Ardennes. This heavily wooded region apparently survived the ravages of the Revolution far better than the Valenciennes area (Douai district) further north, but lack of water communications and poor roads between the two districts limited the access of the Avesnes timber to the dense industrial and agricultural population around Valenciennes. Still, in 1803 the Douai district contained over 13,000 acres of more or less intact woodlands or almost 22 percent of the department's entire forest area, and it produced about 31 percent of the Nord's timber. Production there was concentrated in the state forests of Marchiennes, Flines, Wallers, Saint Amand, Vicoigne, Hasnon, Bonsecours, and Raismes, all with an area of over 1,250 acres.[11] As the names suggest, most of these forests lay on or near the northern edge of Anzin's concessions.

In the early years of the century, Anzin complained to the government about its shameful neglect of France's forest treasures.[12] More practically, it replaced wooden shaft linings below the niveaux with brick, which was cheaper.[13] Little more is known of the company's reaction to the scarcity of wood or of its sources of wood supply until the Restoration. Then a series of circumstances conspired to bring the problem forward.

By 1817 Raismes, the largest of the nationalized forests, was back in the hands of its former owner, the Duke of Arenberg. A descendent of the Cernay family, he was one of the most important owners of Anzin, and, after Desandrouin's death in 1821, a leading member of the company's board of directors. When the duke found himself in financial difficulties, Anzin authorized Scipion Perier to negotiate with him for the sale of Raismes forest.[14] The proposed purchase fell through, perhaps because the forest preserve included l'Hermitage, the ancestral home of the Arenberg family. A happier solution for the duke's plight was arranged in the form of a loan made to him by the company on the security of the forest. Possibly Anzin acquired advantageous timber rights by the terms of the loan, but it was already seeking to achieve the same end, "freedom from complete dependence on the forest proprietors," by a somewhat different path. At the same meeting of the board of directors at which it approved the loan, the company directed Renard to purchase some of the state forest lands then being dribbled onto the open market. This new attempt to secure a sure supply of wood at advantageous prices probably failed also.[15]

Anzin's problem was further intensified in the early 1820s when, on a wave of prosperity and rising industrial production, the price of wood was rising steadily.[16] Local supplies did not suffice. Anzin was forced to raft wood by river and canal from the neighboring department of Pas-de-Calais, from the Somme, and especially the Aisne.[17] It imported Baltic timber through the ports of Amsterdam ("Holland pine"), Dunkerque, and Calais.[18] Hundreds of cartloads of wood followed tortuous roads from the Avesnes region.[19] Anzin's need for Avesnes timber involved it in a running skirmish with the Ministry of Finance. After 1814 French tariff law had forbidden exports of French timber along the Nord frontier. In 1820, responding to the demands of the Avesnes forest owners, who envisaged a rich market in Belgium and especially the Mons mining district, the French administration suspended the export prohibition along the entire northeastern frontier. A series of protests by Anzin were ignored both before and after 1832 when the suspension became permanent. By 1848 at least, Belgian competition had forced wood prices so high along the frontier that Anzin found it cheaper to abandon its local and Avesnes sources altogether and seek its wood supply entirely in the interior French departments such as the Aisne.[20]

Unsuccessful in its attempts to buy up local forests or to shut off the Avesnes area from Belgian rivals, Anzin in the early 1820s under Casimir Perier vigorously set to work to lower its wood costs by improving its purchasing methods. For several years after 1822 the board of directors took upon itself the task of authorizing in very specific terms the amounts, condition, and system of purchase to be used in all major wood transactions. The board felt it necessary to state explicitly that both the quantity and quality of wood purchased should fit current needs and that the prices paid should be as advantageous as possible. At first, the board ordered purchases made only after submission of sealed bids invited by public announcements in the local newspapers, but it later returned to the older system of direct contracts with individual dealers. Similarly, an early attempt to buy wood in an unfinished state and to do some of the lumberman's operations in its own workshops soon gave way once more to the purchase of finished lumber in standard building sizes.[21] The board attempted to circumvent the concerted price-rigging of the local lumber dealers by ordering Renard to establish direct relations with the Amsterdam importers either through their French agents or, if necessary, by traveling to Holland himself.[22] Casimir Perier's report to the board in 1826 pointed up the considerable success of these various measures when he claimed an economy of one-third in wood costs over the past four years. The saving amounted to 136,000 francs per year.[23]

The battle to lower wood costs continued. In 1827 the company directed an employee to conduct a census of all existing timber reserves in the neighboring forests. The purpose of the census was to put the company in a position to challenge the local dealers' "exaggerated pretensions" by threatening to buy up timber rights and produce its own lumber. Once again, the company had initiated a policy that might be described as forced self-sufficiency. At the same time, trials in the pits proved that trash hardwoods such as hornbeam could be employed as propping.[24] Despite these new steps, wood supply remained an expensive problem. The company's consumption of wood per ton of coal produced must have increased as pits went deeper. Prices remained high, although they probably did not increase substantially after about 1826.[25] Thus wood as a proportion of total costs did not fall much if at all before 1833, and its cost remained a serious burden at Anzin. This burden was heavier for Anzin than for most other mining companies on French, Belgian, English, or German fields where local supplies might be cheaply available and where the nature of the seams and the system of working them generally lent themselves to less propping.[26]

After wood, fuel was the most expensive item in the company's supply budget. About one-fifteenth of the coal Anzin mined went to supply its own steam engines and ventilation furnaces. The former alone burned over 16,000 tons in 1825.[27] The company also used considerable quantities to heat company workshops and offices, the barracks in which lived bachelor miners recruited outside the immediate neighborhood of the mines, and the homes of workers, who received "free" coal as a supplement to their wages. In the Etats de Redevance the company put the cost of this coal at its sale value, about one-third more than the cost of extraction. Furthermore, a large proportion of the coal destined for workers' hearths and company offices and shops was in the form of briquettes pressed from slack coal with an extremely low market value.[28] Yet even allowing for these factors, the company's fuel bill ran to perhaps 5 percent of its total costs.

Anzin naturally made numerous attempts to reduce fuel consumption. It has already been pointed out that the improvements in steam engines and their operation in the early 1820s were in large part dictated by expected fuel economies. The Edwards winding engines proved one-third more economical in fuel consumption than the older Perier engines, and changes introduced by Edwards in the company's drainage engines also resulted in a large economy. Overall, the amount of fuel burned per engine per hour fell between 1821 and 1824 from 1.3 to 1.1 hectoliters, a saving of some 16 percent or about 43,000 francs.[29] Other attempts to reduce coal consumption aimed at substituting less marketable coals for prime quality bituminous coals wherever possible. The use after 1797 of briquettes pressed from slack coal was one such attempt that proved eminently successful.[30] The Perier winding engines, like the Newcomen drainage engines, would burn only high quality steam coals in their boilers. Hence the company had transported coal of this quality from the Anzin concession to the pits of Fresnes and Vieux-Condé on the anthracite portion of the coal field. Each concession, however, had used its own coal in its ventilation furnaces.[31] In 1822 when the company was particularly hard pressed to supply the market with the rich bituminous coal mined from the southern seams of the Anzin concession, changes were introduced in the boilers, grills, and chimneys of the new Edwards engines that enabled them to use Fresnes anthracite, then a glut on the market. The adjustments were technically successful, but the cost of transporting coal from Fresnes to Anzin and Raismes was higher than the economy realized, and the next year the company reverted to its former practice. A similar experiment at substituting Fresnes anthracite for Anzin bituminous in the dole to workers of all four concessions proved unsuccessful for the same reason. The company soon ordered its workers at Anzin and Raismes to supply themselves again from nearby pits, although they were to use the semibituminous of the northern canton wherever possible.[32]

For other supplies and services too, the period after 1822 was one of active experimentation to cut costs. In a number of cases, Anzin sought relief by freeing itself from an exaggerated devotion to self-sufficiency. Many of its service and supply operations of long-standing had become unprofitable, and although in some cases the company tried to reinvigorate these operations itself, more often it took the simpler path of contracting with outsiders to provide the needed supplies and services.

Aside from the retreat from the steam-engine business already described, the most important instance of the movement away from self-sufficiency involved the provision of surface transport for coal and supplies. Even after the replacement of horse-powered winding by the Perier engines after about 1800, Anzin had continued to stable large numbers of horses. About 1821 it kept twenty-nine saddle horses for the business needs of its white-collar employees and about one hundred more for the thirty-nine wagons in which it carried coal from the pit heads to the riverside and shifted supplies from storehouses to pits. An employee's horse cost 400 francs, and its replacement cost amounted to 50 francs per year. Even if draft horses were less expensive, the company's total investment in horseflesh would have been at least 50,000 francs and its yearly cost for replacement about 6,000 francs. In addition there was the cost of wagons, fodder, and the wages of teamsters and the considerable number of workers employed in the elaborately equipped stables.[33] Between 1818 and 1821, the average yearly cost of surface transport ran to 151,438 francs or 3.88 francs per horse per day and 1,164 francs per horse per year.[34] Sometime before May 1822 supply transport was contracted out, and on that date coal transport also passed into the hands of local teamsters who furnished both wagons and horses as well as their own services for a stipulated daily fee. Henceforth too, employees received a yearly salary supplement of 1,095 francs per horse out of which they had to furnish their own transportation.[35] In 1826 Casimir Perier estimated the company's savings by getting out of the livery and teamster business at almost 60,000 francs a year.[36]

In some instances, economical provision of supplies and services did not pose the choice between self-sufficiency and outside purchase. For example, by merely transferring its lighting business from suet candlemakers to liquid vegetable oil dealers (who were part of a rapidly growing Nord industry), the company cut its lighting bill in half with a saving of 56,000 francs per year.[37] A little later Anzin made an unsuccessful attempt to transfer the cost of blasting powder to the gallery cutters themselves. The potential saving would have amounted to some 15,000 to 20,000 francs a year.[38]

Although Anzin made a serious attempt after 1822 to abandon some of its less profitable service and supply operations, it was sometimes forced to take on new responsibilities in these areas for technical reasons. It has already been pointed out that at least initially Anzin was forced to manufacture safety lamps in its own workshops because it could not find competent large-scale suppliers anywhere in France. For similar reasons, Anzin built a rope manufactory sometime between 1823 and 1826. A Belgian had developed a process for making flat hemp ropes which wound more smoothly on hoisting drums than the old round style, but no one in the Nord except Anzin itself was sufficiently interested in the new technique to buy the necessary equipment from the Belgian inventor.[39]

Thus, although the general effect of a series of reforms made during the period 1822 to 1825 was to increase Anzin's dependence on outside suppliers and contractors, the policy leading to these changes was not a new theoretical rejection of self-sufficiency, but rather a practical recognition that often a clumsy self-sufficiency meant higher costs. After 1822 efficiency was to be the company's watchword in supply and service operations as in extraction. Investigations proved that Tournelle had badly mismanaged the workshops. Spare parts were not always available when needed or in the proper quantities; the piecework system of payment in the shops was leading to the production of shoddy goods at extravagant cost; and the company's surface installations, including the shops themselves and the engine sheds at the pit heads, were becoming more dilapidated every year.[40] In addition, a thorough inventory of the company warehouses found them overflowing with supplies bought at high prices and in unnecessarily large quantities that bore little relation to the company's current needs. This overstocking not only tied up large sums of working capital, but also often led to the deterioration of perishable materials, which rotted or rusted unused for years on end. Poor supervision and administration had encouraged waste and theft by warehouse employees, kickbacks from suppliers to company purchasing agents, and a general inefficiency and lack of economy in relating need to supply.[41]

Anzin embarked on a broad reorganization of its shops and warehouses. The board of directors ordered the shops "to furnish promptly everything necessary for the regular operations" of the pits and surface installations,[42] but they were no longer responsible for the manufacture of a "host of objects" which could be bought as easily and more cheaply on the market. Henceforth they served as maintenance shops rather than full-fledged "manufactories."[43] The board dismissed Tournelle, replaced him as "mechanical engineer" (shop and machine supervisor) by an "active, capable [and honest?] man,"[44] and gave his functions as director of surface structures to a newly created "architectural engineer." Both of these new posts fell under the immediate supervision of a general director, a position traditionally in the hands of the Mathieu family in the eighteenth century but vacant since the family's emigration during the Revolution. The general director took control of all surface and underground activities concerning production. He thus relieved Renard of an important burden and freed him to concentrate his energies more fully on commercial and legal business.[45]

The appointment of a general director obviously meant far more than a mere shakeup in the administration of the company's supply services. It indicated Casimir Perier's growing dissatisfaction with Renard's performance generally. Here it is important merely to note that the new appointment was part of an overall scheme for the reform of these services. They were better coordinated with production operations, more closely and professionally supervised, and made responsible for their activities on something approaching a cost accounting basis. For it was evident that much of the inefficiency of the supply services had in the past been due to the company's inability to check on their real costs. These costs proved to be discouragingly high when the company adopted a new, more rational, and more thorough system of bookkeeping. The traditional slipshod mode of operations in the shops and warehouses could no longer be justified in the light of the information revealed by the new accounting techniques.[46]

On the purchasing side of the supply problem, a similar tightening occurred. The warehouse inventory of 1822 led to the reduction of supply purchases to the amounts necessary to meet current needs and to the sale at secondhand of immense quantities of scrap iron, nails, bricks, and other useless or overstocked materials.[47] The general director now determined the monthly requirements for operations at each pit and submitted a report of the company's overall needs to the general agent, who in turn took a far more active and personal part in reviewing and verifying those needs. A tighter, more centralized system replaced the lackadaisical method of enrolling newly purchased supplies and of requisitioning supplies from the warehouses. The company dismissed many longtime warehouse workers and appointed a new stores supervisor.[48] It took supply purchases out of the hands of semi-independent buyers working on commission and placed them under an "homme de confiance."[49] It insisted that the general agent, often acting under specific limiting instructions from the board of directors itself, make such extremely large or delicate purchases as the yearly supply of wood props. Indeed, the detailed intervention of the board in supplementing its revised general purchasing policies with minute specifications in important individual cases was a remarkable new feature of the period after 1822. Although Casimir Perier stated in 1826 that the general agent's loyalty and ability were not in question and that Renard too had seen clearly the abuses of the old supply system, the prominence given his new subordinate, the general director, and the closer supervision of his actions by the board made it clear that his former uniquely powerful role in directing the entire operations of the company was sharply diminished by the reforms instigated by Perier after 1822; for Renard alone would not have been able to master the resistance to the new policies evinced by the vested interests of the old system.[50]

The discussion so far of labor and supply costs has avoided a most important problem: how much of this labor and materiel was used to meet day-to-day operational needs and how much was used to create permanent productive facilities? Unfortunately, the nature of the cost figures left by the company makes it difficult to separate operating from capital costs. The detailed lists of yearly expenditures which served as a base for the Etats de Redevance should be the chief source of information for this problem. The problems of undervaluation and of the proportions between cost headings used in these lists have already been discussed. There are other difficulties, however, in using the lists to determine the proportion of capital formation to other costs. Except for an isolated list from 1813, the lists have only survived for the period after 1820. Even for those that have survived, a serious analytical problem is posed by the fact that the headings under which total costs were broken down included various categories of labor and supplies undifferentiated by the function they served. For example, all wage labor was subsumed under a single cost figure regardless of the use to which the labor was put. Shaft sinkers and coal cutters were grouped together as wage labor, although the former were engaged in capital formation and the latter in current production. Workshop labor of all kinds was also grouped into the same general wages category, although some of it was employed in maintenance operations and some in establishing permanent new productive facilities. The one supply item "wood" included both perishable props and construction lumber; some of the latter in turn was used for repair and some for new installations, and the proportions between these two uses changed over the years. Most of the other supply headings, notably for metals, were established on the same simple principle. In part, the clumsy headings were no doubt a result of Anzin's consistent policy of using the Etats to conceal from a tax-collecting government information about its real financial situation. Equally important, however, in the failure of the lists to differentiate operating from capital expenses was the company's own lack of sure knowledge on this important point. Even after the accounting reforms of the period 1822–26, capital expenditures remained hidden in a maze of cost figures organized on a nonfunctional basis.

Nevertheless, suggestive cost estimates can be made for most categories of capital expenditure, especially for the 1820s. Some headings in the Etat lists are functionally composed, and these figures can be supplemented with information from other sources, notably Casimir Perier's report to the board of directors in 1826.[51] For the most important capital item of all, the cost of establishing new pits, information is scarce, and since the cost of a new pit obviously varied somewhat with variations in depth and drainage difficulty and with changes in basic pit size and design over the years, any estimates must be extremely fragile. There are, however, several estimates of pit costs, some of them from non-company sources. These estimates are not entirely limited to the 1820s and are not incompatible with each other.

In 1805 Daubuisson, the visiting government engineer, put the cost of merely reaching the coal measures at 100,000 francs. He cited the pit Bleuseborne, admittedly one plagued by difficulties, as having absorbed 300,000 francs over some ten years before the Revolution without becoming fully operative.[52] In 1821 Clère, the government engineer resident in the Nord, claimed that at Aniche, in terrain similar to that at Bleuseborne, the cost of reaching the coal measures was 200,000 francs.[53] Renard in 1822 and Chevalier in 1828 agreed on the figure of 400,000 francs as a normal one for fully establishing and equipping a pit.[54] Chevalier's figure referred to the Mons area where the terrain was somewhat better, pits shallower, and wages and other costs somewhat lower than at Anzin, and he probably included the cost of a drainage (but not a winding) engine in his estimate. In addition to the cost of the pit itself, there were the allied expenses of explorations, soundings, and unproductive shafts. If 400,000 francs per successful pit is accepted as a plausible figure, including the cost of preliminary expenses and failures, the twenty-eight pits sunk by Anzin between 1800 and 1833 would have represented a total capital investment of over 11,000,000 francs. This total was unequally distributed over these years, since five pits were opened between 1803 and 1806 and seven between 1813 and 1817, while the other sixteen were scattered fairly regularly over the years from 1819 to 1833. Costs were probably lower in the first two periods, since the pits sunk then were generally simpler and shallower than the later ones, and since all but one of those sunk between 1813 and 1817 were located on the shallow seams of Fresnes or Vieux-Condé. The company also spent, however, over 500,000 francs "in sheer loss" from 1804 to 1819 in trying to gain control over the Torrent.[55] Altogether, successful and unsuccessful pit openings to 1819 must have cost at least 4,000,000 francs. While opening sixteen successful new pits during the remaining fifteen years through 1833, Anzin made extensive explorations of its entire territory, opened up new fields at Denain and Abscon, mastered the Torrent at "immense cost," poured funds into sterile Saint Saulve, attempted to renovate the Bois de Boussu, and everywhere deepened existing pits and opened new working areas underground. An estimate of 6,000,000 francs as the cost of all these efforts could hardly be excessive. Over the entire period 1800–1833, 10,000,000 francs would seem the lowest plausible cost for Anzin's basic production facilities with 15,000,000 francs a possible upper limit.

Other capital expenditures amounted to far less. According to the Etats, the cost of constructing roads and buildings averaged only 62,000 francs during the years from 1829 to 1833. For a period when sales and production were expanding rapidly, even double this figure would not be very high, especially if it included worker housing and loading areas at the riverside as well as engine sheds, warehouses, shop buildings, offices, and paved roads from pit head to riverside. Presumably because it was unique and very large, one installation of this sort was listed separately in 1831; a *gare* at Denain was said to have cost 70,000 francs (see table 2). This seems to have included warehouse(s), pavements, dock(s), and perhaps a railed horse tram to the pits to the north. Except in such extraordinary years, however, total annual expenditures for surface installations of all kinds can hardly have amounted to more than 125,000 francs, a very modest sum.

Investment in real estate, lock companies, canal companies, canal boats, and the like was highly irregular and exact figures were not often stated. Real estate purchases were held to a minimum. The company often sold land it no longer needed, and rather than buy land outright, it preferred to indemnify owners for its use.[56] In 1818 Anzin contracted to build several simple locks on the Escaut River near its pits but made no actual investment, since it subcontracted both the construction work and the income from tolls.[57] A year earlier Anzin had taken five shares at 10,000 francs each in a company for the construction of the Sensée Canal, but its only really important venture of this sort was an investment in 1827 of 700,000 francs in a company for the improvement of the Saint Quentin Canal.[58] Such an investment, however, almost certainly came out of reserve funds rather than current income. In 1822 Anzin disposed of the few boats it had built earlier.[59] It is likely that the figure of 162,778 francs for miscellaneous expenses in 1829 conceals some relatively large capital expenditure of a nonrecurring kind (see table 2). Still, it is very unlikely that year in and year out such extraordinary costs can have amounted to 100,000 francs.

On the cost of installing railed haulage underground in the 1820s, the figures are much more certain.[60] Luckily, the period covered by the Perier Report coincides almost exactly with the introduction and substantial completion of the rail system in the existing pits. Between 1822 and 1826 the company spent 188,000 francs on railed haulage, an average of 37,000 francs per year. This figure is of course too high to represent the cost of installing railed haulage in the long run. For the fifteen years after 1826, the projected annual cost of installing, extending, and maintaining the new system was estimated at only 10,000 francs "at the most." Furthermore, even for the initial period of highest cost from 1822 to 1826, part of the sum of 37,000 francs went for maintenance and part merely replaced spending for new installation of the old wooden plank system. Although technically very important, railed haulage did not represent a very significant capital expense.

The amount spent on steam engines and machinery was much higher, second only to that spent for pits. In 1831 the company owned six Watt and six Newcomen drainage engines and had twenty-two Edwards and twelve Perier winding engines.[61] Even if the cost of the drainage engines was not included in the costs of opening and equipping pits, as was assumed above, the Newcomen engines all predated 1800, and the total cost of the Watt engines could not have exceeded 500,000 francs.[62] At an average price of 25,000 to 30,000 francs apiece, the thirty-four winding engines would not have cost much more than 1,000,000 francs and might well have totaled less. Adding a generous 500,000 francs for the cost of engines bought after 1800 but retired from service, of improved boilers, grills, and the like, of pumps and hoisting machinery, and of the steam engine and machines in the workshops, for the entire period the total cost of steam engines and all other machinery would have been at the most 2,000,000 francs, with the greater part of this amount concentrated in the two major engine-buying periods from 1802 to 1806 and from 1822 to 1826.[63]

The picture that emerges from this rough estimate of capital expenditure is somewhat startling, for one might have guessed that Anzin's capital costs would have been burdensomely high. Even in the early nineteenth century, the coal mining industry was known as one requiring extremely large inputs of capital both absolutely and relative to operating costs or profits. Anzin in particular asserted that its growth and its readiness to adopt the best in new techniques led to capital expenditures so high that, together with high operating costs, they merited the benevolent concern of the government. Anzin repeated these claims again and again in its battles with competitors, customers, and government officials, and its enemies tacitly admitted their truth by choosing to concentrate their fire on more vulnerable company policies such as pricing. It was certainly true that the investment totals given here for Anzin were far greater than those its smaller contemporaries could or did spend. Furthermore, although the general belief in Anzin's technical advance has been shown to have been somewhat exaggerated, this belief was largely correct, at least by comparison with most other French mines. The great majority of mine owners had little spare investment capital, their profits were small or nonexistent, and their scale of production was not large enough to warrant the introduction of steam winding, railed haulage, or the other expensive new mechanical techniques. Thus, their capital expenditures were not only lower absolutely than Anzin's, but also probably tended to be smaller in proportion to operating costs.

Yet even at Anzin the amount invested in fixed capital equipment was remarkably low compared to operating costs or net earnings. The total initial cost of all steam engines bought from 1800 to 1833 (ca. 1,500,000 francs) was less than the company's net earnings in any single year after about 1818 and less than its gross costs in any year for the entire period.[64] Other capital expenditures, with the exception of pits, ran far less. For 1822, a year when Anzin was plunging into its most vigorous period of capital expansion, a reasonable estimate of capital expenditures would be as follows: pits, 400,000 francs; steam engines, 92,000 francs; roads, buildings, and other surface installations, 125,000 francs; irregular investments in real estate, etc., 100,000 francs; railed haulage, 37,000 francs; miscellaneous, 60,000 francs.[65] The total is slightly more than 800,000 francs. The Perier Report put gross expenditures for the year at 2,911,173 francs, and net earnings as calculated from Perier's output and expenditure figures were probably about 1,862,000 francs. Anzin's capital expenditures, then, would be only about 28 percent of total costs and 44 percent of net earnings. Capital formation by its very nature proceeds unevenly from year to year and from one period of years to another. The rate of capital formation was possibly higher before about 1820, since operating costs and earnings were lower then, but capital expenditures were also lower for those years and were almost negligible between 1806 and 1813. During the 1820s and early 1830s capital expenditures rose, but so too did operating costs and net earnings. The proportions suggested for 1822 were probably very roughly accurate for the whole period from 1800 to 1831. Thus, it is hardly surprising to find that Anzin paid large "dividends," increased its cash reserve, and financed all capital investment out of current earnings even during the period 1822 to 1826, when sales slumped and the cost of technical renovation was supposedly enormous. As Casimir Perier put it in 1826, "the financial situation of the company has remained reassuring."[66]

Cost per Ton and Productivity

Purchase of Edwards winding engines, introduction of railed haulage, movement away from self-sufficiency, reorganization of shops and warehouses, and a host of other organizational and technical changes in the early 1820s were all intended to lower extraction costs and increase productivity. To some extent Anzin achieved these aims, but its costs remained high and output per man-year low in comparison with the situation in other mining areas.

Between 1815 and 1833, several persons attempted to evaluate the cost per ton of coal extracted at Anzin. Renard stated in 1815 that it was well known that the company's costs were 9 francs per ton.[67] Casimir Perier confirmed this figure for the years 1821–22. He went on to claim a decrease in costs to about 8 francs by 1824–25 as a result of the innovations made since 1821 under his direction.[68] Since Perier made this claim in his report to the representatives of the company's owners in the privacy of a meeting of the board of directors, there is no reason to suspect that his figures were mere propaganda. In 1828 Michael Chevalier, the government engineer who was no great friend to Anzin, independently calculated costs at only 6 to 7 francs per ton.[69] The difference between the figures given by Perier and Chevalier was probably not due to any real decline in costs between 1825 and 1828. Chevalier's estimate was for the cost per ton at the pit head; Perier's figure, like the others cited here, included the cost of transporting the coal to the Escaut River and of loading it onto barges. When these items are added to Chevalier's pit-head cost, his figure for the costs actually borne by most purchasers becomes 7.5 to 8.0 francs per ton or only slightly lower than that of Perier. Far from continuing to fall, costs may have started rising by the early 1830s if two estimates from those years can be credited. In 1832 the Anzin general agent, Mark Jennings, gave written evidence for the great *Enquête* on the coal tariff. While trying to make a case for a protective tariff, he claimed that Anzin's costs were between 8 and 9 francs per ton.[70] A year later the local engineer, Boudousquié, attempting to make a careful estimate based on his own close observation and knowledge of Anzin's operations, set a figure of 8.5 francs per ton.[71] From these varied estimates one fact is certain: cost per ton at Anzin fell about 10 percent between 1821 and 1826 as a result of the economies in technique and organization introduced under Casimir Perier during that crucial period. On the other hand, there is no evidence that costs either rose or fell substantially at any other time between 1800 and 1833.

Although the reduction after 1821 was a substantial achievement, Anzin costs remained comparatively high. Both Chevalier and Jennings compared costs at Anzin with those elsewhere; their figures are presented in table 4.[72] As discussed above, Anzin had certain cost disadvantages. In comparison with Belgium and England, Anzin' supply costs were higher, and in comparison with all the fields listed, Anzin's seams were more difficult to work and hence its operating and capital costs were higher.

Accurate figures for tonnage produced per man-hour or man-year worked are even more difficult to determine than changes in costs per ton. Only the roughest of estimates can be made, but there is

Table 4 Comparative Coal-Mining Costs

Source	*Year*	*Cost per Ton in Francs*							
		Anzin	Mons	Loire	Aniche	Charleroi	Auvergne	Decize, Fins, Blanzy	Newcastle
Chevalier (Pit head)	1828	6.3–6.8	5.7–5.9	2.7–4.5	10.5	10.5	6.8	5.5	. . .
Chevalier (Loaded)	1828	7.5–8.0	6.6–7.2	6.1–7.9	12.3	12.3	7.6–8.6	7.0–7.5	. . .
Jennings	1832	8.9	5.0–6.0	7.0–8.0	. . .	. . .	. . .	. . .	3.0–4.0

SOURCE: AN, F12 2534a, Jennings Report; Chevalier, p. 482.

reason to believe that despite technological advances, improvement was slight. Probably any rise in productivity was less than the 10 percent decline in costs.

The Etats de Redevance supply figures for workers employed both underground and at the surface. There is no reason to believe that these figures are inaccurate; such information was easily determined and a matter of common knowledge. To determine the effective work force as opposed to the number nominally on the company payroll is a more difficult problem. During slack periods the practice at Anzin (as elsewhere) was to put all its employees on a short workweek rather than lay off entirely a part of the work force. There is no way of allowing for such periods of enforced underemployment, although they might last for years, as at Fresnes and Vieux-Condé in the 1820s. If the figures of the output series given in table 1 are divided by the Etat figures for workers underground, the result is a productivity series for tonnage of coal mined per worker per year. The series so calculated swings fairly widely and eccentrically between about 80 and 130 tons. Although the yearly tonnages are generally higher after about 1819, there are important exceptions, which perhaps are due to concealed underemployment, and it would be imprudent to claim that the series demonstrates an improvement in productivity of any particular magnitude.

Clère and Boudousquié made similar calculations for all Nord production (Aniche included) on the basis of their own employment and output figures.[73] The employment figures (allowing for Aniche) are close to those given by the Etats; the output figures are, of course, lower than those given by Michel and Renouard. The Clère-Boudousquié series shows an average yearly productivity per man of about 60 tons between 1816 and 1821 and about 90 tons from 1822 to 1832. Clère and Boudousquié attributed the rise to the introduction of railed haulage, but this explanation is totally insufficient to explain an increase of 50 percent. The number of haulers under the older system constituted only about one-fourth of the men employed underground and one-sixth of all employees, and even after railed haulage was introduced, a substantial number of haulers was necessary. Furthermore, their series does not show a gradual increase of productivity over the years between 1822 and 1826 when the rails were being laid down, but rather a sudden jump between 1821 and 1822, the year just previous to that during which Anzin decided to substitute rails. A better explanation for the jump is a change in the categories used in the calculations after 1821. Previously, "workers employed" seems to have included workers both on the surface and underground; afterward, it included only those employed underground. The outputs calculated on the latter basis range from about 80 to 100 tons, a level of productivity not very different from that calculated on the same basis from the Etats and table 1.

Two final pieces of evidence lend some support to these two series. In 1819 Héron de Villefosse estimated the output of an average Anzin chantier employing approximately 100 workers to be about 40 tons a day.[74] On the basis of 250 working days a year, output per worker per year would have been about 100 tons. In 1828 Michael Chevalier made a similar estimate.[75]

Similar kinds of productivity estimates for other mining areas in the early nineteenth century are no more certain than those for Anzin, but such as they are, they are of about the same or somewhat greater magnitude. All of them have also been calculated by dividing contemporary figures for annual output by workers employed. At Littry in 1822, output per worker was 66 tons, and at Epinac between 1812 and 1821, some 86 tons.[76] Two other French mines show striking increases in output per worker during the 1820s as a result of technical improvements. Blanzy's figure rose from 73 tons in 1812 to 100 in 1824, and Le Creusot's from 110 during the period 1814–21 to 137 between 1822 and 1825.[77] In the Ruhr, output was about 120 tons per worker in the 1820s.[78] All these calculations, except those for Le Creusot, are based on all workers employed rather than merely those underground, and hence the resulting outputs should be raised by an unknown amount to be comparable with the Le Creusot and Anzin figures.[79]

Although most of these calculations are approximations at best, they do invite a general conclusion and raise two important questions. One may conclude that productivity in the early nineteenth century was low everywhere by comparison with the later part of the century. Anzin's output of 217 tons per worker in 1882 was then one of the lowest in the expanded Nord and Pas-de-Calais field, while by 1913 calculations (which included surface workers) gave yearly outputs per worker of 269 tons for Germany, 245 for England, 200 for France, and 155 for Belgium.[80] These national figures included many small, inefficient mines, but, with the exception of those for the Ruhr, the output figures given here for the early nineteenth century were those of relatively large and technically advanced companies.

The two questions these figures raise concern Anzin more specifically and are interrelated. First, why did Anzin, technically advanced as it was, have a level of output per man-year only equal at best to that of smaller French mines such as Blanzy and Le Creusot? Second, why, despite the extensive reorganization and technical improvements of the 1820s, did Anzin productivity rise so little during the same period?

An important factor in the answer to both questions was the system of extraction practiced at Anzin. This system was dictated by geological conditions—deep, wet, thin, twisted seams—and remained essentially unchanged in 1833 except for the new haulage techniques. It involved far more extensive cutting of rock walls and wood propping than was necessary in mines where seams were thicker and flatter and where roofs were stronger. Deep drainage, deep ventilation, and deep hoisting added more labor to the task of extracting each ton of coal. Finally, the difficulties of cutting coal at the face in superimposed tailles on sloping seams also set relatively low limits on output per worker. The system as a whole was admirably designed to solve complex geological problems; it was intricate and sophisticated, but it remained heavily labor intensive despite a number of technical changes.

These changes themselves were not always labor-saving. Output per man-hour was increased by the use of railed haulage and the elimination of bures, and by the board of directors' drift away from the policy of self-sufficiency. On the other hand, more men per ton of coal extracted had to be employed as shafts were deepened and galleries were lengthened and as expansion diverted manpower from the face to the task of opening new pits. Other important developments probably had little effect of any kind on productivity. The Perier winding engine was cheaper than horse-powered winding, and the Edwards engine lowered costs further; both enabled Anzin to work deeper seams than would otherwise have been possible. But steam winding was not labor-saving; in itself it did not increase output per man employed. Improved drainage techniques made possible the opening of the Torrent pits, but once again, an advance that was essential to continuing and expanding the level of output did not lessen the amount of labor involved in production. Expansion did not necessarily increase productivity. Some of the new pits were undoubtedly more productive than the old, but this was by no means true of all. Conditions at Denain were easier than in the old Anzin and Raismes cantons, but those on the Torrent were not, and the relative decrease in the production of the easily extracted coal at Fresnes and Vieux-Condé tended to lower overall productivity.

In general, technical and organizational progress cut costs somewhat or enabled Anzin to work seams that would otherwise have been unworkable, but such progress had little effect on output per man-year. Some changes decreased the labor needed to extract a ton of coal, others led to an increase, and still others had no effect or were two-sided in their effect on productivity. In this light, the surprisingly low ratio of capital expenditures to operating costs was not so much the result of an unwillingness to invest capital as of the continuing high cost of daily extraction. Vital as was the technical progress and expansion of Anzin, especially in the 1820s, it merely enabled Anzin to continue to compete successfully with its rivals despite a generally unfavorable cost structure and an undistinguished level of productivity.

6 Transportation: From Pit Head to Market

The relatively primitive means of moving bulky goods in the early nineteenth century caused transport costs to outweigh production costs in the price many consumers paid for Anzin coal. Furthermore, these transport costs probably fell farther between 1800 and 1833 than the company's production costs. Since the ability of Anzin's competitors to reach a given point of sale more cheaply than Anzin and the ability there of potential buyers to bear the charges of shipment piled on top of the costs of production determined the geographical limits of its market, Anzin was vitally concerned with increasing the efficiency and lowering the charges of moving its output from the pit head to the consumer. Not surprisingly, however, Anzin found its power to influence the many external factors affecting the transportation network fell short of its ability to control the elements determining its internal cost structure. It could improve the connecting links between the pits and the bank of the Escaut River, rebuild and reorganize its loading areas at the riverside, and struggle valiantly, although vainly, to break the coalitions of the boatmen who carried most of its coal away from the riverbank. Although hampered by the relative backwardness of French government officials and entrepreneurs in developing an adequate transport system, it might encourage or even participate directly in several of the schemes to improve and extend the network of waterways over which its coal moved to market. Yet Anzin's most significant influence on transportation patterns was negative: its tenacious, surreptitious struggle to hold off improvement of the stretch of the Escaut that flowed past its own doorstep. Here the peculiar geographical pattern of pits, waterways, and markets dictated the company's anomalous policy. Any increase in speed or drop in costs for coal moving along the river would have benefited Mons producers far more than Anzin. The unimproved Escaut was at once a magnificent natural resource for Anzin and a major obstacle to the supremacy of Mons in the markets of northern France.

Anzin's transport policy naturally reflected the limited possibilities of its age. The early nineteenth century saw a culmination in Europe and America of centuries of development of road, river, and canal for inland transport at the very moment a new transportation age was heralded by the first faint stirrings of the railroad. Despite some interest in Britain's early railroads, the many Frenchmen both in and out of government who were concerned with improving France's inland transportation looked mainly to the earlier British successes in perfecting the old ways of moving goods by water and road. And although the British railroads had been born and proven on the coalfields, French coal companies did not attempt to follow the British example until the 1820s, and even then they conceived of railways, with or without steam locomotives, as rivals of the local roads, mere supplementary feeders for the waterways.

Egged on by the professional enthusiasm of the engineers of the Bureau of Bridges and Roads, the governments of the several regimes between 1789 and 1833 all showed considerable interest in improving French roads and waterways. But judged in the light of past and present accomplishments in England or the grandiose master plans drawn up by the various French governments, results in France were meager. The governments themselves were partly responsible for the failure to accomplish more. In a period of revolution, war, and restoration, no regime consistently supplied adequate funds to do the job. Yet none was willing to free private contractors to do the necessary work as they saw fit. Private capital, in fact, was often repulsed by conditions laid down by governments in the blueprints and concessions; French investors were generally (and often rightly) doubtful about the profitability of transport works. Finally, the strategic aims of the military establishment often distorted or blocked proposed transport schemes. It was only after 1820 that a precarious alliance of government and private enterprise began to make serious headway against the backlog of necessary internal improvements. Consequently, the inland transport system in France lagged far behind Great Britain's as that nation entered the railroad age.[1]

Fortunately for Anzin, a relatively dense transportation network had long served its prime market area in northern France. The roads of the area were numerous, and freight rates were lower than the national average, kept down by the competition of waterways. Yet the roads were unequal to the burden placed on them by the growing traffic spawned by the demands of war and economic expansion. Lack of upkeep during most of the Revolution, poor drainage in some areas, harsh climatic conditions, and the high cost of paving stones were further reasons for the badly degraded character of most of the area's roads during the first third of the nineteenth century. Added difficulties derived from the fact that the commercial freight business was badly organized and ill-adapted to large-scale, long-distance hauling; there was not even one overland freight business at Valenciennes in 1812.[2] Anzin avoided the roads wherever it could, using them only for very short hauls to customers near the pits or to the waterways, which assumed the primary burden of coal transport. In 1808 and again in 1817 it was claimed that Mons mines were sending much of their coal to Lille and Arras by road because their water links to these French cities were so slow and inadequate, but Anzin, with the Escaut River at its doorstep, was never forced to this extremity.[3]

Transported by the Escaut and its tributaries and by canals linking the Escaut basin to Flanders and the Channel coast, Anzin coal in 1800 was able to reach most of the important towns in northern and western Belgium and in the two most northerly departments of France, the Nord and the Pas-de-Calais (see map 4). Yet great difficulties faced shippers over most of this area. The rivers were generally little improved over the state in which nature had fashioned them. Full of curves, boulders, and sand bars, their channels shifting and ill-defined, they meandered over flat, wet terrain. Their tow paths were mostly unfit for horses and often barely passable for men. Much of the summer they lacked sufficient depth to permit the passage of fully laden boats; in winter they froze over for several months; and in spring they overflowed their banks and tow paths and flooded their broad marshy valleys. The canals that linked them shared many of the same deficiencies and had others peculiarly their

Map 4

Major Anzin Markets: French Departments of Nord, Pas-de-Calais, Somme, and Aisne and Western Belgium

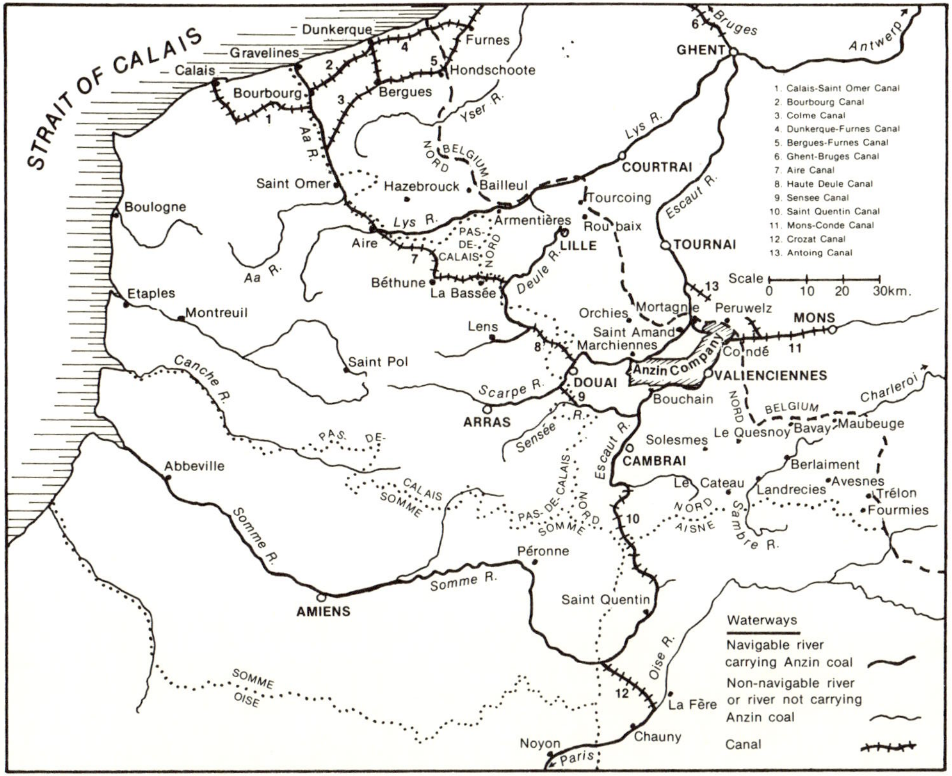

own. They were often of insufficient depth for most boats at the best of times, and many tended to suffer from a permanent shortage of water. The dams and locks intended to regulate the flow of water in rivers and canals and ease the passage from one to another were often bottlenecks behind which traffic piled up for weeks on end until a sufficient head of water built up. By-pass canals around the walls of the larger towns were either lacking or notoriously difficult to navigate. The channel depth in each waterway not only varied with the season but also usually differed from that of the other waterways with which it connected; transshipment was necessarily the normal practice at such points. Navigation was complicated, cumbersome, irregular or intermittent, and even dangerous. Transport by water was very slow, routes were indirect and long, costs were relatively high, and the quantity of goods carried was strictly limited.[4]

Most of these difficulties were attacked and partly corrected between 1800 and 1833. Time, distances, and costs decreased on the already existing waterways. The quantity of coal they carried increased as Anzin's market area was extended in the sense that the company's coal could reach more customers more quickly and cheaply. Even more important, several new canals enabled Anzin coal to reach new markets for the first time or old ones more easily. Southeast toward the Avesnes district, the Ardennes, and the Sambre, Meuse, and Moselle basins, the much-discussed Escaut-Sambre link was not built until much later. Northward toward the French Channel coast a very difficult and roundabout route was vastly improved by completion of the Sensée and La Bassée canals in 1820 and 1826, respectively. The Sensée Canal greatly shortened the route from Anzin to its major markets on the Scarpe and Deule rivers, including towns like Arras and Douai and the industrializing urban complex around Lille. Together, the two new canals gave improved access to Dunkerque and the sea. The most important of the new connections, especially for the future, lay to the south. The Saint Quentin Canal, although officially opened with much fanfare in 1810, was not rendered practicable for heavy coal barges until 1828. Thereafter the Seine-Oise basin with its great industrial and population centers around Paris and Rouen offered a magnificent potential addition to the Anzin markets. The benefits of canal building did not of course fall to Anzin alone. The Mons-Condé (1814) and Antoing (1826) canals helped Mons coal compete with that of Anzin in the whole area from Lille to Cambrai (and eventually Paris) and on the Channel coast. It would be hard to say whether Anzin or Mons benefited most by the improvement of the water network, but it is clear that competition between the two was thereby intensified and that both were able to increase their sales as a result of lowered transport costs and an ability to reach new markets.

From Pit Head to Riverside

From the storage piles near the pits Anzin sold some coal directly by the wagonload to local industrialists and coal retailers who supplied their own transport, but it hauled most of its output to the Escaut River for sale by the boatload. The company itself provided wagon transport to the loading areas until the early 1820s, when it contracted this service operation out to local hauling contractors. Through an employee charged with supervision of transport to the river, Anzin continued to keep close watch over the manner in which the carters performed their task. The new arrangement proved as troublesome as the former, and there were continual disputes between the company and the carters over pay rates, with the latter threatening to "strike" for higher rates and the company denouncing the exigence and greed of the carters. The neighboring Aniche Company, which usually followed Anzin's lead, felt obliged to provide its own carting service, despite its unprofitability, because the local carters would otherwise increase their demands beyond "just limits."[5]

Whether the wagons that carried the coal belonged to customers, the company, or independent carters, the company itself ordinarily built, often with the labor of its own employees and generally on land bought or rented for the specific purpose, the roads that directly connected the pits to public routes or the river. As a sign of proprietorship, Anzin set up stone markers along these roads and even planted their edges with poplars. It was especially active in repairing old and building new company roads in the first few years after the Revolution. Thereafter its road network fell into disrepair until the 1820s, when the opening of new pits and the general renovation of company equipment to meet new market demands led to the construction of a large number of new roads and the repaving of most of the old ones. The local public roads, supposedly maintained by the communes through which they passed, were usually in wretched condition, for agricultural villages were naturally not inclined to keep up and pave roads from their own slim financial resources in order to service the continuous traffic of heavy wagons carrying Anzin coal. Here again Anzin found itself forced to pave and maintain at its own cost or in partnership with the communes the short stretches of local public road over which its coal regularly passed to the Escaut River. And while most of the many bridges over the Escaut were public, Anzin built one for its own use in 1825 to enable it to bring Vieux-Condé coal to the new dock areas on the Fresnes side of the river.[6]

The lack of local paving stone raised construction costs,[7] and maintenance ran high on roads quickly degraded by the constant passage of heavily loaded coal wagons and by a climate where winter invariably meant heavy precipitation together with alternate freezes and thaws. Spring thaws and rains usually resulted in periods of several weeks or more when the public authorities declared the roads closed, and at other times of the year, too, passage even over paved roads was slow and difficult.[8] Figures for the cost to Anzin of road building and maintenance are not available, but at times it must have been substantial, although nowhere near the magnitude of the expenses involved in opening new pits or purchasing steam engines. Particularly during the 1820s, the extension of the private road network throughout the Torrent canton and especially among the pits near Denain and Abscon, in an area where public roads were few and distances greater than on the older pits near Valenciennes or Condé, helped suggest the desirability of alternate means of transport.

Beyond all these drawbacks of high initial and maintenance costs and irregularity of service, there were the heavy ordinary day-to-day expenses of cartage. For some mines elsewhere, the cost of transport from pit head to riverside was as great as the cost of cutting and raising the coal to the surface. This was especially true for the smaller French mines on the hilly edges of the Massif Central or on the flanks of the Alps and Pyrennes, but it also applied to the great Loire field, whose low extraction costs were largely offset by the heavy burden of getting its coal to the Loire or Rhone rivers.[9] The mines of Belgium and the Nord were much better situated, yet until the opening of the Mons-Condé Canal in 1814, two hundred wagons made four weekly trips each to the Condé docks from the Mons mines fifteen to thirty kilometers away.[10] Aniche was only six kilometers from its embarkation point at Bouchain and twelve from Douai, but the cartage costs were 18 centimes per quintal to Bouchain and 30 to Douai, whereas the charges for water transport from these points to the much more distant markets of Lille, Arras, and Dunkerque were only 20–30, 20–50, and 45–65 centimes, respectively.[11] Anzin was excellently placed with respect to its major highway, the Escaut River. Most of its pits, old and new, were much less than three kilometers from the river. Even so, the company paid at least 7 to 10 centimes per quintal for cartage from its pits to the loading areas; this charge amounted to as much as one-seventh of the cost of extraction.[12]

It is hardly surprising, then, to find Anzin proposing to build a tramway in 1810 for transport to riverside.[13] As with the experiment with railed haulage underground proposed about the same time, the company was probably anticipating the expansion of its market to the south and Paris as a result of the expected completion in the near future of the Saint Quentin Canal. But the Paris market remained closed to Anzin, and serious consideration of experiments with rails above or below the surface ceased for another decade. After 1820 Anzin met increased demand by installing railed haulage underground, but at the surface it merely launched a feverish program of road building. This, despite the sensation caused by the publication in 1818 of the government engineer Gallois's description of British tramways, short stretches of wooden or iron rails on which horsedrawn wagons transported coal from pit head to river, canal, and sea.[14] With a very few exceptions, France in 1818 had nothing to match the extensive network of horse-powered tramways on the British coalfields, and the British also led in the practical development of the locomotive with the completion of the Stockton-Darlington and Liverpool-Manchester railroads in 1827 and 1830. Meanwhile, "the first French railroad" began operating in 1827 over the sixteen kilometers between Saint Etienne and the Loire River. Although it was horse-powered, its purpose was identical to that of the Stockton-Darlington; both were built almost exclusively to carry coal from pit head to water. Two longer railroads opened in the Loire coalfield area by 1834.[15] Meanwhile, one of the largest mines in the Mons field, the Grand-Hornu, opened a horse-powered tramway from its pits to the Mons-Condé Canal in 1830. It was rewarded by a rioting mob of carters, who tore up the tramway and pillaged the company's offices and shops.[16]

In the same year, Anzin decided to construct a wrought-iron "chemin-de-fer dit railway" with horse-drawn wagons to connect the three new pits in the Denain area with elaborate new loading and dock facilities on the Escaut.[17] The timing of the decision and the location chosen are easily explained. Certainly the success of the first Loire railroad and the initative taken by the Grand-Hornu played their parts. The Loire and Mons mines were Anzin's great rivals in the Paris market, which was opened to Anzin at last with the improvement of the Saint Quentin Canal in 1828. The Denain docks were closer to Paris by six kilometers of river and several troublesome locks than were the older loading areas near Valenciennes; the Denain pits, producing the best forge coal on the company's concessions, were marked for rapid expansion. Finally, the existing roads were poor and distances relatively great around Denain.

A further motive seems to have been an attempt to obtain leverage in the company's endemic squabbles with the carters. The board approved a new cartage contract cutting standard rates in the same meeting at which it took the decision to push forward with tramway construction. Three years later, when the carters were again pressing for higher rates, three new stretches of tramway were under consideration: a very short line to service the new pit at Hergnies on Vieux-Condé, a more important one to service the Torrent pits, and an extension of the original Denain line to the west where explorations were proving successful and new pits were being sunk.[18] All these lines were short and horse-powered; the initial Denain line was only one kilometer long. But by 1835 Anzin was beginning to link up its isolated stretches of track by building an eight-kilometer line from the old Anzin pits to Abscon by way of Hérin and Escaudin. In 1838 this line was in operation with the first locomotives used in the Nord.[19] Only in 1846 when the Paris-Lille trunk line opened did Anzin's private network link up with common rail carriers serving the markets Anzin had long reached by water. Until then, Anzin's tramways and railroads merely replaced the road as a means of moving its coal from pit head to riverside.

After being carted to the Escaut River, the coal was sold and loaded into boats. Loading arrangements were far from elaborate at the turn of the century. Evidently the loading areas (*rivages*) were merely small open spots on the tamped dirt embankments used to dike the river. The carters probably drove their wagons up onto the embankments and with the aid of a few loaders (*chargeurs*) dumped the coal into the boats. The company owned or leased some of the loading areas; others belonged to communes or private individuals who permitted the company to use them for a fee. Each group of pits had its own loading area, whose location owed as much to the exigencies of river travel and market locations as to the carting distance from the pits. Thus the loading area for Fresnes coal lay almost opposite that used by Vieux-Condé, because navigation through the town of Condé was exceedingly difficult and the major market for both these hard-coal concessions lay downstream beyond Condé. For similar reasons the two chief loading areas for the Raismes and Anzin pits lay downstream from Valenciennes, whose locks and fortifications were also an imposing barrier to navigation.[20]

As in so many other matters, so too in the improvement of loading arrangements the company made minor changes in the first decade of the century, more important ones under Casimir Perier in the early twenties, but only about 1830 did it achieve a decisive breakthrough. It carried out a decision to buy the Fresnes loading area in 1806 at a cost of 10,500 francs, and about the same time it acquired a ninety-nine-year lease for the Vieux-Condé area. The lease with the commune of Vieux-Condé provided for a yearly rent of 300 francs and the rebuilding of the communal hall at an estimated cost of 3,000 francs. In 1810 the company decided to replace the sloping embankment at Vieux-Condé with a stone quai, built according to the specifications laid down by the Bureau of Bridges and Roads at an estimated cost of 8,667 francs. It gave up plans for a new loading area near the Notre Dame gate on the upstream side of Valenciennes when the opening of the Saint Quentin Canal in 1810 did not much increase upstream traffic, and no more is heard about any of the loading areas for more than a decade.[21]

Then, between 1823 and 1826, Casimir Perier led a sweeping reorganization and expansion of the company's loading areas. Their insufficiency had long been evident; boats often tied up for as long as two months while waiting to be charged.[22] By the early twenties, the problem was more serious, if only because Mons producers, able to offer quicker loading as well as coal of higher reputation, were increasingly penetrating the Anzin markets. Even the most loyal of Anzin bulk-coal buyers found themselves forced to warn the company that pressure from their own customers would soon lead them to seek better service from Mons producers. Furthermore, corruption and graft were rife on the waterfront. The coal buyers bribed the minor company officials in charge of the loading areas to load their boats with coal from the pits noted for the high quality of their output. Other buyers, receiving the coal of the poorer pits, unleavened by judicious mixing with better qualities, were naturally indignant and inclined to buy elsewhere.[23] Anzin's enemies pointed gleefully to the long lines of boats waiting to be loaded on the Escaut below Valenciennes as one more reason for the government to take steps to break Anzin's reputed monopoly.

Slow service at the riverside was partly due to insufficient production, especially of top quality forge coal, at a time when consumption was rapidly increasing and Anzin was still struggling to replace exhausted working areas with new pits on the Torrent. But the company could take immediate steps to eliminate graft, tighten discipline, and speed up operations at riverside. The first step was to contract out the job of charging the boats; the loaders, like the carters, were henceforth paid by a contractor, who sold their services to Anzin at piece rates.[24] Work in the loading areas, however, remained under the eye of a company official. This official, a retired naval officer named Tombini, had wide powers and a high salary and status as compared to the weak, bribe-taking employees previously in charge of loading. Tombini was to oversee "transportation from the pits to the loading areas, mixing and arrangement of coal at the areas, loading of the boats, and in a word everything relating to the supervision and good direction of the loading areas." He was informed each day of the number of boats loading and the kind of coal they required, and he kept a register of wagon arrivals at the areas and of boat loadings.[25] By contracting out carting and loading operations while placing them under tightened company supervision, Anzin must have hoped to put a strong double watch over the fractious carters and loaders.[26]

At the same time, the company moved part of its stockpiles from the pit heads to the riverside so as to provide a continuous supply of coal at the point of sale. This important step, plus the expected rapid increase in sales, forced the company to expand the size of all its existing areas and provide rough shelters for its stockpiles. The opening of the Torrent pits and renewed hope of opening a large upstream traffic to the improved Saint Quentin Canal and thence to Paris led to a search for two new sites on the Escaut above Valenciennes. By 1826 each of the loading areas, old or new, was capable of charging twenty or more boats a day.[27]

Even these improvements were soon insufficient to meet rising sales, and in 1830, an especially good year, the company was hard pressed to fill orders made two months earlier, and as many as one hundred boats at a time were lined up below the main loading areas for an average wait of ten to twelve days. Many slipped off downstream to Mons in disgust.[28] Once again, much of the difficulty was due to a sudden expansion of demand beyond the company's immediate production capacity, but Anzin was now determined to provide transport and loading facilities that would suffice for years to come. In 1833 it again expanded all the loading areas,[29] but already a more basic solution was in operation.

Much of the needed increase in output of good forge coal would have to come from the Denain pits, just now coming into full operation and superbly placed to feed the Paris market. In 1830 the company launched construction of new loading facilities at Denain. A hand-excavated canal several hundred meters long led from the Escaut to the village, and a large basin there provided off-river anchorage for dozens of boats. Equipped with spacious docks and an improved system of gravity loading, the new facility was no mere rivage; this gare cost some 70,000 francs. And, a gare in the modern sense, it was linked to the pits by a horse-drawn tramway. The company soon extended the tramway to link other groups of pits with the new loading facilities at Denain. Increasingly, it threw the bulk of the loading onto this single point until through railroads took some of the burden of long-distance transport away from the waterways altogether.[30]

The Organization of River Transport: The Boatmen

Like the carters and loaders, the boatmen (*bateliers*) who transported Anzin coal on the rivers and canals of northern France and Belgium were, from the company's point of view, a troublesome lot, attached to traditional ways and prone to clandestine, illegal organization aimed at enforcing high freight charges and a living wage for all. Water transport was in some ways even more disorganized than land transport. While there were some general boat contractors or boat companies who owned or controlled several boats and with whom a shipper like Anzin could contract for all or part of its carrying business, most boats were owned and operated by their owners with the help mainly of members of their own families.[31]

Almost every good-sized town (and some small ones) on a navigable waterway had its own fleet of boats. Some of them, like Condé, devoted themselves almost exclusively to water transport. Of the 8,000 inhabitants of Condé in 1791, 385 were "master boatmen" (i.e., owners and operators), 800 were master's sons who worked for their fathers, and 700, along with inhabitants of nearby villages, made their living by boat towing.[32] The only industry in Condé was boat building; in 1801 nine boatyards there employed a total of only 30 people. Six more yards were scattered along the Escaut between Condé and Valenciennes. These tiny yards probably operated only during those winter or summer months when ice or low water closed the river.[33]

A great variety of types of boats plied the waterways of Flanders and Hainaut. Most Condé boats were designed for travel primarily on the broad, deep Escaut and carried almost exclusively Mons and Anzin coal. When fully loaded, Condé *nefs* could carry a cargo of about 150 tons of coal, but on leaving the Escaut, the nefs of Condé had to be lightened of half their loads and the excess cargo transferred to smaller boats from other towns with whom the Condé boatmen had rate agreements. The main points for lightening were Mortagne, where the Escaut meets the Scarpe, starting point for the network of waterways leading to French Flanders, and after 1810, Saint Quentin on the canal that connected the Escaut to the Oise and Seine basins. In 1787 Condé was home port for about 350 nefs; in 1812 about 600 operated from Condé and perhaps 200 more from Mons. By 1832 they traveled far from the Escaut with their lightened cargoes of coal on all the waterways of northern France and Belgium, from Rouen and Paris to Brussels and Antwerp.[34]

Although in the eyes of the government the boatmen were small entrepreneurs who were free and indeed were expected to compete with one another in establishing freight rates, in practice they continued to yearn for the restrictive, monopolistic, and cooperative structure of their business as it had existed under the Old Regime. Each town in French Flanders and Hainaut then had its "corporation" of master boatmen with special legal rights to navigation over certain local stretches of water. By far the largest and most powerful of the corporations were those of Dunkerque and Condé. Located at opposite ends of the area's network of waterways, each was strategically placed, and each controlled access to major sources of cargo, Dunkerque of overseas imports and Condé of Mons and Anzin coal. With a combined tonnage equal to that of all their rivals combined, Dunkerque and Condé boatmen gradually encroached on the privileges of neighboring towns.[35] For example, the Condé corporation succeeded in absorbing the much smaller Mons corporation with its privileges on the Haine, the tiny tributary which emptied into the Escaut at Condé and along which most of the Mons pits were located.

Together, the Condé and Mons boatmen had the exclusive right to load coal on the Haine and on the Escaut from Valenciennes past Condé to Mortagne at the Belgian frontier and confluence with the Scarpe. They also had the privilege of establishing the *tour de rôle* on the Escaut and Haine, that is, of dictating the order in which boats might load at any given point, such as one of the Anzin loading areas, or the order of passage at locks and town fortifications. The object of the tour de rôle was partly to distribute the available work evenly among the members of the corporation, but it also placed the boats of other towns at a serious disadvantage. Not only were they forced to return empty to home waters after discharging their cargo in Condéan territory, but they sometimes lost whole months in passage time as nefs belonging to the Condé corporation used their rights to move ahead of them in the long lines of boats that were always waiting to pass the numerous inefficient locks along the Escaut.[36]

The Anzin Company was naturally opposed to the privileges of the Condé corporation of boatmen, but it failed entirely in its long battle to eliminate them. Its failure reveals something of the complexity of administration under the Old Regime, and especially of the relationship between the government and economic activity. Anzin came into existence as a result of a partial defeat for the prescriptive medieval rights of aristocratic seigneurs to the exploitation of the subsoil, but it was unable to destroy the much more recent privileges of a group of petty bourgeois entrepreneurs. Its failure seems to have stemmed partly from division within its own ranks; Croy, a great noble and one of the company's founding fathers, supported the corporation of Condé in his role as haut-seigneur and military commander of the town.[37] Turgot, with his physiocratic conception of economic freedom, abolished the corporations for a short period in the 1770s, but his action seems to have had little practical effect on the rivers of Flanders and Hainaut, and when the Revolution came, the Condé boatmen still ruled the Escaut and the Haine.

The Constituent Assembly abolished the corporations by a law of 17 March, 1791, and two months later, on 12 June, another act specifically threw open inland navigation to all, free of any restrictive agreements or actions. Succeeding governments came under a barrage of appeals from towns like Condé, which claimed free navigation would mean the destruction of the only source of livelihood for the town's population. Under the circumstances of revolutionary turmoil and war, Paris retreated; it issued temporary exemptions from these two laws and delayed their enforcement locally as on the Escaut.[38]

An order for the Nord, issued in 1798 by the Directory at Paris and reaffirming freedom of navigation for all, illustrates the difficulties the government encountered. It provided for recompense to boat owners and freight shippers for property damaged by mobs, broken bridges, or trees felled across stream channels and for enforcement of the order by an "armed force" made up of old campaign veterans stationed at Condé "to protect the liberty of navigation." Mons coal merchants and mine owners greeted the order with public fetes where prizes were distributed to the mining populace amid cries of "Vive le République!"[39]

Yet the government had to continue to concern itself with establishing de facto "liberty of navigation," for the formally abolished corporations continued to operate by tacit agreement or behind the screen of "insurance associations."[40] At least Anzin was now at liberty to take various steps that would give it leverage in its relationship with the boatmen. One measure, long feared by the boatmen, was to build its own boats. In 1801 Anzin proposed to build ten, evidently for its small business with Holland, which it hoped to expand. In 1807 the company decided to "shelter itself from all exaggerated claims on the part of the boatmen" by building "a certain number" of boats capable of navigating on all nearby rivers and coasts, and in 1811 it approved the blueprints for a specially designed boat capable of carrying one hundred tons to Paris by the new Saint Quentin Canal without lightening its load.[41]

One suspects that Anzin's activities as boat builder were in fact very limited. There is no evidence for the number of boats actually built or the number in operation at any given time. That there were some is confirmed by the fact that Anzin decided to sell its boats in 1822. That there were few can be concluded from the lack of evidence in company records, the continuing struggle to obtain satisfactory rates and service from the Condé boatmen, and the failure of the specific commercial projects to which the original construction plans were linked. For the Holland trade was never large and disappeared entirely with the collapse of the Empire, the plans for through navigation to Paris were based on an overly optimistic view of the practicability of the Saint Quentin Canal, and there is no evidence of any kind for company boats on the "rivers and coasts" of northern France. Perhaps as a competitive threat to insure the cooperation of the Condé boatmen, Anzin maintained a small fleet that, given Anzin's resources and the ease with which these canal boats could be constructed, might be rapidly expanded at any time.

Anzin also took steps to provide for a tour de rôle based on the principle of first come, first served. Since the government of the Empire provided "navigation guards" to regulate traffic on the waterways but failed to pay these agents, Anzin stepped into the breach. In 1808 it proposed to pay a yearly salary of 300 francs to the navigation guard at Condé until the government took over the expense itself. Six years later, when the government abolished the guards as an economy measure, Anzin again decided to pay "ces agents de police si essentiels" on the stretch of the Escaut from Mortagne to the Saint Quentin Canal, at least until the merchants and others interested in navigation throughout the area arrived at a permanent agreement for their support. The Restoration government assumed the burden of enforcing freedom of navigation, but the Condé boatmen remained so bitterly opposed to this unwanted freedom and so ready to combine for resistance, that on one occasion in 1824, "the local authorities being powerless," the prefect found it necessary to call out the gendarmerie.[42]

The sale of the company boats in 1822 during the massive reorganization under Casimir Perier was one of many examples then of the company's determination to rid itself of all activities not directly related to the mining of coal.[43] And like many other such steps, it was probably taken by Perier himself who made the decision hurriedly and blindly, without taking sufficient account of the rationale behind the old system or the repercussions of abolishing it. The Condé boatmen probably responded to the sale by making a shortsighted attempt to exploit their restored transport monopoly and by demanding higher rates. Anzin retaliated in 1824, not with more boat building, but with a more subtle and complete, yet less expensive, means of control. At that very moment it was reorganizing its carting and loading operations by contracting those services out to private entrepreneurs who would assume the task of disciplining the fractious carters and loaders. With the boatmen, who were themselves petty entrepreneurs, the problem was somewhat different. To guarantee regular, economical river transport it seemed necessary to break up their illegal but stubborn unity by inducing individual boatmen to bind themselves to exclusive service with Anzin. Choosing those who were financially responsible and could be trusted to live up to their bargain, the company signed contracts with enough boatmen to assure the transport of its entire output. Those chosen received a fixed weekly payment, in return for which they agreed to carry only Anzin coal and to accept whatever load and whatever place in line at the loading areas the company might order. Such boatmen, although they continued to own their own equipment and were not mere wage earners, were very much less independent than they had previously been. Anzin had successfully broken the solidarity of the Condé boatmen and provided for fast, efficient, and regular service at its loading areas.[44]

One serious difficulty remained: its chief customers, the wholesale coal merchants, had to agree to abide by the new system. The company expected resistance from these dealers, for some of them had their own boats, some had contracts like Anzin's with certain boatmen, and the rest had long been accustomed to striking the best bargain possible with independent boatmen. Now the company tried to convince the free-lancing majority of merchants to deal only with Anzin boatmen and to accept the fixed charges paid these boatmen as a regular part of the purchase price for the coal they bought; to persuade those who preferred to contract for transport themselves to do so only with the boatmen favored by Anzin; and to force these dealers and those with their own fleets to accept their places in the loading line with the other favored boats in the order dictated by the company. Perhaps this point of preference in loading was the most sticky one, for a poor place in line might mean the loss of precious weeks while working capital lay idle, customers complained, or the river fell or froze over. Where persuasion failed, Anzin was prepared to secure the recalcitrant merchants' cooperation with the new system by offering them "douceurs particulaires."[45]

The new contract system may never have worked as completely and regularly as Anzin would have liked, but it seems to have served its purpose. In 1827 and again in 1830 the Belgian government paid large subsidies to the mining companies of Mons for the construction of over one hundred boats on the grounds that the Mons coal trade was suffering because of a shortage of boats while Anzin's sales were "très brillantes."[46] Evidently, Anzin had succeeded in its attempt to bind many of the Condé boatmen to its exclusive service by the contract system. This success implies that the terms offered by Anzin were enticing enough to override the boatmen's ingrained habit of seeking economic security through cooperative action with their peers against the mine owners. Their break with the tradition of collective pressure through both legal and illegal associations was no doubt due to the difficulty of maintaining an illegal and purely voluntary solidarity for many years under the combined attack of governmental coercion and Anzin's economic pressure.

More subtle factors, undermining the profitability of the coal-carrying business, may also have played a role, for it is clear that the fortunes of the Condé boatmen had declined greatly in the early nineteenth century as compared with the years of the Old Regime. Already between 1801 and 1808 the number of boatyards around Valenciennes and Condé had decreased from fifteen to six; by 1812 there were none operating, and no new boats had been built in the previous two years.[47] Whatever revival took place immediately after 1815 was limited by the declining profit margins of the boatmen, who were thus unable to replace worn-out equipment. Much of the

Condé fleet was really dilapidated by 1830.[48] Yet the independence of the Condé boatmen was even yet not ended.

In 1832 the boatmen informed Anzin by letter of the formation of the "Insurance and Chartering Society" and asked to open a branch office at Anzin. The company claimed that it "did not want to set up any obstacle to the bettering of the condition of boatmen who have united to form this society, but that it also desired that coal shipments be protected from an excessive rise in freight rates." The Anzin policy was to keep a careful eye on the new organization, which was perfectly legal.[49] So long as the Escaut remained the essential artery of Anzin's trade, the petty entrepreneurs of Condé found new ways to organize legally for illegal ends and threaten Anzin's commercial lifeline by collective action.[50]

Improvement of the Escaut and Other Waterways

Anzin extraction costs were far higher than those of Newcastle and the Loire and substantially higher than those of Mons. Its ability to compete successfully with these rivals in its established markets and to expand its sales into new areas was clearly related to the fact that its markets were more or less sheltered from competitors by high transport costs. The cost of transportation by river and canal fell somewhat in the early nineteenth century, especially after 1815. Nevertheless, the problems of improving the waterway network were so great and government action was so halting that no real breakthrough occurred until the coming of the railroads at mid century. In several places Anzin played a leading role in the improvement of navigation in the Escaut basin. Yet in others, most notably along the stretch of river nearest its pits, it displayed a stubborn reluctance to help push to completion projects that would have aided its Mons rivals more than Anzin itself. Supported by the army's views on protection of the frontier, Anzin succeeded in slowing the development of a really adequate network of waterways in the basin until the 1840s, on the eve of their partial replacement by the railroad.

The archival material dealing with the French waterways is very extensive, but all the evidence shows that the problems faced and the solutions reached were similar everywhere.[51] Therefore the following discussion is focused primarily on an analysis of developments on the Escaut, the most important waterway for Anzin, and its immediate links by canal with other river systems. The major stress is on the technical problems faced, Anzin's role in aiding or blocking improvements, and the effect of improvements on Anzin's competitive position.

Valenciennes, Cambrai, and other towns on the upper Escaut were thriving river ports until the treaty of Westphalia (1648) closed the Escaut outlet to the sea through Antwerp. Thereafter trade on the Escaut fell into desuetude, and by 1770 the upper reaches of the river from Condé to Cambrai (the head of navigation) were no longer passable for through traffic because of the massive fortifications at Valenciennes, and even smaller boats found shorter stretches difficult to navigate. Then, in the early 1780s, the expanding Anzin Company was instrumental in obtaining construction of a canal around the fortifications. Some other parts of the river were straightened and a number of locks installed between Condé and Cambrai. Laurent, the famous civil engineer in charge of these projects, was an uncle of one of the Anzin owners. Anzin pledged to build two "ports" outside the town walls in order to give the townspeople direct access to the new navigational canal but failed to keep its promise;[52] while Valenciennes remained without water transport for all practical purposes, Anzin was in effect provided with a main highway on its doorstep. In 1800 the French portion of the Escaut was used almost entirely for the transport of Anzin coal. By far the greater part of the traffic moved downstream past Condé; movement upstream was limited to a mass of returning empty boats from Condé and beyond and a few laden boats headed for Cambrai. Laurent's system of artificial navigation primarily facilitated the downstream passage of loaded coal boats. But movement in either direction, for loaded or unloaded boats, was extremely slow, laborious, and costly.

The Escaut snaked across its flat, marshy valley floor in great loops, which doubled effective navigable distances, encouraged silting and flooding, and made necessary tedious maneuvers to navigate the innumerable sharp bends. The shallow, shifting, and poorly defined channel was full of rocks and sand bars. Boats foundered at a rate of perhaps one every two years, and each loss stopped the passage of another hundred craft until the sunken hulk could be cleared away. Although the Escaut was well-supplied with water in comparison with many other French rivers, it was nevertheless impassable at certain seasons of the year. From late August to late

October the water level fell so low that all navigation ceased, and in mid-winter the river often froze over for several weeks at a time.[53] But even during the seasons when the river was passable, the water level did not permit the biggest boats to travel more than one-half to three-quarters loaded. Traffic on the river was especially crowded and hence especially slow during November and December, when industries and towns were hastening to replenish coal supplies exhausted during the late summer layoff and to stockpile for the winter months when the river would again be closed.[54] The regular interruption of navigation twice a year for many weeks at a time also tended to reduce the number of annual round trips a boat could make between Anzin or Mons and important markets like Lille. It increased the pressure to speed loading and passage of locks and the importance of a favored place in the tour de rôle. The interruption of navigation by low water or ice thus had repercussions far out of proportion to the actual number of days directly lost.[55]

The low water level and the difficulties of upstream passage were in large part due to an inadequate system of locks. Laurent had installed "simple locks" (*écluses simples*), little more than dams with narrow lift-gates. In the reach behind each lock up to one hundred boats collected waiting their chance to move downstream. Passage usually took place only two or three times a week when the reach was full. Then the gates lifted and the boats first in line shot a narrow rapids on the crest of a sort of cataract. In the reach below they waited again until the river built up a new head of water sufficient to float them through the next lock downstream. The tricky maneuvering at each lock required the use of towing machines and dozens of towmen, who manipulated the boats at the ends of ropes. The services of the towmen alone were said to triple the cost of transport. The difficulties were especially great where the locks lay amidst fortifications as at Bouchain, Valenciennes, and Condé. At Condé, forty men performed acrobatics atop the high walls of the fortifications to maneuver boats through one of the locks. Often the chains, gears, or other parts of the locking mechanisms broke, and traffic along the entire river halted to wait for spare parts and repairs, a problem that grew during and after the Revolution when the entire system of locks fell into disrepair for lack of funds.

A simple lock also failed to provide a really satisfactory water level along most of the reach it controlled. Because of the gradient of the river, the depth of a filled reach might vary from a barely adequate four feet of navigable water just above the lock to less than two feet at the upstream end of the reach. Naturally, the shallower depth determined the effective load a boat could carry. Behind a closed lock silt deposits tended to build up, narrowing the channel, raising the stream bed, decreasing water depths, making passage more difficult, and increasing the danger of floods. The mills that were clustered around each lock for the sake of a head of falling water to move their waterwheels also robbed navigation of precious water. So too did the opening of the Sensée Canal in 1817; the canal diverted so much water from the Escaut that boats could make only one passage per week of the locks between Valenciennes and Condé, and then only by stripping the entire five-mile reach from Valenciennes upstream to Denain of all its water and hence closing traffic in that direction.

Towing was a serious problem not only at the passage of locks and fortifications but also along the entire length of the navigable Escaut and, indeed, along most other rivers and canals in France. The towpath between Condé and Cambrai was treacherously narrow and uneven, a mere track through thick brush or swamps or along the tops of levees. Riverside installations such as boatyards or the Anzin loading docks and workshops forced the towpath to cross the river nine times between these two points. Since horses could not negotiate this type of path, men, women, and children of the riverside communes did the towing along the Escaut. In the hierarchy of transportation workers, towmen ranked at the bottom, far below the boatowner, a petty entrepreneur, and even below the teamster, who had a degree of skill and perhaps his own horses and wagon. Some towmen, like most teamsters, were peasants for whom their transport activity was a by-occupation; others were town dwellers, a swarming proletariat in places like Condé, where the more successful inhabitants were boatmen. Like both their fellow groups of transport workers, the towmen were too numerous to sell their labor at a living wage in an open market. Hence, they formed illegal combinations aimed at forcing up wages, slowing down work, and spreading the available jobs as widely as possible among the members of their miserable group. As late as 1826 neither the gendarmerie nor the regular army units was able to break up permanently the combinations of the Condé towmen.[56] The only possible solutions, it was claimed, were either the building of towpaths good enough for horses or permanent military occupation of the town! But in the early years of the nineteenth century the first solution was not even considered. So long as most loads moved downstream from Anzin and Mons, the current furnished most of the power, and the towing problem was not acute. Depending on local river conditions, four to six towmen sufficed to haul an empty boat upstream against the sluggish current with many long halts at locks and other obstacles. Downstream, six to eight men were needed to maneuver the partly loaded boats. Passages of locks and fortifications in both directions required many more hands, and the few loaded barges moving upstream ordinarily required sixteen to twenty towmen.[57] Only when this upstream traffic became important was there any serious reason to improve towpaths so that a pair of horses could do the work of a score of men. And even when the direction of traffic did change, it was many years before animal power replaced human power on the banks of the Escaut.

In the late eighteenth century Anzin had taken the lead in demanding improvement of the Escaut, and the project undertaken then was directed by a relative of one of its owners, the engineer Laurent. In the early nineteenth century, however, the government Bureau of Bridges and Roads seems to have initiated the plans for further improvements. The project drawn up after 1817 centered on the addition of several locks to break up the longer reaches and the replacement of the old locks. In both instances the new facilities were double-walled locks (*écluses à sas*) of the modern type, in which boats rose or fell in a chamber closed at each end. Such locks operated quickly and without danger; they required little maneuvering and hauling about of the boats; but their prime advantage, of course, was that they eliminated intermittent passage and all the evils associated with it. Boats passed continuously through the chamber locks upstream or downstream, loaded or unloaded, with equal ease; and the water level in the reaches maintained a stable depth. Aside from their intrinsic merits, the chamber locks also had the great advantage of being new and kept in good repair. Besides the replacement and addition of better locks, navigation benefited by a certain amount of channel dredging and straightening and levee raising, which together eased towing, lessened silting, deepened channels, and shortened distances.

Yet the practical results of the work done in the decade after 1817 were not great. Initially, the government engineers had proposed to maintain a navigable depth of 1.80 meters (2 meters of water depth) throughout the distance from Cambrai to the Belgian frontier. This depth would have allowed Anzin's largest boats to pass fully loaded. But along most sections of the channel it proved difficult to maintain a depth of even 1.20 meters at the most favorable seasons. The attempt to deepen the channel further lapsed until after 1836, partly because of technical difficulties and partly because until then the shallowness of the waterways connecting with the Escaut gave no incentive to reach a more ideal solution. More serious yet was the failure to substantially lessen the time of passage by ending intermittent navigation. Although the five new locks built before 1825 worked well, three locks of the older type remained, two of them at Valenciennes near the Anzin loading areas. These three older locks continued to hinder navigation and nearly eliminated most of the advantage gained from the others. They were not replaced until 1836 or later, and it was only in the period 1836–41 that the Escaut was successfully canalized and its navigation made fully artificial by systematic straightening, deepening, and diking. Only then could horse-towed boats move continuously in both directions with a full load.

Meanwhile, movement between Condé and Cambrai past the Anzin pits remained exasperatingly slow. In 1820 it required sixteen days to cover the distance one way, and only six round trips a year were possible. Two years later, under the most favorable conditions, boats took only two days to run the difficult stretch from the Anzin loading areas above Valenciennes to Condé, but eight days was a more normal passage time. In 1832 boats emerging from the Mons-Condé Canal still required from four to eight days to move upstream past the locks at Valenciennes, a trip whose duration would fall to one day, it was said, with installation of the still anticipated three new locks. Four years later traffic on the river had grown enormously, especially upstream, so that passage times had perhaps even increased; the voyage from Condé to Cambrai then took "several weeks" as compared to sixteen days, when the government had first considered the project of rebuilding the river nearly twenty years before.

There can be no doubt that military considerations played a strategic role in blocking the replacement of the three key locks and thus delaying completion of the project to rebuild the river. All three lay among fortifications, two at Valenciennes, the other at Bouchain. All were essential in their old form to the army's plans to defend the frontier by flooding the countryside, as during the invasions of 1814–15. Time and again after 1817, the engineers of the Bureau of Bridges and Roads believed they had successfully overcome military conservatism; time and again at the last moment the objections of the army's corps of engineers halted the plans of their civil counterparts. The clash of military and economic values was heightened by the clash between two proud, bureaucratic corps who questioned each other's professional competence and good faith.

A second reason for the lack of urgency in completing the improvements was that boats traveling to the major markets for Anzin and Mons coal could reach them by routes that avoided the worst bottlenecks on the upper Escaut before 1817, the two locks at Valenciennes. Almost all of the Mons coal passing through Condé moved downstream to the big markets of French and Belgian Flanders. Only very small quantities attempted to move along the upper Escaut past Valenciennes until after the improvement of the Saint Quentin Canal in 1828. Anzin's important hard-coal pits on Fresnes and Vieux-Condé also loaded downstream of Condé and found their way primarily to the Flemish brickmakers and chalk roasters of the Nord and Belgium. Some of the coal from Anzin and Raismes moved upstream toward Cambrai from loading areas well above Valenciennes, but much more entered the river below the Valenciennes locks and moved downstream to join the Mons shipments beyond Condé. The worst stretch was that between Valenciennes and Condé, but this was at least passable downstream, and the stretch above Valenciennes toward Cambrai was the easiest to negotiate in an upstream direction.[58]

The year 1817 saw both the opening of the Sensée Canal and the beginning of improvements on the Escaut.[59] The completion of the canal, in which Anzin played a major role, changed the direction of much of Anzin's shipping. By connecting the upper Escaut just above Bouchain with the Scarpe River, it shortened considerably the distance between the Raismes and Anzin pits and their markets in Arras and Douai on the Scarpe and in the Lille and Dunkerque areas farther north, which were reached by way of the Deule and other waterways. Traffic abandoned the passage from Valenciennes to Condé, and new loading areas built above the Valenciennes locks once again allowed boats to avoid them. It may be significant that the improvements made on the Escaut between 1817 and 1827 benefited primarily the now important upstream passage from Valenciennes toward Cambrai and the Sensée Canal, while the unimproved state of the Condé-Valenciennes stretch of river insured a difficult passage for Mons coal to the upper Escaut.

The roundabout Sensée route had little attraction for Mons producers. They were primarily interested in the passage of Condé and the stretch of river below Condé leading to the Scarpe and thence to Lille and Dunkerque, or to the Belgian Escaut and thence to Ghent and Belgian Flanders. For centuries Mons coal had reached the Escaut at Condé by the Haine or by wagon at loading docks below Condé. Then in 1814, during the Napoleonic regime, the Mons-Condé Canal, roughly paralleling the Haine, was completed.[60] The new canal, which was intended to link the Mons pits to Paris by way of the newly opened Saint Quentin Canal, mainly eased connections with older markets reached by the Escaut below Condé. But the passage from the Mons-Condé Canal into the Escaut past the Condé fortifications remained difficult in 1817:

> The passage of the Condé lock takes place only twice a week. Only fifteen boats pass at one time. The passage is interrupted constantly as a result of floods, and the boats wait one, two, and even three months [sic] to take their turn. The boats can make only three or four trips a year from Lille or Arras to Mons and a hectoliter of coal which sells for 0.75 francs at the pits costs 2.25 francs at Lille because of these delays. Transportation of coal by land is quicker and costs less; [but] the roads have been badly degraded [by the heavy coal wagons].[61]

In the early 1820s the passage of Condé was substantially improved, and traffic on the Escaut below Condé grew rapidly, but this part of the French Escaut, like all the Nord waterways, remained far shallower and less well maintained than the Belgian Escaut and the rivers and canals with which it linked up. After making the short but unavoidable passage of the French Escaut below Condé, Mons boats heading for Lille or Dunkerque preferred to continue down the Escaut into Belgium rather than take the Scarpe, Deule, and other Nord waterways. They then reached their French markets by the very long and roundabout but quicker and easier way of Ghent and the Lys River or the Ghent-Bruges-Ostend-Dunkerque canal system. To avoid paying the regular tariff rate for the short transit on the lower French Escaut and to get around the shallow and intermittent navigation on this stretch of river and the still somewhat difficult passage from the Mons Canal into the Escaut at Condé, the Belgians in 1826 completed the Antoing Canal, which connected the Mons-Condé Canal to the Belgian Escaut and thus eliminated the short loop of French Escaut from their route to French and Belgian Flanders. The number of boats using the French Escaut below Condé fell immediately from nearly two thousand a year to almost none.[62] A decade after the improvement of the Escaut had begun, the pull of the Sensée and Antoing canals eliminated almost all traffic on the Escaut from Valenciennes as far as the Belgian border.

At almost this exact moment, however, another canal was about to open which would enormously increase the importance of the upstream passage from Condé. A water connection between the Escaut and the Seine-Oise basins had been contemplated since at least 1721. By the middle of the eighteenth century the Crozat Canal joined the Somme and Oise, and the passage from Paris north to Anzin was complete as far as Saint Quentin. The gap between Saint Quentin and Cambrai was, however, more difficult to bridge. In two places, underground tunnels were necessary to carry a canal beneath the low ridges of the watershed, and it was difficult to secure an adequate supply of water for such a canal. Various plans under the Old Regime came to nothing. The Revolution revived these plans, and under Napoleon work was actually begun in 1801. Lack of funds and other problems delayed the opening of the Saint Quentin Canal until 1810.[63] Then, hailing the completion of the longer of the subterranean passages as an engineering marvel, the *Annuaire statistique du Nord* announced with satisfaction, "The problem is resolved. The north and south of France are now assured of an uninterrupted communication by water."[64] The Anzin Company itself had expressed interest in the completion of the new water route to

Paris as early as 1798, and Desandrouin evidently based his plans for Anzin's expansion in the first decade of the nineteenth century in large part on the access the new canal would give to the potentially enormous market of Paris and Rouen.[65]

Anzin's expectations and the general enthusiasm proved premature. The canal lay closed most of the year for lack of water, and even when open it was so shallow that only boats of extremely light draft could navigate it. Thus it was necessary to lighten loads at Cambrai and recharge again at Saint Quentin. The Oise and Escaut on either side of the canal were far from ideal as routes for mass transport of cheap, bulky goods like coal. Anzin opened a depot for the reception of coal at Saint Quentin and sales offices there and in Paris; but the company's efforts in these areas remained experimental and the quantity of coal involved remained minimal until after 1828 when the Saint Quentin Canal reopened with a much more satisfactory water level of about 1.20 meters.[66] Thereafter, upstream traffic on the French Escaut grew heavier and heavier as an increasing number of Mons boats attracted by the Paris market joined below Valenciennes those boats carrying Anzin coal toward the Sensée and Saint Quentin canals. By the middle of the 1830s the river had become greatly "encumbered," especially for Mons boats at the passage of Condé, the old Valenciennes locks and Anzin's gare at Denain. Only then was the long overdue replacement of the old locks and major improvement of the channel finally undertaken.[67]

The key stretch on the upper Escaut and the last section improved was that from Valenciennes to Condé. For Mons producers, it furnished access past Anzin to the Cambrai market and eventually to Paris and the basins of the Somme, Oise, and Seine. For Anzin, it was long the only means of reaching all the major markets of the northern part of the Nord department—Arras and Douai on the Scarpe and the Lille and Dunkerque areas beyond. Yet before the practical completion of the Saint Quentin Canal in 1828 opened the passage to Paris, Mons coal had only a minor incentive to use it, and after the opening of the Sensée Canal in 1817 Anzin coal circumvented it. It is not surprising then that the efforts toward improvement of the Escaut navigation between 1817 and 1828 were directed elsewhere, and that Anzin's tacit opposition was an important factor retarding modernization of the Escaut navigation.

Yet Anzin was far from opposing canal and river projects in general or its own participation in them. On the contrary, it was often active in promoting such projects when they favored the expansion of its own markets. Its most important contribution to transport development was the role it played in the improvement of the Saint Quentin Canal. As early as 1801 the Parisian directors of the company were using their influence in the capital to hurry on the construction of the canal.[68] In 1827 Anzin took nearly one-quarter ownership in the syndicate formed for its improvement. The company's interest amounted at face value to 1,000,000 francs, 300,000 of which was actually taken up by the Perier family in its own right and the rest retained by the company as a whole. The actual expenses incurred by Anzin between July 1827 and June 1831 amounted to only 540,974 francs, and the canal by the latter date was already returning a substantial income in tolls to the profit of its concessionaires. Anzin's dominant role is indicated by its strong influence in deciding the terms of the syndicate's conversion into a société anonyme and by the fact that the new canal company's directors probably held their meetings in Anzin's own boardroom at Saint Vaast.[69] Although the Saint Quentin Canal proved a profitable business venture in itself, Anzin's primary motive in improving it was doubtless its desire to open a route to the long-sought Paris market.

Anzin also participated in lesser projects. At the turn of the century, its Parisian directors toyed with the idea of constructing the Saint Quentin canal between the Escaut and Somme rivers.[70] In 1817 it took five shares at 10,000 francs apiece in the company constructing the Sensée Canal in order to have a "deliberative voice" in its affairs.[71] It made at least an attempt to obtain the concession for improvement of the Scarpe in the early 1820s.[72] Both the Scarpe and the Sensée were, of course, essential parts of Anzin's route to the major markets in the north of the Nord department. In 1828 the board instructed Castiau, director of surface works, to visit all the rivers and canals carrying Anzin coal in order to report on their condition. The company was prepared to bid for concessions to repair and improve any inadequate waterways so that such concessions would not fall into the hands of incompetent contractors whose efforts would be likely to interrupt navigation for any length of time.[73] On the whole, Anzin was alert to the commercial possibilities of improved navigation.

Moreover, it was ready to help actively promote certain improvements either to speed construction or to give itself a voice in the very important business of setting toll rates.

Yet despite its other canal and river activities, Anzin took no part in the improvement of the Escaut, its primary road to market. Upon learning of the government projects for the Escaut in 1817, the company discussed undertaking some of the construction itself. Its stated reasons for doing so hardly expressed a far-sighted desire to speed navigation. It hoped to avoid exploitive toll rates set by an unfriendly concessionaire; it wished to prevent any long interruption of navigation while construction took place; and it saw the opportunity of getting the right to dispose of the economically valuable "falls of water" at each lock.[74] In 1818 the board decided to submit bids for the two new locks then planned at Valenciennes, with the intent of leasing out the actual construction to an experienced firm of lock builders.[75] That the bids were ever made or accepted seems unlikely. In any event the locks remained unbuilt until after 1836.

Even more enlightening was the company's petty harassment of the Bureau of Bridges and Roads in its later plans for improvement of the Escaut. In 1827 Renard, who had retired as Anzin's general agent two years earlier and was now a member of the board of directors, protested plans for improvements below Condé in the name of the local commission for the drainage of the Haine and Scarpe valleys. In 1834 he was chairman of the local committee of inquiry which held public hearings on new plans for the straightening of the Escaut between Fresnes and Condé. Still a company director, he was also mayor of Fresnes, a member of the departmental council and owner of a glassworks at Fresnes which used Anzin coal. The other members of the commission included Anzin's current general agent, Jennings, and at least one merchant and one industrialist who did much business with Anzin. The seven complaints reported by the commission were for the most part petty. For example, farmers at Fresnes objected that the improvements would ruin their rich pastures along the river or place them on the other side of the new channel. But the most serious complaint, serious because of the influence of the plaintiff, came from Anzin. The company pointed out that its new loading area at Fresnes would be left high and dry by the proposed changes in the course of the river. The bureau engineers responded that the new channel would only be about five hundred meters farther from the pits than the old and that the reduction in the cost of navigation would far outweigh the slight increase in the cost of transport to the river. Nevertheless, the bureau agreed to permit the old bed to remain in its course over a distance of some sixteen hundred meters for the sake of the Anzin loading docks. Clearly Anzin was not interested in river improvements which meant money out of pocket and might at the same time help Mons coal to pass upstream more cheaply and easily on its way to Cambrai and Paris.

Anzin was, of course, willing to spend very large sums to improve its own access to southern markets by way of the Escaut, while denying them to Mons. According to one engineer of the Bureau of Bridges and Roads, this was one motive behind the company's construction in the early thirties of the railway linking the older pits around Valenciennes with the new loading basin at Denain. The railroad must have cost Anzin a great deal of money, while the replacement of the two Valenciennes locks would have been done at the expense of the government or a concessionaire; the cost of operating the rail line was probably higher than the cost of shipping from Valenciennes to Denain, including tolls. But the railroad put Anzin on the upper Escaut with a clear road to Paris, while improvements on the Escaut would have benefited Mons even more than Anzin.

Anzin could have done much to forward improvement of the Escaut navigation. It had more than sufficient funds and was willing to use them on occasion for other navigation projects such as the Saint Quentin and Sensée canals. It had influence at Paris and knew how to use it, as evidenced by its part in the successful fight to maintain tariff protection against the powerful interests of the Nord industrialists. More directly, Casimir Perier was himself a member of the influential advisory Commission on Roads and Canals during the 1820s. The commission agreed on the need for canal building in France but criticized the government's rigid plans and financial arrangements and wished to see the government turn over all existing and projected canals to private companies, which, as in England, would assume all risks in return for the income from tolls.[76] Anzin was also intimately connected with Honnorey, the canal builder who nearly monopolized concessions for canal and river projects in the Nord area during the Restoration. Anzin seems to have been suspicious of Honnorey when he first bid for the new Escaut locks in 1817, for he was a Belgian, and builder under the Napoleonic regime of the Mons-Condé Canal which had of course redounded to the advantage of Anzin's commercial rivals. But Anzin and Honnorey evidently soon came to a satisfactory agreement when the Escaut improvements and tolls charged there in no way injured Anzin's competitive position. The company almost immediately associated itself with Honnorey in the Sensée project and later in the company for improvement of the Saint Quentin Canal as well. Honnorey was also involved in improvements made on the Scarpe, Deule, and Lys rivers in the 1820s, all of which aided Anzin's commerce.[77] The one great project in which neither the canal builder or the mining company showed any interest was the long-proposed canal between the Escaut and Sambre River to the south. This canal would have opened the Avesnes district to Nord and Mons coal and permitted access to cheap, large supplies of wood, but it would also have opened the Anzin market area to low-cost coal from the Charleroi area and would have given the Charleroi mines a shorter route to Paris than Anzin's by way of Saint Quentin. Anzin, it was claimed, had scuttled this projected canal once in the 1780s, and its influence in the early nineteenth century again helped postpone its completion until after mid-century.[78]

In the eighteenth century, the canalization of the Escaut had occurred with Anzin's approval, indeed at Anzin's instigation and under its virtual direction. The key improvements in the nineteenth century did not take place until after 1833, although they had long had the support of the Bureau of Bridges and Roads, Mons producers, Nord industrialists, and coal consumers of all sorts, actual or potential—in fact, of all persons living along the entire water route from Dunkerque to Paris and Rouen. It says much for the influence of Anzin and the army corps of engineers that they could fight off so long improvements so obviously in the common economic interest of all northern France.

Although the Escaut and its connections, the Sensée, Saint Quentin, and Mons-Condé canals, were the most immediately vital waterways for the commerce of Anzin and Mons, the other waterways of the Nord and the Pas-de-Calais (and later those of the Seine-Oise basin) were hardly less important. They suffered from much the same deficiencies as the Escaut and in greater degree because their channels and water supply were smaller. Improvements were made after 1800 and especially in the 1820s by Honnorey, but even the building of modern locks and new canals and the canalization of river channels left most of the Nord waterways in a far from satisfactory condition in 1833.

The Nord department had one-sixth of all the improved river and canal mileage in France at the beginning of the nineteenth century, but most of the waterways were adapted only to serving a local, rural economy. These waterways permitted the movement over short distances on tiny boats of small quantities of agricultural produce and such farmers' supplies as fertilizers. For example, the very dense and ancient network of tiny canals on the Flemish coast was merely a substitute in those wet lands below sea level for the ill-paved local roads used in other parts of France.[79]

Other waterways in the Nord had a more general importance, and some had long carried Anzin or Mons coal. Among these, the Scarpe, a tributary of the Escaut, was one of the most important. Along it were located the important coal markets of Saint Amand, Douai, and Arras. Despite improvement of the Scarpe by Honnorey in the late 1820s, the ascent of the river to Arras from its junction with the Escaut at Mortagne still took eighteen days for partly loaded coal boats in 1833. Yet the Scarpe was heavily traveled because it was a necessary part of any all-French route from the coalfields to French Flanders. Mons boats, however, avoided the Scarpe as a through route by using Belgian waterways, and after 1820 Anzin replaced the long upstream haul on the Scarpe with a very short one downstream by the opening of the Sensée Canal.[80]

From the Scarpe at Douai a canal led to the Haute Deule, on which Lille was located. Both as canal and river the Haute Deule suffered excessively from a shallow channel and seasonal lack of water. Nevertheless, it carried a heavy traffic consisting almost exclusively of loaded coal boats moving from south to north and empty boats returning in the other direction to the pits of Anzin and sometimes Mons.[81]

Beyond Lille, the Nord waterway system broke down badly. The Basse Deule met the Lys at Deulemont on the Belgian frontier. The Belgian portion of the Lys, to the east, was well maintained. It was the major route, via Ghent and the Belgian Escaut, by which most Mons coal reached Lille. To the west, however, on French soil, the Lys was barely passable.[82] At one time, Anzin coal had used it to reach Aire, the head of navigation, and even beyond by canal to Saint Omer. This route had become impracticable by 1813, and Anzin then supplied these towns by wagons from Arras.[83] The opening of the La Bassée Canal in 1826 connected the Haute Deule to the Lys at Aire and thus greatly shortened and improved navigation to this point by eliminating a part of the Haute Deule, the difficult traverse of Lille's fortifications, the Basse Deule, and a long upstream stretch of the wretched Lys.[84]

Unimportant in itself, Aire was not far from the channel coast and theoretically in communication with it by canal to Saint Omer, the Aa River to Bourbourg, and the Bourbourg Canal to Dunkerque. In fact, these three waterways were almost impassable to all but local traffic.[85] Their difficulty helps explain Anzin's lack of success on the channel coast in competition with Mons, which reached Dunkerque and Ostend by the much longer but better maintained canal system along the Belgian coast west of Ghent. As late as 1835 navigation between Dunkerque and the French interior was closed entirely for eight months, and Lille had to obtain by road the foreign raw materials its industrialists imported at Dunkerque.[86]

The port of Dunkerque itself was wholly inadequate. Nearly abandoned by commercial shipping during the years of naval blockade under the Empire and with few natural advantages as a port, it was in 1815 almost cut off from both the sea and the canals of the interior by sand bars. Ships of as little as 250 tons could enter its port only on the highest tides, its quays were in ruins, its colonial trade gone, and its water connection to the interior nearly impassable. But as the chief outlet for the Nord and as a naval station, it received much attention and money from the Restoration government. The Ministries of War, the Navy, and the Interior together spent over 7,000,000 francs between 1820 and 1830 in reestablishing and improving the port of Dunkerque, its facilities, and its communications with the sea and interior waterways.[87]

Between 1820 and 1833 the government and private companies had done something to improve all the essential waterways of the Nord and its neighboring departments. Coal prices in towns on these waterways fell somewhat in response to lower transportation costs, and Anzin correspondingly enlarged its market by adding new customers, who were able to buy at lower prices (see table 5). The

Table 5 Transport Costs from Condé, 1825–29 (In Francs per Metric Quintal)

Year	Arras	Lille	Saint Omer	Dunkerque	Compiègne	Paris	Rouen
1825	1.05	0.85	1.20	1.15	1.80	2.85	2.85
1826	1.00	0.85	1.10	1.00	1.90	2.85	3.00
1827	1.00	0.85	1.10	1.00	2.00	3.25	5.25
1828	0.80	0.75	0.90	0.85	1.70	2.80	2.85
Jan. 1829	0.85	0.65	0.85	0.85	1.20	2.00	2.10

SOURCE: AN, F12 2534c, Coupin, customs official at Condé, "Mémoire . . . sur la demande faite d'une diminution des droits . . .," 20 Jan. 1829.

NOTE: The figures apparently include navigation duties. The cost of transporting Anzin coal to these markets from the loading areas at Valenciennes would have been somewhat different than those cited here for Mons coal emerging from the Mons-Condé Canal at Condé. Obviously, too, the total transport costs for Mons shippers were somewhat higher than the ones given here, since the cost of transport from their loading areas to Condé is not included.

Saint Quentin Canal had opened one potentially enormous new market in the Seine-Oise basin. Yet most of these improvements had not come until the end of the 1820s or even later, and most of what was needed to provide a really efficient transport network, even at the existing level of technology, remained to be done. Railroads were as yet totally unused for long-distance transport, and the waterways remained far below their potential in carrying capacity, time of transit, and economy of cost. Transportation improvements continued to lag behind the growth of industry and its increasing consumption of coal. The existing waterway system in 1833 was just barely adequate to serve the growing needs of the coal industry and its customers. Anzin did something to improve the system at certain key points, notably the Sensée and Saint Quentin canals, but on the whole it tended to accept the existing network and to work for expansion of its market by more conventional and less expensive means than canal and lock building.

7 Anzin Markets: The Consumers of Coal

One historian has remarked that "the economic history of the Nord region . . . 'took off' under the Empire."[1] The production of coal, small metal goods, and the traditional textiles—woolen and linens—suffered during the Revolution, but forges, tileworks, oil and flour and especially cotton-spinning mills, breweries, and distilleries expanded rapidly with the establishment of financial and political order under the Empire.[2] It was especially after 1820, however, that contemporaries noted a general industrial advance. An "immense takeoff" seemed to be occurring, and as one local observer put it, a "crowd of factories," most of them using coal, suddenly "sprang from the ground" in all branches of industry.[3]

Nevertheless, although the Nord department developed much faster industrially than most other areas in France and although coal played a greater part in this development than anywhere else, coal was still not fully integrated with the industrial life of the Nord in 1833. Of the three industries on which the wealth of the Nord in the later nineteenth century was based—coal, metal, and textile—only the textile industry was really significant in terms of the revenue directly produced or the population directly supported. And although parts of the textile industry were undergoing rapid changes in the first third of the nineteenth century, many of its branches continued to be organized in the traditional ways. Only the fragile beginnings of coal-based industry around Valenciennes and of textile concentration in the Lille-Roubaix-Tourcoing area were evident in 1833, and the industrial connection between the two areas was still tenuous.

The Anzin market area was roughly bounded by Tournai, Lille, Arras, and Cambrai in a great half-circle thirty to sixty kilometers from the Anzin pits near Valenciennes. The eastern half of the circle around Valenciennes was blocked off by the Belgian tariff frontier, the Mons coalfields, and the lack of water transport between the Escaut and Sambre watersheds. Beyond the radius of the half-circle to the north, west, and southwest, the length of the water haul put coal prices too high for all but a few users and discouraged the growth of new consumers. Although Anzin coal reached Dunkerque, Paris, and Rouen in small quantities before 1833, Lille, Arras, and Cambrai roughly marked the points at which Mons and other coals or traditional fuels like wood could begin to exclude Anzin coal. Even within this half circle, of course, Anzin was in constant competition with other fuels and other coal producers. In the following pages special emphasis is given to the market around Valenciennes in order to assess the company's growing impact on the local economy.

Domestic Consumption

The extent to which coal was used in homes for heating and cooking is difficult to determine with any accuracy. Nevertheless, as late as 1822 one-half of Anzin's sales in the Nord were to householders.[4] In northern France, the cold, damp winters made some kind of heat a necessity for all levels of the population during several months of the year, and the combination of a bad winter and high coal prices caused widespread suffering among the poorer classes of the towns.

The domestic market had special characteristics that made its extent and growth somewhat different from other markets. Of the three possible domestic fuels—peat, wood, and coal—individual householders made their choice mainly on the basis of price but also on the basis of convenience, smokiness, and smell of burning. Those who were able to afford a comparatively gracious style of living, especially the bourgeoisie of the towns, usually bought wood even when it cost more than the two less pleasant fuels. The mass of the population, however, was forced to buy the cheapest fuel available. Over great stretches of the Nord and neighboring departments, especially in rural areas, the cheapest fuel was peat; in a few places in this relatively deforested region it was still wood; but within about a ten-mile radius of the pits of Anzin and Aniche and in and around many towns located on navigable waterways, the cheapest fuel was coal. As wood continued to become more scarce and expensive and as improved and extended waterways lowered coal costs at older points of consumption or enabled entirely new areas to be reached cheaply, domestic coal consumption increased. On the other hand, coal often forced its way into industrial use in areas far removed from the coalfield long before it was accepted as a major household fuel. Industries could often absorb high transport costs when coal offered technical advantages over traditional fuels.[5]

Paris may serve as an example of an area where the consumption of coal for heating and cooking made little progress in the early nineteenth century. The cost of the transport of coal was high on the inefficient waterways that connected the capital with the Loire, Nord, and Belgian fields, but wood floated in huge rafts from the nearby forested areas to the south and east.[6] Parisians in 1764 credited their woodfires for their freedom from the chest ailments, jaundiced complexions, and soiled laundry that were supposedly the common lot of Londoners as a result of their domestic use of coal.[7] The revolutionary governments campaigned vigorously for the use of coal instead of wood, then temporarily more scarce than usual,[8] but in 1814 Paris still impressed an English traveler with the "stupefying" transparency of its air.[9] Two years later, it was said, domestic use of coal was still in its infancy in the capital.[10] Chevalier noted in 1828 that the small quantity of smokeless coke made available as a by-product of gas manufacture was in great demand in Paris for domestic heating, but that ordinary coal, even when mixed with wood for easier kindling and cleaner burning, was still little used, although he thought that it was bound to become more important in the near future when the opening of an improved water route to the Nord and Belgium lowered its price.[11] His prediction proved premature, for even in 1840, twelve years after the improvement of the Saint Quentin Canal, only some five thousand of the more than one hundred thousand tons of coal shipped to Paris each year were sold for domestic use.[12] As early as the Revolution, however, a great number of metal, textile, and other craft trades in and around Paris had already come to depend almost exclusively on coal fuel, and this dependence had continued to grow between 1800 and 1833.[13]

The situation in much of the Nord was the reverse of that at Paris, for the domestic market probably did not give way to industry as the major consumer of Anzin coal in the Nord until the late 1820s. In the Valenciennes area, coal was cheap and wood dear as early as the late seventeenth century when wagon loads of coal arrived regularly from Mons for use in the homes of "the poor."[14] By the middle of the eighteenth century, Anzin coal had become the standard fuel for households in the vicinity of the company's concessions and had reached the grates of the lower strata of the population in more distant towns such as Lille.[15] So accustomed to coal heating was the population of these areas that a particularly cold winter, during which Anzin, it was claimed, would take advantage of its power to raise prices, was a physical and financial disaster for the poor.[16]

By 1822, despite a considerable extension of the Anzin industrial market during the past fifty years, one-half of the company's sales in the Nord were for domestic consumption.[17] Most of these sales were within a few kilometers of the pits, where Mons coal could not compete and where transport in small quantities by wagon did not raise prices sufficiently to price even the poor out of the market. In this limited area Anzin had a monopoly on the domestic market for coal. and that market was large since even the wealthier inhabitants were now accustomed to coal fuel. In Valenciennes itself, all but two of the twenty coal merchants who bought from Anzin operated on an extremely small scale (less than seven hundred tons per year and mostly about two hundred) and thus were undoubtedly engaged primarily in retailing coal to householders.[18]

In such towns as Lille, Arras, and Cambrai, at the outer limits of Anzin's effective reach, coal was an important domestic fuel, but industry in the towns and surrounding countryside took an even larger proportion of coal sales. Anzin coal competed there with that of Mons, and wood and peat were still serious competitors for the domestic market. By 1829 there were forty-one coal dealers in Lille, and several of the seven wood dealers and three charcoal dealers also sold coal.[19] Lille's industries absorbed large amounts of coal, but most of the coal dealers must have been small retailers selling to households. Yet in 1837, the domestic market for coal in Lille was still capable of considerable expansion.[20]

Great deposits of peat were found in the bogs of the half-drowned valleys on the upper reaches of the Somme, Authie, Maye, Scarpe, Sensée, and Escaut rivers as well as throughout the wet lowlands of Flanders, especially around Bergues. The Pas-de-Calais and Somme departments, both part of the Anzin market area, were the largest producers of peat in France. About 60 percent of the Nord peat production was cut near Douai in the valleys of the Scarpe and Sensée, two rivers that carried a large part of Anzin's coal. Most of the rest came from the areas around Lille and Bergues. Peat cutting was usually a small-scale summer operation, often a by-industry of peasants who dried and cut or shaped the peat into briquettes. Despite extensive flooding of peat deposits during the Revolution when the Old Regime's regulatory system disappeared, Nord production around 1800 totaled over one billion briquettes, which sold for 2 to 3 francs a thousand. Peat was burned in a variety of local industries, such as brickworks, breweries, and sugar-beet refineries, but it was mainly burned in the homes of those who could not afford a less repulsive fuel. Peat was too bulky and low in value to be worth transporting any considerable distance from the point of production. Each producer's market was highly local, often comprising no more than the producing commune itself. Nevertheless, it was found so widely that it was "the exclusive fuel of a great number of communes in the Nord," particularly in the Douai area where it severely restricted the Aniche Company's natural market for domestic fuel.[21]

Only in the immediate vicinity of the pits in and around Valenciennes did Anzin have a monopoly of the coal market, and coal a near monopoly of the domestic market. Elsewhere in the Anzin market area, several factors could help expand the domestic market. Improved fireplace chimneys and stoves better adapted to burn ordinary coal or cleaner-handling coal briquettes were the object of considerable inventive effort, although none of it by Anzin itself.[22] Improvements in living standards, a continued rise in wood prices, and exhaustion of the better peat deposits would have affected the domestic market only slowly, and none could be speeded up by company action.

A more important factor was population growth.[23] The Nord, with 765,001 inhabitants in 1801, was the largest department in France. Except for the Seine (Paris and vicinity), it was also the most densely populated department. Two other departments within reach of Anzin coal, the Pas-de-Calais and Somme, also ranked among the largest ten. By 1831 the population of the Nord had grown some 29 percent, about 50 percent more than France as a whole. The Pas-de-Calais department grew at about the same rate as the Nord; the three other departments in the Anzin market area, the Somme, Aisne, and Oise, grew much less rapidly. Thus the potential domestic market for Anzin coal was large in 1801 and considerably larger in 1831.

A final possibility for increasing the domestic market was to lower the cost of coal by improvements in water transportation, which would bring immediate and substantial results by expanding the geographical and social reach of coal for domestic use. Furthermore, it was within the power of the company to do something toward achieving this goal. Anzin's efforts in this direction have been described in the previous chapter. That they were only partially successful is evident from the fact that householders continued to burn peat and to a lesser extent wood in many areas where coal was the primary industrial fuel.

There was certainly much room for expansion of the domestic coal market in France as a whole. In 1802 a member of the Council of Mines estimated that only one-sixth of French coal production was used for domestic heating.[24] Since production was still quite small, it is evident that wood was overwhelmingly the most important domestic fuel. Later the famous Chaptal was only somewhat in error when he wrote in 1819 that "wood is still the only fuel used in our domestic hearths," and the situation had not radically changed by 1833.[25] It was still only on or very near the coalfields that coal was the preferred fuel at most social levels. A short distance from French coalfields, even along rivers and canals or behind the port towns which welcomed English colliers, coal was not an economically viable fuel at least for domestic use.

The Limeburning and Construction Industries

One of the Anzin Company's larger markets was the limeburning industry. Heating limestone in a kiln drives off carbon dioxide gas. The friable residue is quicklime, which has a variety of important agricultural and industrial uses. The hard coal from Anzin's Fresnes and Vieux-Condé concessions was ideally suited, indeed essential, for the limeburner. It burned extremely hot and long without flame or smoke, and its high kindling point mattered little in the long, continuous operations of the limeburner.[26] Cheaper than charcoal and technically an improvement over wood or peat, hard coal began to replace these traditional fuels in the limeburning industry from the moment the Fresnes deposits were discovered in the early eighteenth century. Since the Mons field did not produce any coal of the proper type, by 1800 Anzin hard coal had gained a monopoly in the limeburning market throughout the company's sales area.

This market was widely scattered among many small-scale, mainly rural producers; limeburners sold mostly to local farmers and builders, and limestone deposits near the surface or in outcrops were abundant throughout northern France and northwestern Belgium. Anzin sold large amounts of hard coal to limeburners in the Somme, Aisne, and Pas-de-Calais departments as well as the Nord.[27] In the Pas-de-Calais there were seventy-nine limeburning establishments in 1813; all but four employed fewer than ten workers.[28] The industry was organized on a somewhat larger scale in Belgium. The center there was the vicinity of Tournai on the Escaut, and a lesser concentration was located near Vervik and Menin just north of the frontier beyond Lille.[29] Anzin's shipments of hard coal to Belgium for limeburning accounted for the bulk of all French coal exports.[30]

This Belgian market was highly prized; its loss during the 1820s due to a retaliatory Belgian tariff was regarded as a calamity. Until then Anzin's monopoly had enabled it to sell its hard coal at a price slightly above that charged for its softer coals, which had many more uses but were in sharp competition with similar types produced by Mons. The stability of the limeburners' demand for hard coal was an important factor in the company's prosperity before the industrial expansion of the 1820s. Output at Fresnes and Vieux-Condé was at prerevolutionary levels during the first two decades of the nineteenth century, when Raismes and Anzin bituminous production remained below 1789 output.[31] During the 1820s Anzin sought with some success to expand the French market for its hard coal by finding new uses for it. It had always been used, for reasons of economy, in the domestic hearths and stoves around the Fresnes and Vieux-Condé pits as well as by Nord and Pas-de-Calais limeburners and brickmakers. Because of its long-burning quality and because its price had been lowered after the loss of the Belgian market, some blacksmiths, textile dyers and bleachers, brewers, and even sugar refiners found it economical to mix coal from Fresnes and Vieux-Condé with their usual fuel.[32] Nevertheless, the limeburning industry remained the most important market for Anzin's hard coal.

Most of the burned lime sold in the Anzin market area was used in agriculture as a soil conditioner.[33] In 1802 a Belgian claimed that agriculture profited more from the expansion of the coal mining industry than either commerce or industry. This exaggerated claim certainly had some truth to it for areas where agriculture used advanced techniques. One such area was the Nord: "It would be difficult to add anything to the perfection of cultivation of the soil in French Flanders."[34] There and elsewhere in northeastern France and the Lowlands, farmers depended heavily on soil conditioners such as marl, coal and peat ashes, barnyard manure, and town wastes, but burned lime was the most important.[35]

The agricultural demand for burned lime was very large. The heavy population densities in the departments where Anzin sold most of its coal reflected primarily a vigorous agricultural economy together with the local commerce and craft industries dependent on it. The rural portion of the population of the Nord in 1836 was 61 percent; in the Pas-de-Calais the proportion reached 78 percent, slightly higher than for France as a whole, and the figure was even higher in all the other neighboring departments and Belgian Flanders.[36]

The agricultural demand for burned lime, and hence for Anzin coal, was not, however, easily or rapidly expandable. No extensive new tracts of soil were opened to cultivation in an area where commons and fallowing were almost unknown; farming was already intensive, and the use of lime was a long-established practice almost everywhere. Thus it seems likely that the rise in output at Fresnes and Vieux-Condé after 1830 was mainly due to the reopening of the Belgian market, the result of political rather than agricultural changes.[37]

Burned lime was also an industrial raw material. Hydrated or slaked by adding water, it formed the basis for cement, mortar, plaster, and whitewash. Slaked lime was a chemical base whose caustic properties were useful in tanning, pulp and paper manufacture, and sugar-beet refining. A building boom and the rapid growth of sugar refining in the 1820s may have lessened somewhat the impact of the lost Belgian market on Anzin's sales to limeburners during that decade, but the amount of lime needed by the building and sugar industries was relatively small. Like the domestic market, the limeburning industry grew slowly and mainly in response to population increase. One must look elsewhere for the causes of Anzin's rapidly increasing sales after 1815.

One possible cause was a spurt of activity in the building industry. In the Anzin market area, especially on the heavy clay soils of Flanders, buildings were mostly made of brick, and the area's brickmakers were held in high repute.[38] Like the limeburners, they found the coal mined at Fresnes and Vieux-Condé exactly suited to their needs. The brickmaking industry was second only to the limeburning industry as a consumer of Anzin hard coal.[39] In the Nord alone in 1813 there were 122 brickyards with an average of ten workers each.[40] The production of other materials used in building also involved the consumption of coal, both soft and hard. Population growth, the postponement of construction after the Revolution, and particularly the growth in the number of people with middle-class incomes and aspirations were probably responsible for something of a building boom in the 1820s and 1830s, especially between about 1820 and 1826. This increased construction helped take up the slack in Anzin's hard-coal market.[41]

The Agricultural Industries

In the Anzin market area, where agriculture was diversified and market-oriented, several industries that processed agricultural products became important coal consumers after 1800. The old emphasis on the production of wheat, cattle, sheep, and flax for the basic needs of the local towns and countryside was giving way to a broader spectrum of products, many of which required considerable processing before they entered the market.

By far the most important of the new processing industries was sugar-beet refining. The sugar refining industry is an intensive user of fuel, since the juice of cane or beet requires several successive boilings.[42] From the beginning, sugar-beet refining was organized capitalistically and used coal on a large scale; a sizable refinery producing 4,000,000 francs' worth of sugar per year used over one thousand tons of coal.[43] With the metal, textile, and glassmaking industries, sugar-beet refineries had become one of the four largest industrial users of Anzin coal by 1833.

When the revolutionary and Napoleonic wars, and especially the British blockade of the continent, cut France off from colonial sugar supplies, the growing and processing of sugar beets was undertaken in France.[44] This development initiated the Napoleonic regime's most famous experiment in wartime synthetics and one of the most important early successes for industrial chemistry. Destined for a great future throughout the northern European plain from Paris to Moscow, beet sugar had its first success in the Nord and Pas-de-Calais area after 1812. The infant industry collapsed in 1815 when England dumped colonial sugar on the French market.[45] It slowly recovered after 1818 and began to expand rapidly by the mid 1820s, although the real boom did not develop until the next decade.[46]

As beet sugar became important, the refining industry developed new centers inland in competition with the cane refineries of the ports. At least one beet refinery, using Anzin coal and producing 150,000 pounds of sugar a year, was operating in the Valenciennes area by 1827,[47] and in the next two decades several private bankers emerged at Valenciennes who owed their success to profits from the sugar-beet industry.[48] Refineries using Anzin coal were also important in the Pas-de-Calais and Aisne by 1833, although the center of the industry in the early years was the area around Lille.[49] A spokesman for Anzin regarded this rapidly growing industry as a chief cause of the company's prosperity and expansion in 1829–30.[50] By 1837 the thirty-four beet refineries in the Nord were consuming some two hundred thousand tons of coal, equal to about one-third of Anzin's total output, although many of the refineries used Mons coal.[51]

Meanwhile, other industries that processed local agricultural products had also become extensive users of coal. Like sugar-beet refining, at least two of them owed much of their development to war shortages. Ground, roasted chicory root was introduced in the Napoleonic period as a substitute for the tropical coffee bean. Either the ovens used in roasting or the steam engines used in grinding might use coal.[52] At the same time, the shortage of animal fats led to increased use of vegetable oils for lighting and soapmaking. Anzin's own replacement of suet candles by oil lamps has already been noted. The oil was pressed from the seeds of a variety of plants including flax, hemp, and rape. The smaller oil mills operated their presses by water, animal, or even wind power,[53] but the bigger ones employed steam engines, which might use as much as three-fourths of a ton of coal a day or about one-fourth as much as an average sugar refinery.[54] Oil from the Nord and Pas-de-Calais had a wide market. It was sold in Paris and Rouen and even as far away as Marseilles.[55]

The brewing industry, which required large amounts of heat, showed a marked increase in coal consumption after 1800 as town breweries expanded their production at the expense of village breweries. Still, many village breweries survived, and their small size did not necessarily prevent them from using coal.[56] Certainly the six breweries near Fresnes, employing a total of fifteen workers, and the four near Anzin employing twelve used local coal in producing beer for a thirsty mining population. Altogether, there were twenty-four breweries in the Valenciennes area in 1827.[57] Dupin claimed in 1818 that the Nord, a stronghold of French beer production, burned over forty thousand tons of coal to produce over one million hectoliters of beer.[58] Much of that coal was Belgian, and Dupin's figures may be exaggerated; yet forty thousand tons would have represented well over one-tenth of Anzin's total output, and breweries in the Pas-de-Calais and other neighboring departments also bought Anzin coal.[59]

Distilleries had long used coal,[60] but they too increased consumption after 1800, especially those in northern France and Belgium, where gin had become the common drug of the working man.[61] Before the Revolution, there were only three distilleries in the Nord, all at Dunkerque. By 1802 the department had seventy-two, mostly in villages near the sources of their raw material.[62] Some fifteen years later, eighty-seven distilleries produced 25,505 hectoliters of gin and consumed about seven thousand tons of coal.[63] There were nine in the Valenciennes area alone, one of them at Anzin.[64]

Blacksmithing deserves mention as a coal consuming trade closely tied to agriculture, although blacksmiths were found everywhere, in town and country alike. As a shoer of horses, the blacksmith was one of the most important customers of the makers of nails and bar iron, who became important consumers of Anzin coal in the 1820s. Especially in the rural areas, the blacksmith was also an expert in many areas of metal working; he made and repaired agricultural implements as well as the looms and other tools and machines used by local craftsmen. It is difficult to estimate the amount of coal burned by the blacksmiths' forges in the Anzin market area. Even in country districts, however, away from the waterways, most blacksmiths seem to have preferred coal to charcoal, at least for some of their needs. Indeed, in many rural areas where there were no industries that burned coal and where peat was the chief domestic fuel, the blacksmith was the only customer for Anzin coal.[65]

Except for brewing and blacksmithing, these agricultural industries were essentially new ones. They represented a clear increase in Anzin's potential market. Oil refining and distilling made most of their impact before 1820; like domestic consumption and limeburning they helped shore up the Anzin market in the years of relatively weak demand. But only sugar-beet refining, which, after 1820, became one of the major consumers of Anzin coal, offered much hope for a strong immediate increase in coal sales.

The Glass Industry

The Nord industry tied most closely to coal mining was glassmaking. The relationship between the two industries had its beginnings in the late seventeenth century when a technique was developed that permitted the first of two fusions to take place in a furnace fired by coal. During the eighteenth century, cheap coal pulled the industry out of the woodlands and onto the coalfields. Unlimited supplies of cheap fuel, the technique of coal fusion in large furnaces, and the growth of the market for glass bottles and window panes, led at the same time to concentration of the industry in relatively large capitalistic units.[66]

By 1789 there were six glass factories in the Nord. Two, in the Avesnes area, still used wood exclusively. Two of those that used coal were located near the coal pits at Aniche, one was on the Hardinghen field near Calais, and the other at Fresnes. The Fresnes firm was owned by the Desandrouin family. Jean-Jacques Desandrouin had founded it, like the Anzin Company itself, as a result of the Treaty of Utrecht, which had separated French Flanders and Hainaut from both the Charleroi glass industry and the Mons coalfield. With the discovery of the bituminous seams south of Fresnes and the formation of the Anzin Company, the Desandrouin family's interests shifted decisively toward coal mining, but this early linkage was symbolic of the continued close ties between coal and glass.[67]

The Nord glassmaking firms were hard hit by the reunion of the Belgian provinces with France during the Revolution and under the Empire. Most of them, including the Fresnes firm,[68] suspended operations for at least part of this period because of overwhelming competition from Belgium, where the cost of all the materials was lower. After 1814, the separation of Belgium and France and the passage of a law prohibiting the entry of foreign glass cleared the way for rapid expansion of the Nord glass industry.[69] A similar set of circumstances stimulated Anzin output, which went in increasing proportions to feed the furnaces of new glassworks established near its pits.

In 1815 there were only two glass firms operating in the Douai-Valenciennes area, one at Aniche, which employed 34 men, and Desandrouin's at Fresnes, which employed 65. Using entirely local materials, the Desandrouin works produced 256,920 panes of window glass and 88,655 bottles in that year. Desandrouin and Dorlodot both opened new glassworks on the Anzin pits in 1819. By 1822, although Desandrouin had died, his new Anzin works, which had two furnaces and 108 men, was nearly as large as the expanded Fresnes operation, which had 130 men. Dorlodot's firm employed only 26 men at one furnace, but he added another furnace in that year and made application to start another operation at Aniche. One other firm began operation at Anzin in 1822 or 1823. About that time, Anzin's Renard claimed that the company furnished coal to five glassworks: three on its own concessions, one near Cambrai, and another at Dunkerque, although the last two apparently used Aniche or Mons coal as well.[70]

By 1828 the three Anzin firms had five furnaces among them, employed forty, forty, and ninety-six men, and produced as many as 1,294,000 bottles and 587,000 window panes a year. There were now two operations at Fresnes (the old Desandrouin unit being still the largest in the Nord), one at Bruai, and one at Saint Saulve.[71] A total of at least seven glassworks thus depended on coal supplied by the Anzin Company at a time when the annual coal consumption of an "ordinary" glassworks was estimated at about thirty thousand tons.[72] Furthermore, it is clear that an industry that could successfully locate at Hardinghen, Aniche, or Fresnes did not require high-grade bituminous forge coal. By using a coal for which demand was low, glassmakers helped relieve Anzin of one of its most severe marketing problems.

The glass industry reached a temporary peak in the Nord about 1828. Thirteen new firms, banking on an exclusionist tariff policy after 1815 and the prosperity of the wine and spirits industry and the construction industry, entered the market during the Restoration. By 1834 twelve firms were active in the Nord, nine of them near Anzin or Aniche. The department's glass production, valued at 1,580,700 francs, sold in Brittany, Paris, and Rouen, as well as the Nord and its neighboring departments.[73] The same nine firms, with some changes of ownership, were still operating around Anzin and Aniche in 1840, but their production had almost doubled.[74] Glassmaking continued to consume a large share of the Anzin output into the 1870s.[75] The process of concentration on the coalfields (and of integrated ownership) had occurred elsewhere too.[76] Charleroi firms almost monopolized Belgian window and bottle output,[77] and glass produced on the Loire and Nord fields together accounted for almost three-fourths of French output by the late 1830s, a proportion almost equal to the part these two fields held in the coal industry itself.[78] In the words of a jingle current in the Loire field:

Dans la houille et dans la verrerie,
Ces diables enrichés menent joyeuse vie.[79]

The Iron Industry

As in glassmaking, the 1820s saw develop around the Anzin Company pits an iron-working center, which owed its existence in large part to French tariff policies. Except for two large-scale refineries, the Valenciennes iron industry consisted mainly of handicraft or putting-out enterprises that made small finished products for an agricultural population. Yet both modern refineries and traditional craft enterprises furnished a significant new market for Anzin in the 1820s, and their establishment near the Anzin pits foreshadowed the immense future development of metallurgical production of all kinds on the Nord coalfields.[80]

Two major technical triumphs in the iron industry were achieved in England during the eighteenth century but made little headway on the continent until after 1815. The success of coke smelting and puddling depended in part on the quality of the coal and ore used; many firms found that methods used in parts of England did not work because raw materials were different and technical expertise was lacking in France. An abundant charcoal supply, the separation of ore and coalfields, and the dispersion of the market for finished products also slowed the adoption of "la méthode anglaise" on the continent. Isolation from England during the wars of the Revolution and Empire and a genuine spirit of conservatism in the iron industry further contributed to technological backwardness.[81] Finally, iron smelted or refined with coke or coal was generally regarded, with or without justification, as inferior to iron smelted or refined with charcoal. This preference was reinforced by the fact that the type of iron most in demand then was the high quality bar iron worked by smiths or other craftsmen into a wide variety of objects requiring different characteristics in the metal used. In the face of a preference that often amounted to prejudice on the part of its best customers, the iron industry in France was perhaps not so mistaken in its aversion to iron made with coal.[82] Lower costs were not as important as proper quality.

Hardly surprising, therefore, is the fact that Anzin and the iron industry went their separate ways until after 1815. The Nord and neighboring areas depended until the fall of the Empire on the Liège-Charleroi region for most of its iron products. Then the French tariffs of 1817 and 1822, which affected imports of all kinds of iron goods, stimulated development of the rudimentary Nord industry along the Belgian frontier, a development partly along traditional lines using charcoal but partly on the basis of the new coal-using techniques recently adopted by much of the Belgian industry.

The smelting branch of the industry made the least progress in the Anzin market area. The only iron ore deposits in the Nord were at Trélon near Maubeuge in the Avesnes area and isolated from the Anzin coalfield by the lack of water transport.[83] At the same time, the rich "blackband" ore, or ironstone, intercalated in the coal seams of the English Midlands attracted much attention in France, for it demonstrated to the highest possible degree the great advantage of a close proximity of ore and coal. The government mining engineers were convinced that French coalfields, if examined closely, would reveal the same fortunate formation.[84] Despite a number of notices in the *Journal des Mines,* this optimism was largely unfounded, but the engineer Daubuisson had recognized deposits of ironstone in the coal seams at Anzin as early as 1805. An assay of the ore showed a mediocre iron content of 26 percent.[85]

Nothing more is heard of Anzin ironstone until 1817. The date is suggestive, for Anzin was then launching its project for completion and expansion of its foundry and iron-working shops and making arrangements with Edwards for the production of his steam engines on company premises. In March, on the suggestion of the company's iron founder, the first to suggest using the ore, the board of directors put in motion plans for the erection of a blast furnace. Presumably, they expected it to burn local coal coked by the company. A location for the blast furnace was to be selected, and an estimate of construction costs was to be submitted at the next regular meeting of the board in September.[86] In June the Bureau of Mines in Paris received a specimen of Anzin ore.[87] Successful tests were reported on somewhat doubtful authority sometime in the next year.[88] And here the story ends, leaving many questions unanswered: Was the result of the assay or of an estimate of ore reserves discouraging? Did the construction costs prove too high? Did the Periers' purchase of Chaillot remove some of the incentive for setting up a thoroughly integrated operation from ore to steam engine at Anzin, or were there second thoughts concerning the wisdom of a first-rank coal company's attempting to become a novice iron manufacturer? Or were there technological problems concerning the coking of Anzin coal? Until the discovery of the rich Denain seams in the late 1820s, Anzin coal was regarded as much inferior to that of Mons for coking. Anzin coal was not used for smelting until 1835, when a furnace was opened on the Denain seams, and the Valenciennes smelting industry ran a poor second to that around Maubeuge on the Trélon ore until the 1880s.[89]

The secondary metallurgical processes, the founding of cast iron and the refining of wrought iron, had better success with Anzin coal. Iron founding had early adapted itself to the use of coal fuel, and the first foundry in the Nord began operations in 1734 between Valenciennes and the village of Anzin. The date clearly indicates that the foundry was intended to use the coal discovered there by Desandrouin in that same year. The foundry was fed by scrap iron from old war materiel, fueled by Anzin coal, and operated mainly by English workers drawn from the English royal arsenal at Woolwich. It disappeared sometime before 1789, although there were then two small casting firms at Dunkerque using Mons or Anzin coal and scrap iron to cast cannons and cauldrons for the navy during the Revolution and Empire. Another foundry with a single furnace opened on the Anzin pits at Raismes in 1796. It cast cauldrons and pots of all sorts as well as "wheels for machines" and some cannon pieces from scrap. It employed only six workers in 1801 and four, seven years later. In 1804 the Anzin Company set up a foundry in its own workshops for the casting of some of the larger pieces used in constructing its Newcomen and winding engines. As noted above, Anzin considerably expanded this operation in the 1820s. By 1844 the local government engineer claimed that there were fifty such small casting units operating in the Nord as auxiliary workshops of most of the larger industrial firms.[90]

In contrast to the many scattered foundries, the few forges in the Nord had a more general market and tended to locate near the Anzin pits. Thus, the most important and most modern firm in the whole Nord iron industry was a refinery founded by a Belgian named Renaux on the Anzin coalfield at Raismes in 1823–24. This refinery had three coal-fueled furnaces in which it puddled and rolled scrap and charcoal pig iron from Champagne into wrought iron by the English method. It hammered and slit most of the wrought iron into bars but could also process it into rolled shapes such as boiler plate. A foundry completed the firm's ability to perform any of the major secondary iron processes.[91] A forty-horsepower steam engine that burned about twenty-five tons of Anzin coal a day powered the hammer, rollers, and slitting and plate mills.[92] The firm was fitted out by the famous English firm of Manby and Wilson, which did much in the 1820s to popularize both puddling and the steam engine through its machine shops at Charenton near Paris and its participation in the integrated Le Creusot operation.[93] Like the Raismes foundry opened almost one hundred years before, the new refinery initially employed "principally" English workers.[94] A firm where English workmen processed Champagne pig iron by an English technique on equipment probably built at Paris by Englishmen had located at Raismes in order to take advantage of the cheap, abundant coal from Anzin pits.

This so-called "vaste établissement" was one of Anzin's most important customers. The initial work force of 110 soon grew to 150. It was especially noted by contemporaries that the entire force was gathered in one spot in a single set of connected workshops and that it worked year-round, in contrast to putting-out operations, whose employees were scattered over the countryside and often worked only a few months of the year.[95] In 1828 a smaller refinery, also using modern puddling and rolling techniques in the production of bar iron, separated from the Raismes works and set up between Trith and Saint Leger on a bend of the Escaut between Denain and Valenciennes. It too used Anzin coal to process charcoal pig-iron, which was brought all the way from Brittany by sea.[96] Production at Raismes grew from fifteen tons of finished iron products a year to two thousand by 1827 and, including the Trith output, to twenty-five hundred tons in 1834.[97] The remaining owner at Raismes, Dumont, like many other French industrialists of the period, went to England to study methods in the iron industry there, and he returned to build the first coke-fired blast furnaces in the Valenciennes area, in 1835. The coke came from coal mined by Anzin in its new seams at Denain, where Dumont built his furnaces.[98]

Nailmaking was the one tertiary metallurgical process which, while maintaining its handicraft technique and putting-out organization, became an important consumer of Anzin coal.[99] Like the cast-iron founders, nailers had long been accustomed to using either charcoal or mineral coal fuel, mostly the latter in Belgium and the former in the Avesnes region, the two areas from which most of the Nord department drew its supply of nails until 1814.[100] Then between 1814 and 1816 three Belgian nailmakers, pushed out of their native land by English competition, crossed the border to the Nord, where tariff protection and the abundant coal around Valenciennes promised a prosperous future. Settling down around the Anzin mining villages, they brought a large new industrial population to the coalfield. In 1822 the three new naileries employed 588 workers, of whom all but 37 were Belgians who had emigrated with their employers.[101] As Renard expressed it, the Anzin pits "bristled" with nailers' forges, all of them erected since the fall of Napoleon.[102]

Unlike the Raismes refinery, these new industrial consumers of Anzin coal did not represent any fresh access of metallurgical technique or centralized production. Removal from Belgium represented simply a migration of classic handicraft and putting-out firms due to the fortunes of war. "These nails are made by hand without the aid of a steam engine, water wheels, or machines. These manufacturers or merchants have some workshops at their places of residence, but their workmen are scattered in the [nearby] villages."[103] By 1824 the total nailmaking population numbered about one thousand in the Valenciennes area. The largest of the four "manufacturers and merchants" was the same Renaux who had founded the Raismes refinery in that year; his nailmaking operation employed three hundred workers at forty scattered forges.[104]

Napoleon's bad fortune had thus become Anzin's good luck in more ways than one. Aside from the stimulation that Restoration tariff policy had given to the glass industry in the Nord, it had directly led to the movement of Belgian nailmakers to the Anzin pits and then to the establishment of a modern refinery there to supply them with the cheap bar iron they needed. The Nord coal industry owed its inception to the enterprise of a restless Belgian and to the use of the English Newcomen engine in the early eighteenth century. Now, a century later, Belgium and England contributed ideas, capital, and men to the establishment of an infant metallurgical center on that coalfield.[105]

The Textile Industries

Since the Middle Ages the Nord had been a major center of the linen and woolen industries. Together with the rapidly expanding cotton industry, they still constituted by far the most important industry in the Nord during the early nineteenth century. All three industries contributed something to Anzin's prosperity.[106]

The linen industry as a whole was declining in the Nord, but it still employed about 30,000 female hand-spinners, who worked in their homes throughout the countryside during the 1830s.[107] Thus, in the early years of the Empire, "almost all the wives and young women in the village" of Fresnes were occupied in spinning flax. In fact, there were an estimated 500 women listed as so employed in the same report that put the number of men and boys employed in the Anzin Company pits there at 544. These figures indicate that flax spinning at Fresnes was probably the usual occupation of the female half of the Anzin mining population. Since the same situation existed in the Raismes and Anzin communes, it is obvious why so few women were employed in the Anzin Company pits, although the practice was fairly common in other coalfields, including nearby Mons.[108]

In the mid-1820s, three large putting-out firms at Valenciennes organized the local batiste and lace industries. They employed about four thousand women in the surrounding rural communes and produced about eighteen thousand pieces each year, and at Valenciennes alone there were twenty-six "merchants" and six "bankers" connected with the trade and financing of the batiste and lace industries.[109] All this activity was of only indirect significance for Anzin. Its coal was not needed in any of the steps in the making of linen cloth,[110] but the company's labor policy might well be affected by the long-term fall in the demand for fine laces and batistes and consequent unemployment within its miners' families.

The woolen and worsted industry held an intermediate position between linen and cotton with respect to its rate of expansion, its degree of mechanization and urbanization, and its dependence on coal. Carding or combing, spinning, weaving, and finishing were all tending to move from the huge expanse of rural communes stretching from Lille to Abbeville and to settle in the Lille-Roubaix-Tourcoing area where machine spinning, the Jacquard loom, and steam finishing were replacing the more primitive rural technology and increasing the usefulness of steam engines and the tendency toward factory organization. This industry by 1833 had become a minor customer for Anzin coal.[111]

The Nord cotton industry might almost be described as a boom industry. Despite frequent setbacks during commercial crises, the spinning of cotton yarn by jenny and then mule had experienced periods of very rapid, even speculative expansion since the Revolution. The first use of the steam engine to move spinning machinery in the Nord was apparently at Roubaix in 1802 and the next not until 1818 at Lille, but by the end of the 1820s cotton spinning was almost wholly mechanized and steam-powered, and the industry was concentrated at Lille and Roubaix.[112] It was variously estimated that a steam engine sufficient to drive the machinery in an average-sized spinning mill used three to six tons of coal a day.[113] Such finishing operations as bleaching, dyeing, and printing were also becoming large consumers of coal in the form of steam heat. Anzin was forced to turn its attention to the problem of supplying such semi-distant markets as Lille with the type of coal best suited for boiler firing at a price competitve with its Mons rivals.

Steam Engines and Steam-Heat Boilers

Almost no steam engines were at work in the Nord before 1820, except those employed for pumping and winding in the mines, but by 1825 at least twenty-three other engines were in use elsewhere in the Nord. Cotton spinning mills had either fourteen or twenty-one engines, oil refineries probably had five, and flour milling, nail making, cannon founding, and chicory grinding each had at least one.[114] During the next decade the number of steam engines in the Nord more than quadrupled. In 1835 the Nord had 297 engines, far more than any other department. Four of the other leading departments—Seine-et-Oise (surrounding Paris proper), Seine-Inférieur (Rouen), Aisne, and the Pas-de-Calais—were part of the Anzin market area.[115] Together with the Nord, they contained nearly half of all the steam engines in France. Much the same industries then used steam power in the Nord as in France generally. The French spinning, mining, sugar refining, and metal industries ranked in that order as using the most engines, although the 266 engines employed in mining supplied more horespower than the 404 used for spinning. In the Nord, the spinning mills used 103 steam engines, the sugar refineries used 83, and the coal mines used 58, but again the rated power of the engines used in mining was considerably greater than the power of the engines used in either spinning or refining. Other industries accounted for little more than one-sixth of the total. Among the more important were various metallurgical operations, cotton weaving and finishing, and vegetable oil refining, each of which had eleven engines.[116]

The amount of coal used by the mines to fuel their own steam engines was probably no more than 5 percent of their total production, but these engines accounted for over one-third of the steam power produced in the department. One must therefore acknowledge that as late as 1835 the Nord market for steam coal was not overwhelming. And in 1828, when there were certainly less than half as many engines as in 1835, when the Paris-Rouen market was barely beginning to open, when Mons was competing vigorously with Anzin in the Lille area, where so many of the Nord's engines were found, and when only the new and undeveloped Denain field produced highly regarded steam coal, the Anzin sales to engine users must have seemed merely a promise of bigger things to come, of much less immediate consequence than such markets as householders and the limeburning, metal, and glass industries.[117]

Sugar refiners and textile manufacturers bought large amounts of coal to produce the steam heat necessary for sugar boiling and various cloth-finishing processes. The steam-heat boiler was first employed in France in 1803 but, like the steam engine, came into general use only after 1820. By 1836 the Nord had eighty-nine boilers and ranked third, well behind the Gard and the Seine, while other departments in the Anzin area, the Pas-de-Calais, Aisne, and Somme, ranked fifth, ninth, and tenth, respectively. (Like the Gard, most of the other leading departments were located in the Rhone valley and used small boilers to prepare silkworm cocoons.) Aside from the silk industry, sugar refining and textile manufacturing together accounted for the largest number of boilers used in both France and the Nord.[118]

Types and Quality of Anzin Coal

Almost to a man, the French users of coal believed that English, Mons, and Loire coals were of better quality than those marketed by Anzin. The company leaders never devised a successful stratagem to overcome the low repute of most of its coal. There was no attempt, for example, to convert low quality coal into more marketable form by processing it into briquettes, coke, coal tar, or gas, although each of these products was known by 1833 or earlier. True, the company cleaned the coal of most stone fragments, hand-separated it by lump size, and did some mixing of grades either at its customers' requests or, as its enemies claimed, in order to hide the particularly poor coal from certain pits, although here it went no further than the normal practice of the day. Not even Casimir Perier's insistence on better cleaning and more judicious mixing could overcome the harsh fact that most of Anzin's own customers regarded its product as less desirable than that of its rivals. In this light, the concentrated drive to pass the Torrent and seek out new seams beyond the old workings takes on a fresh meaning. It was a desperate search for resources that would meet the qualitative as well as the quantitative demands of coal users. Only after 1830, with the opening of the excellent seams at Denain, could Perier and his colleagues feel confident of Anzin's future competitiveness. Until then, the widespread use of Anzin coal owed more to the protection of its markets by transport costs and tariffs than to the intrinsic quality of the coal itself.

Long before the efflorescence of specialized coal uses and more accurate scientific measurement of the widely varying mechanical and chemical properties of different coals in the latter part of the nineteenth century, producers and consumers separated coals into a number of grades on the basis of lump size and burning qualities.[119] The bulk of the coal sold by Anzin and its rivals was of middle size (*moyen*), that is, within a wide range between slack, or small (*menu*) and the large (*gros*) lumps known as *gaillettes*.[120] Aside from lump size, consumers considered four major qualities in choosing a coal for a specific use: caloric value; ease of kindling and time required for complete combustion; length and type of flame; and degree of agglutinization or caking of the burning lumps.

In all these respects, it was a common article of faith held by most French coal users that English Newcastle coal was generally superior to any continental and especially French coal. This antinational prejudice, as it was termed by the Bureau of Mines engineers, may have been reinforced by the almost total inability of most French consumers to obtain English coal because of blockades and tariffs imposed during the Empire and Restoration. Although French coal improved in quality over the years with more careful handling and cleaning and more restraint in mixing,[121] the very inaccessibility of English coal made it seem more desirable. Nevertheless, three types of coal readily available in France gained high reputations for certain uses: Saint Etienne and Rive-de-Gier (Loire) field coal for metal forges; Mons field flenu coal for boilers; and Anzin semianthracite hard coal from Fresnes and Vieux-Condé for roasting limestone and baking bricks and tiles.

The special nature of Anzin hard coal has already been pointed out.[122] It was the only type mined by Anzin whose quality was unanimously regarded as superior to that of its rivals for any particular purpose. The coal from Anzin's northern canton on the Raismes concession was semibituminous with a rather low caloric rating, a relatively high kindling point (though lower than that of Fresnes and Vieux-Condé), a clear flame, and little caking ability. The combination of these qualities made it a mediocre coal, poorly adapted to any major industrial use. Aside from that used for domestic heating, where almost any kind of coal could be and was used if transport costs were low, it was often mixed in small quantities with the richer bituminous coals from the more southerly Raismes or Anzin pits.

Far more serious for Anzin's commercial position was the very ordinary reputation of its coal from the older cantons on the southern faisceau of bituminous coal, which supplied the bulk of its output throughout this period. The pits of southern Raismes, "old Anzin," and the Torrent furnished a rather hard bituminous coal.[123] Its short, unsteady, smoky flame burned hot, slow, and long, but it was relatively hard to light and tended to form a hard crust of caked coals. Boiler operators preferred a coal that ignited quickly, then burned with a long lively flame, quickly raising steam as the flames played up around the boiler. Forge operators required a coal that settled down well on the grille and held its heat by forming a partial crust that inhibited the play of the flames. Obviously Anzin bituminous coal fitted the requirements of the forge better than those of the boiler, but forgemen complained that its crust was too hard, and stokers in all industries did their best to convince their employers to avoid the crusty Anzin coal.

Numerous comparative tests held under government or private auspices did not seem to confirm this reputation to any significant degree. The bias and scientific accuracy of these tests were usually suspect, but in general Anzin bituminous coal seems to have held its own with that of Mons, Saint Etienne, or even England. Anzin naturally enough made much of these test results, but it is doubtful if they shook the ordinary customer in his long-established notions of what coal worked best for his own needs. It is significant, too, perhaps, that Anzin, never shy at blowing its own horn when there was the slightest excuse to do so, never went further than to claim that its standard-grade bituminous was as good as that of its rivals. By implication, Anzin accepted the common opinion that Mons flenu coal and Saint Etienne forge coal were the standards against which all boiler and forge coals should be measured. The coal from the new Denain field, of little commercial importance until after 1830, was much closer in its characteristics to Mons flenu than that from the older pits to the east. Even those who regarded as a misnomer Anzin's designation of its Valenciennes area output as "forge coal" grudgingly admitted that Denain coal was "similar" to flenu since it burned quickly with a long flame and little crust. After 1830, without qualms, Anzin sold the Denain output as "steam coal."[124] Notions about the inferiority of Anzin standard-grade bituminous coal did not, of course, prevent consumers from buying Anzin coal even for forge and boiler operation; but, along with the claims of insufficient production and monopoly prices, it was a staple argument for those who fought in the name of the public interest for lower coal tariffs or the formation of rival concessions.

Several conclusions emerge from this review of the major consumers of Anzin coal. One is the limited importance of new technology in expanding the market for coal in the early nineteenth century. Of course some new techniques did so thoroughly revolutionize an old industry that they really created a new one based on coal; English puddling revolutionized the manufacture of wrought iron and the steam engine made possible the rapid growth of cotton spinning in an area without adequate water power. Technological innovation did help create some wholly new industries in the Anzin market area. The establishment and expansion of distilling and sugar and vegetable oil refining between 1800 and 1835 was partly based on the exploitation of known techniques in combination with an advanced agriculture; but it also depended on an extensive use of steam engines and heat boilers fueled by Anzin and Mons coal. In fact, the use of steam as prime mover or source of heat in a variety of industries was the most important single creator of new uses for Anzin coal in the decade after 1820. Most consumers of Anzin coal, however, were little affected by technological innovation. One technical revolution in glassmaking had run its course in the eighteenth century and another would begin after 1833, but in the early nineteenth century the rise in the number of glassmaking firms and in output per firm was due to the increase in consumer demand. The end of a long period of war and the separation of Belgium from France led to a tariff policy that was directly responsible for the establishment of nailmaking around the Anzin pits, and the transplantation of this industry across the frontier in no way affected its traditional forms of organization and technique. Limeburners and household consumers together may well have accounted for over half of Anzin's sales as late as 1833; yet neither was affected much, if at all, by technical innovation.

A second general conclusion concerns the limited importance of heavy industry in the expansion of Anzin markets. In the early nineteenth century, even those coal users whose production was centralized in workshops or mills on a capitalistic basis were remarkably small by present-day standards. Although the Raismes refinery was considered a "vaste établissement," it employed only 150 workers in the late 1820s. The nearly 600 nailworkers around the Anzin pits worked at dozens of scattered forges. The largest glassworks in the Nord employed only 130 workers, the others considerably fewer. Brickworks, lime kilns, foundries, breweries, and sugar refineries seldom employed as many as 30 workers before 1833, and the average spinning mill used a single twenty-horsepower steam engine. Furthermore, most of Anzin's industrial customers did not sell semi-finished goods to other industries; they produced either goods for sale to the immediate consumer like beer, sugar, lime, and cast-iron pots, or goods like nails, window glass, and bar iron that were only one step removed from final consumption.

A final conclusion concerns the relative unimportance of the industrial town as a market for Anzin coal. In the company's primary market area, Lille was the only urban center that ranked among the twenty-five largest towns in France;[125] it was the only one in the area that grew to an appreciable extent in the early nineteenth century;[126] and it was the only one that depended heavily on industry for its prosperity and growth. Together with its satellites, Roubaix and Tourcoing, and its rapidly growing suburbs,[127] Lille was a center for sugar and vegetable oil refining, cotton spinning, and cloth-finishing, as well as a host of lesser industries dependent on coal.

The other major towns in the area—Dunkerque, Cambrai, Valenciennes, Douai, and Arras—had important commercial, administrative, religious, or military functions, but they were not industrial centers,[128] and their populations virtually stagnated between 1800 and 1833. The mining communes on the Anzin Company concessions showed a high growth rate during these years, but their total population remained small, ranging from about 500 to a little over 4,000. Here where the impact of mining and related industry can be clearly measured, the total increase in population during the first three decades of the nineteenth century amounted to only 4,454 as compared to an increase of 37,384 for the Valenciennes district as a whole. Since the towns in the district—Valenciennes, Condé, and Saint Amand—together lost population during this period, by far the greater part of population growth in the district boasting the largest mining company in France must have been rural.[129] Some of this rural growth was undoubtedly attributable to small-scale industry, but most of it was simply due to a high birth rate among peasant families and a consequent growth in density of the rural population.

During the Restoration, Anzin coal sales approximately doubled. Most of the increase occurred in the late 1820s when improved transportation, expanded production, and an aggressive sales policy enabled Anzin to satisfy the growing demand for cheap fuel despite severe competition from Mons and the mediocre reputation of most of its coal. The increased demand was only partly the product of new technology and new industries. The slow surge of industrialization was the chief stimulus to the transport improvements that opened new areas to Anzin coal. Nevertheless "industrialization" in the early nineteenth century did not mean primarily the creation of a vast, interconnected, urban-centered complex of heavy industries based on the use of mineral fuel. In spite of the important changes that took place after 1815, the Anzin market remained closely linked to a prosperous, highly developed agriculture, to a growing rural population, and to a variety of coal-using industries that supplied the domestic fuel, lime, and building materials needed by the countryside and that processed the products of its labor.

8 Commercial Organization: Prices, Markets, and Profits

Whether in serving its traditional customers or in supplying a growing industrial market, Anzin in the early nineteenth century showed little commercial flair or originality. Mons competitors dictated its pricing policy, and it neither invented nor adopted improved sales techniques. The striking geographical expansion of its market toward the end of the period owed more to new or improved waterways than it did to the initiative of company salesmen, advertising, or the improved processing of its product. An immunity to commercial booms and busts stemmed mainly from a lack of critical ties to the forces behind the trade cycle and shielded Anzin from the possible consequences of failure to control the market through innovative commercial policies. Anzin focused its undoubted managerial talents and expansive zeal on the supply side, on exploiting its coal reserves more fully and efficiently, for it was the lack of good forge coal rather than coal buyers that set the practical upper limits to sales. Even Casimir Perier would do little more than exploit aggressively the opportunity provided by the opening of the Seine basin, and with all his aggressiveness and drive for control, he never proposed that the company assume the role of its wholesale dealers by replacing them as essential middlemen, the chief salesmen for Anzin coal. Like Desandrouin before him, he insisted on limiting the volume of coal sold by maintaining prices that reflected the company's quasi-monopolistic advantages in the marketplace. The brilliant profits that resulted from this policy reflected the owners' consistent preference for a high rate of return over a maximum volume of sales.

Sales Prices and Payments

Although Anzin sold an essentially uniform product, significant differences in coal types, sizes, and methods of sale complicated its pricing policy. The company mined coal of two basic qualities, the hard coal of Fresnes and Vieux-Condé and the soft coal of Anzin and Raismes. It sorted both types into three grades according to lump size: large, middle-sized, and slack, or small. Finally, it sold each of these varieties, with the exception of slack, at either wholesale or retail prices. The possible permutations and combinations among these various categories produced a list of ten prices.

The demand for soft coal rose faster than that for hard coal; by about 1830 the former outsold the latter by more than two to one. The combined proportion of large lumps and slack sold was never more than perhaps one-eighth that of the middle-sized grade. And at least 80 percent of sales was at wholesale prices. Thus, by 1830, 63 percent of all the coal Anzin sold was of the middle-sized grade at wholesale prices, and about two-thirds of this, or 45 percent of all sales, was soft coal.[1] Thus the wholesale price of the middle-sized grade soft coal may be taken as the norm in tracing the Anzin pricing policy for the period 1800 to 1833. On the basis of fragmentary evidence, the evolution of sales prices in francs for a heaping hectoliter of Anzin middle-sized soft coal at wholesale can be summarized as follows: 1800–11, 1.30; 1811–17, 1.25; 1818–29, 1.30; 1829–33, 1.20.[2] Evidently, if these figures are approximately correct, formal prices for this grade of coal did not change often or by very much over a period of a third of a century.[3]

Anzin's enemies, in order to give substance to their charges of monopoly against the company, said that in fact its real prices rose considerably as a result of fraudulent practices. They claimed that, in shifting from an older measure to the smaller hectoliter about 1800, the company had not changed prices proportionately; that another shift in measures from the heaping to the level hectoliter had occurred in 1818 without any corresponding change in prices; and that Anzin had increasingly managed to sell bad coal at the price of better by mixing the two grades. Altogether, they claimed, these abuses had resulted in a hidden price rise of some 0.15 francs or over 10 percent. What they pointed to as "a notorious fact," the company simply denied.[4] Whatever the truth of these charges, it is clear that Anzin sale prices ought to have included the 0.10 to 0.25 francs paid for transport, measurement, and loading, since the company insisted on performing these operations itself and since these charges were far higher than the actual cost of the operations to the company.[5] For soft coal sold at the wholesale price at riverside, these charges fell from 0.250 to 0.175 francs between 1811 and 1829. The total sale price of this type of coal, including these charges, thus fluctuated over the period 1800 to 1833 from a high of about 1.600 francs to a low of 1.375 francs, with the lowest price occurring at the end of the period.[6]

This slight fall in Anzin prices was the result of competition with Mons. As will be shown in a later section, Anzin prices were not closely linked to cyclical movements of the French economy. Nor did they fluctuate seasonally, although the company dealers might adjust their retail prices to changes in seasonal demands. Anzin fixed its prices in response to changes in the price of Mons coal.[7] Mons prices in turn were set by competition between some thirty-five to forty producers. The French coal tariff and lowered extraction costs for the larger producers forced Mons prices down in the 1820s and especially in 1828–29.[8] Anzin met the threat by dropping its prices slightly and by forcing its dealers to operate on a lower profit margin.[9] Shielded by the tariff and transport costs, Anzin was in a powerful position with its dealers and continued to sell at a large profit, while many of the smaller and disunited Mons firms, which could not agree on a common marketing policy, operated at a loss.[10]

The real prices paid by wholesalers fluctuated far more often and to a somewhat greater degree than the formal ones quoted above and listed in table 6. In fact, the company issued new price lists in 1811 and 1829 in part to rationalize the chaotic multiplicity of prices which had resulted from the extension of commercial favors to individual coal dealers in previous years. No sooner did the new rates become standard, however, than new deviations in a downward direction began. The favors received by one merchant elicited demands from others, and once again formal list prices became mere points of departure for the calculation of real prices.

The two major types of commercial favors consisted of rebates on the formal price and the extension of credit terms. The degree to which the company favored a given customer depended on a

Table 6 Sales Prices at Anzin in 1811 and 1829 (In Francs per Heaping Hectoliter)

Types of Coal	Retail	Wholesale
	1811[a]	
Anzin and Raismes		
Large	2.90	2.75
Middle-sized	1.55	1.50
Fresnes and Vieux-Condé		
Large	2.90	2.75
Middle-sized	1.65	1.60
	1829[b]	
Anzin and Raismes		
Large	2.35	2.175
Middle-sized	1.50	1.375
Slack	1.00	0.975

SOURCE: Registres, 6 June 1811, 16 May 1829.
a. Includes 0.15 francs for measuring and loading plus, for wholesale prices, 0.10 francs for transport to loading areas.
b. Includes 0.10 francs for measuring and loading plus, for wholesale prices, 0.07 francs for transport to loading areas.

combination of four factors: the amount of coal he bought; his competitive situation; his willingness to attempt to open up new markets; and his loyalty, or willingness to buy coal from Anzin exclusively. Since most dealers received no favors unless they were considered loyal, and since few new markets were within reach of most dealers, the first two factors tended to be decisive. The more coal a dealer bought and the more threatening the competition of other dealers in his area handling Mons coal, the more extensive a dealer's rebates and credit terms were.

Of the several types of rebates, the most common was the standard percentage reduction on formal sales prices, the type shared by most wholesale merchants.[11] Often the reduction was at a flat rate (usually 3 percent) for all dealers buying more than a given amount of coal, but frequently the company granted rebates on a sliding scale, the percentage increasing with the amount of coal purchased. In 1807 the sliding scale allowed a rebate of 1 percent for every 10,000 hectoliters (ca. 1,100 tons) up to a maximum of 10 percent for all purchases over 100,000 hectoliters.[12] Since the average dealer in 1810 bought perhaps 50,000 hectoliters, the average rebate in this case amounted to about 5 percent.

When the flat rate system was in operation or when the sliding scale did not sufficiently protect or encourage certain buyers, the company might grant special rebates above the ordinary level. Such favors tended to become general as word of them passed along and nonfavored buyers put pressure on the company for equal treatment. Probably the company's largest buyer in 1810 was Soyez of Cambrai.[13] In 1808 he was already receiving the 10 percent rebate accorded those taking more than 100,000 hectoliters, but because of the price war Mons was then pursuing at Cambrai, the company granted him an additional 2.5 percent rebate on condition that the size of his purchases proved his "attachment" to Anzin. In 1810 Mons had "redoubled its efforts" to seize the Cambrai market, and Anzin raised Soyez's rebate to 15 percent at a time when it apparently expected his purchases to amount to about 200,000 hectoliters. The next year it revoked all rebates and lowered list prices but immediately granted Soyez a 3 percent commission on his gross sales. The uproar from the company's other dealers must have deafened the board, for at its next meeting four months later, four other merchants demanded and received identical commissions. At the next meeting after that, Soyez had yet further bad tidings for Anzin; the Mons dealers' price cutting had virtually eliminated his profits, including the commission. The company took an unprecedented step by granting him outright an "indemnity" of 250,000 francs on condition that he buy at least 50,000 hectoliters that year.[14]

Anzin seldom had to resort to such extensive and varied inducements in order to meet competition and hold its dealers, but the Registres are full of lesser favors granted above the standard rebates. Thus the company regularly found itself with three sets of prices: the formal price list, the standard rebate, and a variety of special rebates on the standard rebate. It was then that the company was likely to cancel all rebates and issue a new formal price list, which had the effect of establishing uniform prices for all at a level about equal to that paid by buyers previously receiving the standard rebate. Almost immediately, however, outcries poured in from those buyers who had previously been most favored, and the process of special favors, which in turn widened into general favors, began again.

Such a sequence seems to have occurred in the late twenties, if the figure given in the Etats for value of sales and rebates for 1826 through 1829 can be accepted as proportionately if not absolutely correct. The average rebate on wholesale purchases in 1826 through 1828 was about 5 percent. In 1829 the company put a new formal price list into effect for the first time in several years. It lowered the price of middle-sized coal by about 7 percent, but in that same year rebates still amounted to about 2.5 percent.[15] The new price list thus lowered real prices to wholesalers by 5 percent in order to meet lower Mons prices but at the same time brought formal and real prices into closer agreement.

It is noteworthy how quickly some dealers nevertheless reestablished their advantage. The company granted, to the few who bought in very large quantities, large rebates of a new type fixed at a flat sum in francs per hectoliter for all purchases over a certain base amount. In 1832 the rebate amounted to 0.115 francs (about 8 percent of the formal list price) for each hectoliter bought over the base of 55,000 hectoliters.[16]

The second type of commercial favor extended to wholesale buyers was commercial credit.[17] A complete list of the firms to whom the company extended credit exists for 1810. It states the maximum amount of credit in francs that each of seventy-five firms could draw on against the company books. The amount varied enormously, of course, ranging from 600 to 280,000 francs in a total of 2,100,000 francs.[18] The total certainly increased absolutely and probably also increased relative to cash sales after 1817, when the company tendered the privilege of wholesale buying to local industrialists as well as commercial wholesalers.

Like rebates, the amount and term of the credit extended varied from dealer to dealer depending on such circumstances as quantity of coal purchased, competitive position, and solvency. These short-term commercial loans ordinarily ran for at least three months, never for more than a year. The interest charged was invariably 0.5 percent per month or 6 percent per year throughout the period 1800–33.[19] There was a strong tendency to cancel interest rates by rebates, and so rebate and credit terms often counterbalanced each other and made the formal list prices real prices.

Anzin sometimes insisted on security for its short-term credit, especially when the dealer involved was new to the company's list of buyers.[20] Its established customers, however, simply received book credit in an amount sufficient for the purchase of the quantity of coal they expected to sell in the next year. The company did not always guess right; it was occasionally forced to write off bad debts. Most of the bad risks were small dealers, new customers located in small towns, about whom the company had evidently known little, but on whom it had nevertheless decided to take a risk without requiring security, on the chance of opening a new market. Occasionally, however, a large established customer failed. The largest of these failures was Jean of Valenciennes, whose drawing account on the company books in 1810 was 145,000 francs, second only to that of Soyez of Cambrai. Yet Jean was bankrupt and 200,000 francs in debt to Anzin by 1816.[21] Jean's debt was only the largest of many uncollectable accounts Anzin had been carrying for years when Casimir Perier took over the company leadership. Between 1824 and 1828, in addition to settling with Jean, Anzin wiped off its books over 450,000 francs of old debts incurred before 1822. Thereafter, it moved more prudently in extending credit and more vigorously in dunning customers who were slow in settling their account.[22] Uncollectable debts incurred between 1822 and 1833 amounted to less than 50,000 francs.[23]

The company also faced a contrasting problem. Some of its dealers were so liquid that they turned their credit account with Anzin into a savings account. They deposited commercial paper or cash with Anzin far in excess of the sums needed to balance their accounts. Instead of using this balance to buy coal, they left it on deposit at the regular interest rate of 6 percent. Somewhat annoyed at being used as a savings bank, Anzin ruled that henceforth such dealers would find the funds they left with the company placed in two separate accounts. The one used for coal purchases would draw interest at 6 percent; the other, for excess funds, would receive only 5 percent. The excess cash balances were doubtless soon put back into circulation.[24]

Anzin customers settled their accounts with the company in a variety of ways. Those local purchasers who bought by the wagonload at retail prices paid cash on delivery; cash payment was part of what was meant by the phrase to buy at retail (*en détail*). The wholesalers who bought by the boatload and on credit were scattered in towns throughout northeastern France and western Belgium.

Upon filling their orders, the company billed them directly, and they paid by commercial paper—bills of exchange or drafts—drawn on their accounts with private bankers in the chief towns of the area, such as Cambrai, Arras, and Lille.[25] The bills and drafts received in the mail every day by the company's general agent were passed along to the head bookkeeper, who entered them on the company books before returning them to the general agent.[26] What happened next is not entirely clear. Anzin may have had accounts with local private bankers who would cash drafts and discount bills, but there is no direct evidence to prove it.[27] The company may also have held some bills to maturity before settling them locally. It seems, however, that the normal procedure was to send commercial paper to the Perier private bank in Paris where Anzin maintained a current account. The Perier bank then discounted the bills on the Paris financial market or cashed the drafts with the Nord bankers' correspondents in Paris. The receipts were credited to the Anzin account, which drew interest at a rate that fell gradually following reduction in the general discount rate. The dates of changes in the rates are uncertain. Apparently, an interest rate of 6 percent around 1800, when the discount rate stood at 9 percent, dropped to 5 percent sometime in the next few years as the improved commercial and financial climate of the Empire led to a reduction of the discount rate to 6 percent. In 1820 the interest rate fell again, to 4 percent, because the discount rate had fallen to 5 percent. The interest rate may have dropped again in the late 1820s, for the Periers were claiming in 1826 that they could now barely pay 4 percent, given the prevailing discount rate.[28] Thus the Periers kept their banking and mining interests separate. Although they performed a service for Anzin by facilitating the settling of payments with its customers, they did so without sacrificing the usual profit associated with their discount business.[29]

Anzin does not seem to have formally drawn up written contracts with its customers stipulating the terms of sale and exact quantity to be sold over a fixed period of time, although such contracts were normal practice in the Nord by the middle of the century.[30] Mining companies then agreed to deliver regular amounts of coal at regular intervals, such as twice a week, and accept payments made by the month. The contracts were usually for a one-year term at prices fixed for the contract period, but with old local customers closely tied to the mines, such as cokeries, they might be for as many as twelve to fifteen years.[31] Fixed contracts of this sort with dealers in distant towns were difficult to carry out before the coming of the railroad, because of the uncertainty and slowness of water freight. It is possible, of course, that Anzin signed contracts with some of the local industries that sprang up around the pits in the 1820s, but it is more likely that some of these industries simply saw to it that they were regularly supplied with the wagonloads of coal they needed without making any formal arrangements with Anzin. Their wagons would be expected on a certain day of the week at an accustomed pit, but there was no obligation for them to buy or for Anzin to supply any fixed quantities of coal over an agreed period of time. All the many items in the Registres that mention the quantities of coal to be bought by customers are stated in a contingent style: this customer is to receive this rebate and these credit terms *if* he buys this much coal in the next year.[32] Anzin remained free to manipulate prices and terms of sale with each customer and for each order placed.

The Distribution of Sales and the Search for New Markets

None of the Anzin Company's own commercial correspondence or accounts has survived. In the previous section, it was necessary to reconstruct prices and sales techniques mainly from a large number of miscellaneous and fragmentary references found in the Registres of the board of directors. There is, however, one source of information in the Registres that can be used for a closer analysis of the Anzin market at one particular moment. In February 1810 the board of directors approved the list of credits for that year, as submitted by the general agent, in what was presumably an annual occurrence. But at the next meeting, in October, the list itself was for some reason inserted into the Registres, the formal record of decisions taken by the board. The list, taking up three full pages, gives the name of each of the seventy-five firms to whom the company extended credit, their place of business, and in the last three columns, the amount of credit in francs that each firm might draw on for the purchase in 1810 of Anzin (and Raismes), Fresnes, and Vieux-Condé coal, respectively.[33]

Given the prevailing prices, it is possible to convert the money values of the list into quantitative ones in hectoliters or tons. As always when dealing with Anzin statistics, there are great technical difficulties in making the conversions, and the results are only rough approximations.[34] In addition, of course, the listed values are merely credits for coal yet to be bought and not records of actual purchases, although it is certain that most if not all of the credits granted reflected the solid ground of previous years' purchases. Some weaker firms may not have received as much credit as they requested or enough to cover all the coal Anzin expected them to buy, and other firms certainly did not use up all the credit granted them by year's end. Nevertheless, the list of credits for 1810 is the sole comprehensive source in the entire period from 1800 to 1833 that makes possible a detailed analysis of Anzin's most important customers and the distribution of its markets at a given moment. Even allowing for errors as great as 20 percent more or less in converting money credits to tonnage credits, or in assuming tonnage credits to approximate tonnages purchased, the list is invaluable.

Of course, some of the Anzin output escaped the 1810 list because it was not sold at all but fed the company's steam engines and ventilation furnaces and heated its offices and shops or went to employees as a supplement to salaries and wages. Anzin also claimed that it stood ready to give coal free to any "unfortunates" too poor to buy it.[35] Much of the coal used in these ways was slack coal, which was difficult to sell commercially.

Retail sales were also almost excluded from the 1810 list. In dividing sales into those made *en détail* and those made *en gros,* Anzin was making a much broader distinction than would strictly be expressed by the words *retail* and *wholesale.* Wholesale purchases were naturally those made in large quantities and at a reduced price, but they were also those made by the boatload on the Escaut River, for nonlocal markets, and on credit. Retail sales, for cash, by the wagonload, at the pit head fed the purely local market of Valenciennes and the neighboring communes. Large numbers of the local population, mostly peasants and villagers, one assumes, went directly to the Anzin pits to buy coal from the storage dumps there and carry it away in wagons, farm carts, and even wheelbarrows. Some tiny industrialists, such as village brewers, may have supplied themselves in this way, but most bought for their own domestic use or for sale to a few neighbors, while a few undoubtedly made a business of making the rounds of a limited area with their wagonload of coal. The total amount of coal involved in these sales was "minimal," although it was presumably as much greater than that sold in the better-known Valenciennes market as the population of the countryside was greater than that of the town.[36]

The eighteen dealers who supplied the domestic market of Valenciennes in 1810 bought by the wagonload and at retail prices, but they were sufficiently large and established to receive credit. The amounts ranged from 500 to 6,000 francs apiece and totaled just over 25,000 francs, a figure that indicates a total of about 2,000 tons or an average of some 100 tons per dealer. All bought soft coal exclusively, since it came from pits adjacent to the town and had better burning qualities for household use than hard coal did.[37]

Industry, particularly glassworks and metalworks, provided much the most important of the local markets in the long run, but as has already been shown, few such industries existed in 1810. Only one is clearly identifiable on the credit list for that year, the Fresnes glassworks owned by Desandrouin, but unfortunately it is the sole firm on the list whose credit columns were left blank. Like other glassworks under the Empire, it may have temporarily shut down, or, because of its owner's position in the Anzin Company, it may have commanded indefinitely large credits.[38]

Except for this special case, until 1817 the company treated all purchases by local industries, however large the amounts, as retail sales at retail prices simply because it delivered them by the wagonload. Then the migration of nailmaking firms from Belgium, preparations for the opening of new glassworks, and direct competition from outlets for Mons coal at Valenciennes led Anzin to set new, more favorable terms. Stating that "big factories" and the military establishments at Valenciennes and Condé bought their coal from Mons suppliers because Anzin charged them retail prices, the board ruled that henceforth large buyers by the wagonload would pay the same wholesale prices, or nearly so, as buyers by the boatload.[39] By the early 1820s, the local industrial market was growing rapidly, and as new firms established themselves, they soon demanded not only wholesale prices, long credit terms, and the customary rebates, but special rebates as well. Favors granted one firm were quickly demanded and received by all its competitors. The local industrialists, least favored under the Empire because they had been virtually nonexistant, probably, as a group, shared the most advantageous sale conditions and prices of any Anzin customers in the 1820s.[40]

Important though this new market was, its size ought not to be exaggerated. Together, retail and wholesale purchases by wagon amounted to only 22 percent of all sales in 1830, and 25 percent in 1831.[41] Since retail sales in Valenciennes in 1835 totaled only 5,000 tons (or about 1 percent of the 1830 output) and rural retail sales were probably only somewhat larger, most of the 22 or 25 percent of wagon sales must have gone to the new local industries.[42]

Anzin sold almost all of its output in 1810 and 75 percent of it by 1831 in boatloads to large dealers scattered throughout the larger towns of the Nord and nearby areas. These men almost monopolized the credit list of 1810 (see table 7). They bought at wholesale prices

Table 7 Credits Extended to Anzin Dealers, 1810 (By Percentage)

Dealers	Credits
Lille	20
Cambrai	18
Western Belgium	16
Arras	12
Valenciennes (wholesalers)	10
Smaller towns	10
Dunkerque	5
Douai	4
Valenciennes (retailers)	1

Source: Registres, 4 Oct. 1810.

and paid on credit. They sold Anzin coal to householders, retailers, and industrialists, to town and country. From their dumps at dockside in the river towns, where their headquarters were invariably located, they supplied wide areas by wagon and often transshipped coal to smaller towns. They mixed coals to meet special needs.[43] Each one's market ended where another's began or where transport by wagon raised prices too high to compete with other dealers or other fuels. Together they blanketed an area that included all the

Nord department and most of the Pas-de-Calais as well as parts of the Aisne, Somme, and (for hard coal) western Belgium. Each of these firms was a single-ownership or partnership concern; although some may have been sociétés en commandite, they were probably of the simpler partnership form as opposed to sociétés whose capital was divided into negotiable shares.[44] None was a corporation. Like most sellers of foodstuffs or fuel, they still operated under a certain amount of local regulation, and Paris had permitted them to form a trade association (syndicat). It was they who contracted for river transport with the boatmen and paid the tolls at locks and their own town's entry taxes; it was they who knew their own market, ordered accordingly from Anzin, and arranged for sale to the local consumer. Information about commercial conditions, Mons competition, and other matters vital to Anzin's welfare normally passed from the dealers to Anzin rather than the reverse. They were not only Anzin's most important customers but also in effect its salesmen, shippers, and informants. Anzin depended heavily on the loyalty and ability of the owners of some fifty wholesale firms scattered through its prime market area.[45]

The most important duty of Anzin's general agent was to keep up an extensive correspondence with the company's customers. Each of the three men who held the post of general agent between 1798 and 1839 was ignorant of the technical side of the mining business when first appointed. All three were chosen primarily for their presumed abilities as salemen and managers. The forced retirement of the first general agent, Renard, in 1825 and the dismissal of the second, Paulze d'Ivoy, in 1827 were both in part a result of their failing to live up to the exacting standards of the Periers. The tendency to stress the commercial side of the general agent's duties was accentuated after Renard's retirement by the appointment of a works director to relieve the general agent of day-to-day supervision of coal mining and to permit him to concentrate his attention on the company's commercial fortunes.

In the 1820s Casimir Perier and his brother Joseph kept up an almost daily correspondence with their general agent. In writing to Paulze d'Ivoy between 1825 and 1827 they emphasized that he was to seek out new customers "without respite" and regain any that had been lost without alienating dealers who had remained loyal through Anzin's earlier, more difficult years.[46] He was to force dealers to meet their financial engagements on time while dunning them so tactfully as not to lose their business.[47] He was to make certain that the employees at the riverside loading areas did not favor some dealers above others by giving them a preferred place on loading schedules or a better quality of coal. He was to make certain that there was sufficient coal on hand to fill all orders when navigation resumed on the Escaut after the summer drought or winter freeze. If coal was in short supply at these crucial times, "the loss of money is certainly an annoying matter, but what is worse to my mind is the bad effect it will have on the public."[48] From Paris, the Periers asked constantly for information on the current level of sales. They compared current sales levels with those achieved during the same period in past years, and if the comparison was not encouraging, they insisted on knowing the reason why. Was Anzin's coal not "assez beau"? Were the dealers "mal servis"? Was extraction lagging or was there favoritism at the loading areas?[49]

By far the largest of these dealers, according to the 1810 list, was Soyez of Cambrai. He received a book credit of 280,000 francs, but the suspicion is strong that he actually bought nowhere near that amount. The huge credit was probably part of the extravagantly favorable treatment granted him because of the price war he was waging with Mons dealers, and he and Anzin may well have assumed that the opening of the Saint Quentin Canal in that year would result in a tremendous increase in sales for Cambrai dealers who would be involved in the new trade to Paris. The Paris trade did not materialize, and in 1812 Soyez received a large indemnity from Anzin on condition of buying at least 50,000 hectoliters, the equivalent of only some 75,000 francs' worth of coal.[50]

The ill-fated Jean of Valenciennes, bankrupt by 1816, was the second largest dealer in the list with a credit of 145,000 francs. Six other dealers received 70,000–90,000 francs, eight 50,000–69,999 francs, thirteen 30,000–49,999 francs, and four 18,000–29,999 francs. These thirty-three large dealers accounted for about 95 percent of all credit sales in 1810. The remaining forty-two firms on the credit list included the eighteen Valenciennes retailers, the largest of whom received only 6,000 francs, and a variety of smaller dealers granted 12,000 francs or less apiece. The range in size even among the top thirty-three was obviously great. Soyez's credit was fifteen times greater than that of the smallest, who received enough credit to buy a little over 1,500 tons as compared to well over 20,000 for Soyez. Apart from Soyez and Jean, the range was considerably less: thirty-one dealers with credits of 18,000 to 90,000 francs.

Sixteen of the largest thirty-three further dominated Anzin's roster of dealers through their control of the three major consuming and distributing centers of Lille, Arras, and Cambrai. Nine large dealers at Lille obtained credits of 418,000 francs, about 20 percent of the total authorized by the 1810 list. They served the most populous and most industrialized part of the Anzin market and included within their sales area the growing industrial towns of Roubaix, Tourcoing, and Armentières. Yet 40 percent of their credits were for hard coal, a fact that indicates the degree to which domestic, building, and especially agricultural needs still dominated the Anzin market.

In contrast to Lille, Arras and Cambrai stood at the heads of navigation of their respective waterways in 1810 and served somewhat more indefinite areas as centers for the diffusion of coal along the shifting frontiers of the Anzin market area. Arras in the Pas-de-Calais had four large dealers who received credits amounting to 254,000 francs, or about 12 percent of the total credits for 1810. Like Lille, Arras took about 40 percent of this amount in hard coal. At the head of navigation of the Scarpe, Arras was a dead end for water transport and never seems to have attracted the kind of price war Anzin and Mons engaged in at Cambrai at the northern terminus of the Saint Quentin Canal. Still, Anzin supplied only about one-half of the coal consumed in the Arras area, a proportion that was probably near that found at Lille and similar towns some distance from Anzin and within fairly easy reach of Mons. From Arras, coal moved in wagonloads to numerous points to the north, west, and south in the Pas-de-Calais and even as far as the Somme department. It circulated especially in the vicinity of Saint Pol, and from there, again loaded onto boats, reached as far as Montreuil on the Canche River.[51]

Cambrai's credits, taking those of Soyez and two other larger dealers there at face value, amounted in 1810 to 373,000 francs, or about 18 percent of the total, a proportion which, if correct, placed Cambrai second only to Lille among Anzin's individual markets. Some reasons for regarding this figure as inflated have already been pointed out. It may be added that the location and organization of

Anzin's loading areas and the difficulty of upstream navigation to an unindustrialized town slightly smaller than Valenciennes indicate that Cambrai was not nearly so important a market as Lille before 1810. After that date it probably served as a transshipment point for coal embarked on the tortuous Saint Quentin Canal for the upper Somme and Oise valleys. Still later, when the canal had been improved, Cambrai again received attention as a way depot for coal intended for Paris.[52]

Together, the towns of Lille, Arras, and Cambrai received half the credits extended by Anzin in 1810. Although Lille was incomparably the largest town in the Nord–Pas-de-Calais area, Arras and Cambrai were about the same size as three others towns that received considerably fewer credits. One of these, Dunkerque, was a special case and will be discussed as such. Another, Douai, was a limited market for Anzin and Mons coal alike because of its proximity to the Aniche pits, for which it was the major outlet. Nevertheless, Anzin's two dealers there received 90,000 francs in credit, of which 70,000 francs was for the purchase of hard coal, a grade not produced by Aniche. Finally, as related above, eighteen small retailers supplied the domestic needs of Valenciennes, but the town also harbored two large wholesale dealers who together obtained 200,000 francs in credits, or almost 10 percent of the total. It is possible these dealers competed with peasant carters for the custom of the local countryside. More likely, they sold by the boatload to larger towns, and the predominance of hard-coal credits may indicate that they sold especially to the limeburners of western Belgium. If they were indeed wholesalers buying for sale at a distance, they broke with the normal pattern, according to which Anzin's wholesale merchants were located in the areas where they sold their coal.[53]

Other, smaller towns each with a single Anzin dealer received substantial credits: Marchiennes, 62,000 francs; Condé, 50,000; Bailleul, 50,000; La Bassée, 70,000; Saint Omer, 30,000. In all but one of these towns the credits were overwhelmingly for the purchase of hard coal, which implies that it was for domestic and rural rather than industrial consumption. Each was on a waterway and some distance from the major population centers like Lille and Arras. Together these towns, filling in the chinks in the Anzin market area, accounted for 10 percent of the credits listed.

The final important market as revealed by the 1810 list was the limeburning industry of western Belgium. Here Anzin sold its coal to some of the burners directly, although smaller burners may have bought fuel from larger ones or from wholesale dealers of the usual sort.[54] Seven firms at Tournai obtained 215,000 francs in credit, four in Menin and Vervick received 76,000, and one firm each in Merlbeck, Lynaerde, Lepighem, Ghent, and Bruges totaled less than 50,000. Together, the limeburners of western Belgium accounted for 16 percent of the 1810 credits. All of their credits were, of course, for hard coal, and they constituted almost 30 percent of the Fresnes–Vieux-Condé market.

All the markets discussed thus far, with the partial exception of local industry, were traditional ones dating from the eighteenth century. In these areas, Anzin's role in selling was essentially passive; it accepted orders for coal of a specific grade and attempted to fill these orders as quickly as possible. Sometimes, however, the company made a direct effort to stimulate sales by taking a personal hand in marketing its product in order to reach a new geographical area beyond the range of its regular dealers. Most such attempts proved abortive; markets that it could not serve in the usual way also proved beyond the company's own ability to open and keep open. And Anzin gradually turned over to dealers of the traditional type those new markets where it did manage to establish itself and returned as quickly as possible to its passive role of initial supplier.

Anzin did try in at least three isolated instances to open new markets by direct demonstration of the quality of its coal. The first attempt, evidently associated with the opening of the Saint Quentin Canal, occurred in 1810 and 1812 when Anzin tried to introduce the use of Fresnes coal to the brickmakers and limeburners of the Rheims area. The company went so far as to build a lime kiln there at its own expense in order to demonstrate the advantages of its hard coal over the wood fuel used previously. "The military events of 1814 and 1815" intervened, and in 1819 the company decided to write off as a loss its expenses of some 13,800 francs.[55] The second such attempt took the form of an extraordinary 20 percent rebate to Scipion Perier in 1810 for his experimental glassworks at Romesnil, in large part no doubt because of his position in the company, but also, it was said, "to give an example to other glassworks in the area."[56] The third involved the Periers' hopes of creating a market for Anzin soft coal in Paris. In 1825 they ordered "some lumps" of the coal extracted at the Torrent pits sent to Paris in a wagon that was carrying cast-iron samples to their Chaillot works. If the coal from the new pits proved more acceptable for forges and coke ovens than that which Anzin had previously tried to sell in Paris, "we will be able to have a considerable market here."[57]

The Periers' call for coal samples in 1825 was merely a minor incident in Anzin's continuous effort to open the Paris market after 1810. The direct-selling techniques that Anzin used in the lower Oise and Seine valleys had to a large extent been worked out in the earlier unsuccessful attempt to place its coal in the ports of the Channel and the Atlantic. Already, in 1798, the company had proposed to extend its production in order to try to supply "or at least compete with . . . the irreconcilable enemy [England]" in Normandy and Holland.[58] At the end of 1801 the scent of peace in the air (and hence freedom from naval blockade and seizure) plus prohibition of English coal imports set the stage for a vigorous campaign to invade coastal markets. It was then that Anzin decided to build ten boats to open the Holland trade via the Escaut. Much more important were plans for shipping coal to Ostend and Dunkerque. There Renard himself would try to find merchants who would take Anzin coal on consignment to sell for a commission in the many ports with which they maintained commercial relations. Renard was to draw bills for Anzin on the Perier banking firm in Paris, where a sort of office for the new trade would be set up. The Perier office would be a clearing house for correspondence. The staff there would discount Anzin's bills and out of the receipts, amounting to as much as 100,000 francs, pay expenses, such as freight charges and merchants' commissions. The Periers' payments were backed by Anzin's bills, which would balance on the Perier books the sales receipts of Anzin coal in Le Havre, Nantes, La Rochelle, Bordeaux, Rochefort, and other ports. For its services, the Perier bank was to receive a 2 percent commission on gross sales.[59]

After the signing of the Peace of Amiens on 25 March 1802, "the export of coal from the pits of Anzin and Mons to the ports of the Channel and the Ocean was of the greatest consequence" for the shippers of Dunkerque, Ostend, and Antwerp.[60] Since September of the previous year, Anzin had expedited 10,000 tons of coal to various ports from Dunkerque and Ostend. Then on 31 May the French government lifted the prohibition on English coal imports and imposed in its place a tariff, heavy at Dunkerque but progressively lighter as one passed west and south along the French coasts. The nascent Anzin (and Mons) trade along the coast was nipped in the bud by the reappearance of English colliers in such French ports as Cherbourg, Nantes, and Bordeaux.[61] Anzin, which had "profited from the freedom of the seas . . . to extend [its] markets," now found itself in September with 12,000 tons of coal lying unsold in distant ports and with a loss for its venture into distant markets of 53,000 francs.[62] Renard protested in vain to the government, pointing out the three-to-one advantage the English had over Anzin in costs of extraction and the ease of transport to the sea by English canals as compared with the "tortuous rivers" of northern France.[63] In desperation, both he and Scipion Perier went personally to Ostend to establish a depot there for Anzin coal already awaiting shipment or to convince the now cautious shippers to take the coal either on commission or for their own account.[64]

No more is heard about this venture, since with the resumption of war with England in 1803 all questions of tariffs and English competition became academic. Coastwise shipping again became nearly impossible. In 1813 only thirty-five ships carrying cargo of any kind left Dunkerque legally, and only nine of them carried any coal (most of which was probably from Mons), all of it destined for either Channel ports or Saint Gilles in the Vendée.[65] According to the credit list for 1810, three Dunkerque dealers received 111,000 francs, or about 5 percent of the total, but most of this amount must have been spent for coal sold to local consumers. Appointment of a salesman working on salary plus commission "to place coal . . . in Holland" in 1810 seems to have had no particular success, although it is one of the few times Anzin used company salesmen in the early nineteenth century.[66] After 1815 the Dutch market was wholly absorbed by the Belgian mines, now part of the new Kingdom of the Netherlands.[67]

At the same time, peace and a very restrictive tariff on coal imports by sea freed the coasts once more to French colliers, but although English imports were minimal and Dunkerque's coal shipments grew during the Restoration, especially after the rehabilitation of the port,[68] Anzin shared in the coastwise traffic only to a minor extent, probably because the French waterways leading from the Escaut to Dunkerque remained much poorer than the Belgian route. Anzin's chief dealer in Dunkerque, the firm of Carlier and Mirland, received extraordinary commercial favors in the mid-1820s when it was fighting for the business of supplying the Royal Navy in its Atlantic yards. In 1825, for example, Carlier and Mirland was to receive a rebate of 0.30 francs per hectoliter for all coal sold to the navy if it "formally engaged" (by written contract) to take, at regular prices and conditions, 30,000 hectoliters for sale to local consumers.[69]

By the early 1830s, Anzin was dealing directly with merchants in distant ports who had firm contracts to supply the navy at Rochefort, Brest, Lorient, and Cherbourg. Again, transportation costs were so unfavorable to Anzin that it was forced to extend extremely large rebates.[70] Even so, by 1835 only about 10,000 tons of Anzin coal were heading out from Dunkerque each year or about 2 percent of its total production. Even more striking was the loss of the local market around Dunkerque on the coast of French Flanders, where Anzin sold only 5,000 tons.[71] Never able to establish itself firmly on the high seas, Anzin now found itself virtually ousted from its long-standing, though feeble, outlet in the northern district of its own department.

Anzin's efforts to open a market in the Seine-Oise basin ran parallel in many respects to its attempts to penetrate to the ports of the Channel and the Atlantic. The key to both markets rested in large part on the improvement of inland waterways. In both markets Anzin faced competition from neighboring Mons and from another mining region, England or Saint Etienne, which had long dominated the intended market. The distance, difficulty, and cost of transport to these markets forced Anzin to adopt what were for it unusually aggressive and direct sales techniques outside the usual dealer pattern. And both markets raised false hopes and encouraged plans for expansion of output before proving unworth the effort expended on them. Only the final results were different, for eventually the Seine-Oise basin did become one of Anzin's most important markets.

Until 1810 there was no water route connecting the Escaut and Seine-Oise basins. Hence most of the Seine-Oise region was little accustomed to the use of coal fuel. The Paris coal market itself was relatively small and limited mainly to small-scale or consumers' goods industries, which required a quick, hot, even grate fire. For this purpose, the forge-grade coal of the Saint Etienne part of the Loire field had a preeminent reputation, and along with other coals from the Midi, it monopolized the Paris market, which it reached after a very long haul on the Loire, the Seine, and the canals in between.

The opening of the Saint Quentin Canal in 1810 enabled Anzin and Mons coals to pass for the first time south of the Escaut basin to the valleys of the Somme, Oise, and Seine. The distance from Saint Etienne to Paris by river and canal was 552 kilometers, but the water distance from Mons and Anzin to Paris was only 340 and 323 kilometers, respectively, and they had an even greater advantage in distance throughout the populous valleys of the Oise and especially the Somme. In fact, Saint Etienne coal had never been able to penetrate much beyond Paris and the Seine valley proper as far as Rouen. Paris, henceforth, would serve as a sort of watershed between the market areas of the southern and northern fields and as their chief point of direct competition.

The water route from Anzin to Paris was complicated. From Valenciennes on the Escaut the route led upstream to Cambrai at the head of the Saint Quentin Canal. On the upper Somme at the southern terminus of the new canal was the important and growing industrial town of Saint Quentin. There followed a short downstream traverse of the improved Somme to the Crozat Canal, which in turn emerged on the upper Oise at Chauny. Below Chauny the Aisne River joined the Oise at Compiègne, and the Oise itself struck the winding Seine not far below Paris near Pontoise. All these points were distribution centers in their own right as well as transshipment ports for Paris, but Saint Quentin was the most important since it controlled the exit from the Saint Quentin Canal, the entrance to the Crozat Canal, and the headwaters of the Somme River and was itself an increasingly large consumer of coal.

Not surprisingly, then, Casimir Perier proposed in 1808 to set up in his own behalf a depot for Anzin coal at Saint Quentin. The company was so eager to penetrate this new market that it agreed to assume Perier's losses. Perier, however, quickly abandoned the idea of operating the Saint Quentin depot himself and opted instead to set up in Paris, which was not only potentially the most lucrative of the many markets to be opened but also his personal residence and the center of his banking operations. The Saint Quentin depot was taken over by the company, which appointed a manager, Wattelez, and a clerk to handle its business there. Meanwhile, Anzin bought land for new loading areas at Valenciennes, upstream from its troublesome locks, and at Chauny on the Oise at the southern end of the Crozat Canal. It made its arrangements with the dealer Soyez of Cambrai at the northern end of the Saint Quentin Canal and approved blueprints for a boat capable of carrying a full load of coal to Paris without transshipment. Thus in the two years before the Saint Quentin Canal opened, Anzin had secured a position at most of the key points on the water route to the capital.[72]

Unfortunately, the Saint Quentin Canal and, to a lesser extent, the Oise proved unnavigable for fully loaded boats at any time, and much of the year low water closed them to all traffic. In addition, other long delays were incurred during loading on the Escaut, transshipping at various points between waterways, and passing the innumerable locks along each of the waterways linking the Nord and Paris. A round trip between these two points by a boat carrying a half-load of coal south and returning empty could easily take as much as a year.[73]

As sales of Anzin coal at Paris languished, Perier once again divested himself of direct responsibility for opening a new market, and Wattelez became company agent in Paris at double his former salary and with an incentive plan which would pay him one-third of the profits on all the coal he sold in the capital. Twenty months later, Anzin cut all direct ties with the Paris outlet, and Wattelez became an independent dealer, assuming for his own account all sales of Anzin coal in the entire Seine-Oise basin. He received favorable credit terms and the usual 3 percent rebate, with the special stipulation that the total amount of the rebate over a year must in any case equal his former salary of 2,400 francs as a company sales agent. The company apparently hoped for sales of at least 80,000 francs in the area, a figure which would have represented, however, not much more than five thousand tons of coal purchases some three years after the much heralded opening of the Nord-Paris waterway.[74]

Anzin's position in the Seine-Oise basin and especially in Paris continued to be discouraging. Wattelez gave up his Paris dealership almost immediately because of poor health. His successor there,

Lhuillier, had to be given not only the rent-free use of the premises of the Paris outlet, still owned by the company, but also its office furniture "as a pure gift" and one year's work from each of the two laborers employed there. His credit with the company was only 20,000 francs.[75] When Wattelez died in 1815, Lhuillier also succeeded him as company manager of the Chauny loading area and as independent dealer for Anzin coal in the surrounding area. His credit at Chauny was 120,000 francs, "on sufficient guarantee," with a 3 percent rebate until the Saint Quentin Canal was "perfected."[76] Clearly Anzin was succeeding far better on the upper Oise than in Paris. "The regions neighboring the Saint Quentin Canal use coal, whose consumption increases every day," but in Paris Anzin coal continued to meet with indifference.[77]

Between 1815 and 1818 Lhuillier spent 11,389 francs out of his own pocket in setting up an expanded sales outlet in Paris, but by 1821 he had run out of money. Out of consideration for his "efforts" to "introduce and propagate" the sale of Anzin coal there, the company granted him a remission of interest on his debts. His credit for the Paris outlet continued to be only 20,000 francs, a sum sufficient to purchase less than 2,000 tons.[78] Lhuillier passed from the picture soon afterward, and on his death in 1830 the company wrote off a long-standing debt of some 35,000 francs, which he had incurred in the process of being the first dealer to make a real attempt to sell Anzin coal in Paris. Casimir Perier apparently took charge of the Paris depot again in the early 1820s, but its operations remained small and sporadic. The company did rent storage space for 150,000 hectoliters at Compiègne in 1824.[79] Compiègne was the next important point on the Nord-Paris route below Chauny. The company's operations in the direction of Paris continued to expand much faster than its sales in Paris itself. It seems likely that with the failure of Lhuillier, the Paris outlet was allowed to mark time under the nominal direction of Perier until the impending improvement of the Saint Quentin Canal and the Oise River enabled the coal from the new pits on the Torrent to reach the capital.

Although Anzin met with almost total failure in Paris for nearly two decades after the opening of the Saint Quentin Canal, Mons gradually whittled away at Saint Etienne's erstwhile monopoly there. By 1818 Mons sales in the capital amounted to fifty thousand tons, a figure that continued to increase in the next decade. The Mons share of coal sales in Paris rose from nothing in 1810 to 26.67 percent in 1818 and to 54.09 percent by 1825, despite the steady deterioration in the condition of the Saint Quentin Canal during those years.[80]

The construction company organized by Honnorez and heavily financed by Anzin succeeded by 1828 in greatly improving navigation of the Saint Quentin Canal by halting seepage and maintaining a higher water level. Transit times and freights between the Nord and Paris fell one-third between 1827 and 1829.[81] Because of lower transport costs, Anzin and Mons could sell their coal in Paris at prices well below those Saint Etienne could afford to charge despite its lower extraction costs. Table 8 summarizes the situation in 1829.

Table 8 Factors Affecting Coal Prices from Pit Head to Paris, 1829 (For Forge-Grade Coal)

Factors	Mons	Anzin	Saint Etienne
1. Water distance from field to Paris, in kilometers	340	323	522
2. Cost of extraction, in francs per heaping hectoliter	0.63–0.65	0.70–0.75	0.30–0.50
3. Cost of transport, pit head to riverside, in francs	0.10–0.15	0.125	0.38
4. Sale price at riverside, in francs	0.85	1.375	1.06
5. Cost of transport from riverside to Paris,[a] in francs	3.25[b]	2.57	3.82
6. Total of 4 and 5	4.10	3.945	4.88
7. Sale price at Paris, in francs	4.25	4.25	4.66 [sic]

SOURCE: Chevalier, p. 482.

a. Includes freight charges, canal and lock tolls, Paris entry tax, and charges there for unloading and measuring.

b. Includes 0.33 francs for customs duties.

While the Saint Etienne wholesalers sold at a loss, coal from the northern field continued to edge its way into a dominant position in the Paris market. From 54.09 percent in 1825, its share of the market rose to 63.80 percent in 1830, and the gain in tonnage sold was even greater as total sales increased rapidly after 1828.

In addition to a new advantage in transport costs, northern coal profited from changes in the structure of the Paris market. The small craftsmen and workshops demanding coal for hot grate fires were joined by householders and especially users of steam engines and heat boilers and manufacturers of coal gas, all of whom found cheaper northern coals more satisfactory for their purposes than those of Saint Etienne.[82]

For some years after 1828, however, it was mainly Mons rather than Anzin that benefited from faster, cheaper transport and the changing market. Anzin's Paris outlet remained small, and sales in Paris in 1829 were said to be still "very variable" from year to year.[83] In that particular year, Anzin sales included twenty boatloads of Fresnes coal for limeburning and a "little" soft coal for domestic heating.[84] At least occasionally during these years, two large Parisian firms also bought Anzin soft coal by the boatload as steam-engine fuel.[85] Both, however, had personal and business connections with the Periers, and both received extraordinarily high rebates. Evidently, Anzin hoped to use the patronage of these firms to help overcome the prejudice in favor of Mons flenu that still prevailed in the Paris market. Total sales for Anzin in and around Paris in 1830 still amounted to only 20,000 tons as against 70,000 for Mons and 67,000 for the Midi.[86] To spur sales in the Paris-Rouen region, the company asked Casimir Perier to take the unusual step of hiring a traveling salesman, a sure sign of commercial difficulties.[87]

Although Anzin sold only some twenty thousand tons in the immediate Paris area, its sales throughout the valleys of the Somme, Oise, and Seine amounted to a respectable one hundred thousand tons.[88] After Paris, the largest single potential market in the region was the Rouen area, 100 kilometers farther than Paris from Anzin, Mons, or Saint Etienne. Before the Revolution, it had drawn its coal supply in equal parts from Newcastle and from Littry in western Normandy. Both sources were cut off by the Revolution, and Saint Etienne replaced them.[89] Attempts at supply from Mons and Anzin about the time of the Peace of Amiens were frustrated by high transport costs.[90] With the opening of the Saint Quentin Canal in 1810, Anzin succeeded in temporarily placing a little coal on the lower Seine through two dealers, one of whom had migrated there from Condé at Anzin's expense and had received extremely generous credit and price terms in recognition of "the distance and difficulty of transport" and his "zeal and attachment" to the company.[91] As in Paris, the foothold was soon lost, and after 1815, Mons and

Newcastle shared the market with Saint Etienne to the virtual exclusion of Anzin. After 1828 Anzin made some progress, but Mons was the chief beneficiary of lower transport costs in Rouen as in Paris. As late as 1835, Anzin sales in Rouen amounted to no more than ten thousand tons against only four hundred tons for the southern French fields, fourteen thousand tons for England, and forty-seven thousand tons for Belgium.[92]

Anzin's progress along the Somme River was more considerable. Before 1828 some Anzin coal had reached the Somme valley by wagon from Arras, and some had probably penetrated the barely navigable upper Somme from Saint Quentin. After 1828, improvement of both the Saint Quentin Canal and the upper Somme led to a brisk trade in such towns as Péronne, Amiens, and Abbeville.[93] By 1835 Anzin sales in the Somme basin amounted to thirty thousand tons.[94] Nevertheless, it is clear that Anzin's chief market south of the Escaut basin in the first years after the improvement of the Saint Quentin Canal in 1828 lay along the Oise with the chief distribution points located at Compiègne and especially Saint Quentin.[95]

Although the Anzin market in the Seine-Oise basin as a whole was nonexistent in 1810, negligible in 1820, and still modest in 1827, by 1830 it may have accounted for as much as 28 percent of the company's sales and thus have absorbed most of the company's very rapid increase in production in the late 1820s (see table 9). Apparently Anzin had benefited more than Mons by the improve-

Table 9 Anzin and Mons Markets by Area, 1830 (In Tons)

Market Area	Mons Sales	Anzin Sales
Escaut basin	250,000	260,000
Seine-Oise-Somme basin (Minus Paris and vicinity)	80,000	80,000
Paris and vicinity	70,000	20,000
TOTALS	400,000	360,000

SOURCE: AN, F12 2534d, Pichault de la Martinière, "Mémoire."
NOTE: Figures are based on the official records of interior navigation. For somewhat different figures at slightly later dates, see Lentacker, p. 1402, and Lévy-Leboyer, *L'industrialisation*, pp. 303–13.

ment of the Saint Quentin Canal, and although Anzin still sold far less than Mons in the immediate vicinity of Paris, it had caught up in the rest of the basin. The tariff on Belgian coal and the cost of transport from Mons to Valenciennes naturally weighed heavier on the consumers of the upper Oise and Somme than on those of more distant Paris, where such costs shrank in relative importance beside the still greater transport costs to the capital. Then too, many of the consumers in the nearer parts of the basin were either new coal users who were less prejudiced again Anzin soft coal than the older Paris industries or were consumers who required the hard coal of Fresnes–Vieux-Condé on which Anzin held a monopoly.

The division of sales reached in 1830 represented only a temporary equilibrium between the company's old and new markets. While sales in both areas increased rapidly in the next five years, the proportions changed. Table 10 indicates roughly the amounts of

Table 10 Anzin Markets by Area, 1834 (In Tons)

Market Area	Sales
Consumed by the company	30,000
Escaut basin[a]	300,000
Exports to Belgium (hard coal)	20,000
Shipments from Dunkerque for French coast	10,000
Seine-Oise-Somme basin[b] (including Paris and vicinity)	140,000
Total	500,000

Source: AN, F12 2534b, Isoard, "Informations recueillies," 1835.
a. Includes 5,000 on Nord coast and 5,000 at Valenciennes.
b. Includes 30,000 in Somme and 10,000 at Rouen.

coal sold by 1834 in the more important sectors of the Anzin market and especially the growing relative weight of the market below the Saint Quentin Canal. Table 11 demonstrates in greater detail the striking geographical extension of the Anzin market in 1833 as compared with the range of markets revealed by the credit list of 1810. The list is not entirely accurate: it is obviously incomplete for the Nord, and sales in many of the towns listed outside the Nord were undoubtedly quite small and perhaps in some towns intermittent, while sales in Belgium and elsewhere are excluded as

Table 11 Anzin Markets by Town and Sales Prices, 1833 (For Middle-Grade Soft Coal)

Department	Town	Local Price to Consumer (In francs per Metric Quintal)
Nord	Valenciennes	1.67
	Lille	1.90
	Roubaix	2.20
	Cambrai	1.90
	Douai	. . .
Pas-de-Calais	Arras	2.36
	Saint Pol	3.30
	Saint Omer	2.50
	Aire	2.40
	Montreuil	4.14
	Calais	3.70
	Boulogne	4.25
	Saint Vevant	2.40
	Lens	2.60
	Béthune	3.36
	Bapaume	2.80
Aisne	Saint Quentin	2.30
	La Fére	2.50
	Laon	. . .
Oise	Compiègne	2.35
	Creil	2.65
	Beauvais	. . .
Somme	Ham	2.40
	Peronne	2.50
	Amiens	2.95
	Abbeville	3.20
Seine	Paris and vicinity	3.80

SOURCE: AN, F12 2534c, "Tableau de la répartition et des prix de vente," [1833].

are those to Rouen and the naval establishments of the Atlantic coast. Nevertheless, the range of towns included suggests the importance of the Saint Quentin Canal in expanding the Anzin market area.

The Course of Sales and the Business Cycle

In preceding chapters it has been pointed out how in the long run, and especially during the Restoration, both industrial expansion and improvements in transportation were more important in stimulating sales than anything Anzin itself did or could do. It remains to be seen how short-run factors, and in particular the business cycle, affected the evolution of Anzin's sales. A close correlation between the general business cycle and the Anzin sales cycle might be expected to help explain the timing of important managerial decisions such as the introduction of railed haulage, the search for new concessions, or price changes. In fact, there is a demonstrable lack of correlation between the two cycles, which reflects the fact that neither the coal industry nor Anzin itself were well integrated parts of the general French economy in the early nineteenth century. One must look elsewhere for an explanation of short-run changes in the course of Anzin's sales.

The course of the French business cycle is fairly clear for the first third of the nineteenth century, owing especially to the labors of Chabert and Gille.[96] There were roughly five periods of recession or depression: 1798–1801; 1806–7; 1810–14; 1817–21; and 1825–32.[97]. Their causes are still in dispute among economic historians,[98] but fortunately this lack of agreement does not matter in a study of Anzin, for it is generally agreed that the coal industry played no role in touching off economic downswings before 1833. Thus one can regard the Anzin sales cycle as either consequent to the general cycle, or unrelated to it, but not a cause of it, even in its own market area.

The data needed to reconstruct the Anzin sales cycle are sadly lacking in quantity and quality.[99] The materials available for this purpose are essentially the same production, cost, and price figures previously described. Thus it has been necessary to use output figures as a substitute for sales, although the two might obviously diverge fairly widely in any single year or over a short period of years.[100] Aside from the missing sales figures, the most serious gap in the data concerns unsold stocks of coal on hand, both their absolute quantity at any given moment and changes in their relative size from year to year.[101] Also lacking are reliable, detailed figures on movements of productivity and prices from year to year, data that would be extremely useful in analyzing the real meaning of changes in sales.[102] Nevertheless, this section compares the course of Anzin's sales from about 1800 to 1833 with the cycle for the French economy as a whole.

After a rapid and steady recovery from the nadir of 1794, Anzin production reached a temporary peak in 1799. There followed a period of severe deflation and commercial stagnation between 1798 and 1801, and sales slumped off in 1800, 1801, and 1802. Thus the board complained in 1801–2 about "mediocre sales" and growing stockpiles due to Belgian competition and the high prices charged by the Condé boatmen.[103] Among a large number of remedies proposed were the granting of rebates and the unsuccessful project to expand the company's commerce in the ports of the coast. The period 1801–5 was one of growth and prosperity for the French economy as a whole, but while Anzin output slowly rose, it did not quite attain the level reached in 1799, itself far below that of 1789–91. Then the short, light slump from October 1805 to July 1807 lowered Anzin output in 1807. Once again, unsold stocks at the pit head soared, and the company proposed large rebates to revive sales in competition with Mons.[104]

From the end of 1807 to the end of 1810, the Empire experienced its most prosperous years. Anzin output leaped upward by more than 20 percent in 1808 and continued to rise slowly for the next two years. Although unsold stocks remained relatively large, rising sales forced the company to open another working area in the spring of 1810, and production that year reached a new postrevolutionary peak.[105] The opening of the Saint Quentin Canal raised hopes high, and projects for exploration, new pits, and new concessions abounded. In May 1810 a severe depression, however, had begun in France; it would last, with a slight improvement in 1812, to the spring of 1814. Anzin granted an unprecedented number of rebates in 1810 and 1811, it lowered the formal price list in 1811, and cut hard-coal prices once again in 1812.[106] The board adjourned salary increases because "circumstances" required the company to practice "the greatest economy" in all its expenditures.[107] Nevertheless, production fluctuated around the 1810 level, and following a slight fall in 1812, output in 1813 rose to approximately prerevolutionary heights.

The period from May 1814 to June 1816 marked a "powerful flight" of the whole French economy.[108] Yet Anzin production fell off badly in 1814 and continued until 1818 to fluctuate well below the level of 1808–13. This weakness was undoubtedly the result of local troubles and uncertainties caused mainly by the invasion of the Allies, the fall of the Empire, and the restoration of peace and the Bourbons. The advance units of the Allied armies arrived on the Nord frontier in January 1814. Defensive flooding for the protection of fortified frontier towns, including Condé and Valenciennes, brought navigation on the Escaut and other rivers to a halt. Even land transport became impossible as enemy patrols scoured the roads.[109]

More important in the long run was the lengthy negotiation of the new frontier, a process which threw all Anzin's calculations for the future into confusion for three years. In 1814 Louis XVIII and Talleyrand hoped to conserve at least some of the Belgian departments absorbed during the Revolution. French retention of Mons would naturally have worked to Anzin's continued disadvantage. The Allies refused, however, to allow France to keep any of the territorial gains won after 1792, and they reestablished the Nord frontier along its prerevolutionary line between the Anzin and Mons pits. Napoleon's return from Elba and his subsequent defeat once again threw open the boundary question. This time, the Prussians were determined to weaken France irrevocably by including parts of the northern departments in a *cordon sanitaire* along its eastern frontiers. Anzin was threatened by the direst of disasters; it might well find itself a part of the new Kingdom of the Netherlands, separated from its French market by an international frontier, and forced to compete directly with Mons for the Belgian market. But the treaties of October 1815 maintained the frontier of 1814 without serious revision.[110] Anzin remained French, and before the end of 1816 it obtained some shelter from Mons competition for the first time since before the Revolution by a protective tariff—the "Anzin tariff" as it was sneeringly referred to in the Nord. The Anzin owners had held their breath for three years while the old regime passed and the new one bumblingly came into being. The majority stockholders of the Paris group were closely identified with the Empire and could expect no favors from the returning Bourbons. Thus even the confidential Registres are extremely meager for the period 1814–17; they include no new projects and no reference to the war or political events. Only in March 1817 was it possible to breathe a sigh of relief; sales now seemed assured "so long as the French frontier here remains unchanged."[111]

At the very moment when Anzin's future seemed to brighten, France was entering a commercial and financial crisis that soon slipped into a prolonged industrial recession reaching bottom in 1820 and lasting until 1821–22. But once again Anzin output, and presumably sales, followed an inverse course from that taken by the French economy as a whole. Although output dipped slightly in 1817 to the lowest level since 1807, it leaped in the next year by about 50 percent and surpassed easily the highest previous figure for output, reached in 1790. In the depths of the country's depression in 1819–20, output at Anzin remained near this new peak, then rose slightly to new record levels in 1821–22, years of general recovery in France. The best years of the Restoration for French business were 1823–25. Anzin output dropped in 1823, but it rose sharply in 1824 and, after marking time in 1825, rose again in 1826 to a level 22 percent above that of 1790. In fact, the years from 1818 through 1826 were for Anzin a single period during which output was remarkably stable and failed to reflect in its minor fluctuations the much greater swings in the general French economy. It achieved this stability despite growing imports from Mons, which, leaping over the tariff barrier, averaged well over two-thirds the quantity mined by Anzin at the beginning of the period and one-fifth more than the Anzin output by 1825.[112] Evidently, the new glassworks and ironworks near the Anzin pits, the revived sugar-beet industry, and other local developments, including the tariff, more than outweighed the general atmosphere of stagnation from 1818 to 1822, but the loss of the Belgian hard-coal market in 1820 and the growing difficulty of supplying expanding industry with sufficient good-quality soft coal prevented Anzin from taking full advantage of the general upswing after 1821.

"The good years of the Restoration were over by 1825."[113] Following a financial crisis in 1825, sector after sector slumped until by 1827–28 a general depression had set in. The Revolution of 1830 only worsened the economic situation, and real recovery did not get underway until 1832. Yet Anzin output during those years was generally rising. While the rest of the economy was foundering in 1827, Anzin output exceeded four hundred thousand tons for the first time and was more than 20 percent above that of the "good year" 1823. The next two years were also good ones for Anzin, but in 1830 an amazing, absolute increase in output of almost one hundred thousand tons jumped production almost 20 percent over the previous year and sent it over five hundred thousand tons. The opening of the Saint Quentin Canal in 1828, a small price cut in 1829, and the beginning of mass production at the new Denain pits in 1830 sent sales rocketing upward. Although the July days themselves had "no unfortunate consequences" for Anzin sales,[114] the general weakness of the economy finally caught up with Anzin's soaring production in 1831 and 1832 when output slumped slightly, only to more than recover the lost ground in 1833. With the return of good times in 1833–34 and boom times after 1835, output rose steadily year by year until the business cycle turned downward again after 1839.

The paradox of Anzin's experiencing such unprecedented prosperity during the years of prolonged slump in the French economy was noted by Villeneuve-Bargemont, prefect of the Nord, in one of his monthly reports to the minister of the interior. He pointed out in July 1828 that while the Nord's traditional, rural lace industry was declining, while many Nord boat haulers were without work since the opening of the Antoing Canal, and while Lille's modern textile industry had cut production and laid off workers in the past year, Anzin sales were booming.[115] Anzin itself affirmed that at the end of 1829 and beginning of 1830 the coal trade had never been more prosperous.[116] This prosperous growth was achieved despite the fact that Belgian coal exports to France continued to rise, and year in, year out approximately equalled Anzin's own production,[117] while Anzin hard coal continued to be excluded from Belgium until 1831. It was achieved despite the rapid fall of Mons prices since 1823–24, a fall that left many Mons firms operating at a loss by 1829.[118] But Anzin met Mons prices easily by judicious rebates, partial price cuts, and the small general price cut in 1829.[119]

Anzin's success in the face of a price war, severe competition, and a slumping economy was due to factors that ran counter to the prevailing bad times. On the supply side, the triumph over the Torrent and then the discovery of the rich Denain seams enabled Anzin to meet its customers' needs for growing quantities of good soft coal. On the demand side, many of the biggest coal-using industries in the Anzin market area continued to expand despite the general stagnation. Born in the first years of the Restoration, the industrial nucleus around the pits was still growing. Textile firms suffering from high costs and falling prices mechanized and bought steam engines. Sugar-beet refineries sprouted throughout the northern departments. And the improvement of the Saint Quentin Canal by 1828 opened huge new domestic, agricultural, and industrial markets south of the Escaut basin.

It has been remarked that until 1848 the Nord coal industry "remained in large measure a favored, 'sheltered' sector."[120] There are three main reasons why Anzin's growth was so little affected by short-run business fluctuations: the business cycle hit Anzin later than other firms in other sectors of the economy; it affected Anzin less profoundly; and its effects were less important than other factors, even in the short run.

Even when the general cycle seems to have affected Anzin and even allowing for the slower response to the cycle of the output figures used here in place of sales, the Anzin sales cycle usually followed the general cycle by one or more years. For example, the general cycle turned downward in 1798, late 1805, early 1810, and 1825, but Anzin output did not fall off until 1800, 1807, 1811–12, and 1831. The lag between 1825 and 1831 was particularly pronounced. Each time, there is descriptive evidence to link the fall in output with falling sales and to link falling sales with generally contracting commercial horizons. Obviously coal did not set the industrial pace, let alone the general economic trend. Falling coal production and sales reflected belatedly the contraction in other sectors. This lag in response can be explained by the fact that the initial shock to the economy most often occurred in the form of bad harvests or a financial crisis which in turn affected industry and finally Anzin. Only when falling prices, closing markets, and tightening credit forced the manufacturers of textiles, glass, iron, beet-sugar, and other products to cut production did Anzin sales fall.

Furthermore, recovery on the upswing was relatively easy, rapid, and early for Anzin. The recovery was due in part to the fact that Anzin sales did not fall as far as those of many other firms, and in part to the fact that there was a general buoyancy in the coal industry during the early nineteenth century. Anzin was the last in and first out of depressions, and its slumps, so far as they reflected a generally weak economy, were comparatively short.

Anzin's slumps were also comparatively mild. The downturns in the four instances already cited were relatively slight and thus easily made good at the earliest signs of recovery. Even today, the demand for coal is rather inelastic; it does not fall or rise much in the short run. It was much more inelastic in the early nineteenth century when a much larger share of the market depended on domestic and agricultural consumers, whose purchases of coal could not be put off either because of rising coal prices or falling incomes. As a household and farm necessity, Anzin coal was assured of a large, stable market in good times or bad.

Even the industrial portion of the Anzin market was relatively stable. It consisted of a great variety of industries, not all of which were likely to be hit by falling sales at the same time. Although "local" until after 1828, the Anzin market area was relatively large and varied; industrial conditions at any given time were likely to differ widely over the several departments and in the many towns supplied by the company. The great number of individual firms using Anzin coal also insured the company against sudden loss of a large part of its market. Although the most volatile of industries, textiles, was as yet only loosely connected to the coal industry, its crises, by emphasizing the need to cut costs, might well lead to the adoption of new machinery and hence to increased coal purchases. The inelasticity of demand in half of the Anzin market and the variety of industries, places, and firms included in the other half gave to Anzin sales a stability on the downward side that insured it against the catastrophic drops in income so often suffered in an economic downturn by such other producers as farmers and cotton spinners.

Finally, Anzin was never vulnerable to the drastic credit contraction that was sometimes one of the causes of an economic crisis and always a potent factor in its spread and duration. Because Anzin was eminently solid financially, its credit with its suppliers was always good, and it could pay in cash if the supplier himself felt pinched for ready money. More important, its own customers did not lack for credit in a crisis so long as their fundamental position was solid, for Anzin itself supplied them with the credit needed for their purchases. Anzin's great size and financial reserves kept open the possibility of sales when other firms, depending on a stricken money market to finance their clients' purchases, found themselves obliged to contract their operations because their customers could not pay cash or obtain credit.

The business cycle also affected Anzin sales less than those of most firms because its effects were often wholly or partly cancelled by other factors, operating either in the short or long run, which ran counter to the direction of the cycle. The importance of these other factors is most evident in the entire period from 1814 through 1830. From 1814 to 1816, 1817 to 1822, and 1825 to 1830 Anzin sales moved in the opposite direction from the business cycle, and between 1822 and 1825, when the Restoration economy as a whole had its most expansive years, Anzin sales were nearly stagnant and production dropped in three of the four years. Eleven of the years between 1815 and 1833 were "bad" ones for the French economy; in seven of those years Anzin's production rose.

Various factors helped counter the effects of the cycle. In 1814–16, political uncertainties temporarily retarded sales in a period of general expansion. In 1816 politics in the form of the protective tariff provided Anzin with a long-run stimulus, although Belgium's own tariff in 1821 perhaps came close to canceling out its effects, since they were mostly felt in the form of higher profits per ton sold rather than increases in the quantity sold. The relative stagnation from 1822 to 1826 in the face of rising demand reflected an insufficient supply of good soft coal, and expanded sales were possible after 1826, despite a downturn in the business cycle, because the insufficiency was made good. But the overriding factor that caused the slow and relatively steady upsurge in production and sales after about 1818, even in the face of periodic general depressions, was the growth of the Anzin market.

The Anzin pricing policy did not follow the business cycle even as closely as its sales or production did. Ordinarily, coal prices, like sales, fall before production. Anzin's general price cuts in 1811 and 1829 did precede output drops in 1812–13 and 1831–32, but these overall cuts and the company's timing of rebates throughout the period were both responses to price cuts by Mons rather than direct adjustments to the business cycle. Anzin's full complement of rivals discouraged increases in coal prices when other prices were rising, and Mons' prices, which determined Anzin's, were generally falling during the whole period. Anzin never made an attempt to improve its competitive position by lowering its prices unilaterally, for such a policy would have lowered profits. General changes in coal prices, such as those made by Mons and met by Anzin, had little effect in stimulating an inelastic demand in the short run. They were thus a very ineffective solution to the problems posed by the business cycle, and were primarily short-run reflections of desperate competition in the long run among the many comparatively small producers in the Mons field.

Most of the important managerial decisions affecting production were also not motivated by the business cycle, but rather by problems or opportunities special to Anzin. The most important instance was of course the massive and varied series of innovating and expansionist policies carried through between 1822 and 1825 under Casimir Perier. Technical innovations often occur during periods of commercial difficulty in an effort to cut costs per unit of output by raising productivity. Lowering costs was certainly one of Perier's aims in the early 1820s, but increased output of a better product, good soft coal, was more important, because Anzin's commercial difficulties were the result of its own special circumstances (and even in part self-created by earlier neglect) rather than the result of bad times.

Thus an analysis of Anzin's commercial fortunes and their relationship to the business cycle confirms Labrousse's remark that the French coal industry in the early nineteenth century was "si résistante aux crises."[121] In contrast, the modern sectors of the textile industry expanded very rapidly during the same period, but by fits and starts, suffering periodic moments of overproduction marked by severe contraction of sales, prices, profits, production, and employment. Both the expansive élan and the susceptibility to swift catastrophe shown by the textile industry were basically the results of rapid technological changes that drastically lowered costs and hence prices. The market thereby enormously increased, but by the same token the structure of the industry remained highly unstable. Anzin was less affected than the textile industry by rapid technological change, less subject to wild swings of consumer demand, and more sheltered from price competition. Thus its expansion was not so much the effect of its own efforts as it was the reflection of the slow, tentative, but cumulatively powerful surge of the Industrial Revolution.

Profits

The amount and rate of profit earned by Anzin was a carefully guarded company secret.[122] The secrecy stemmed not only from the ingrained sense of privacy typical of the French business firm but also from Anzin's realization that open avowal of its immense profits would only add to the cries of monopoly that furnished its enemies with their most effective weapon. Nevertheless, a few figures are available to serve as bench marks for an estimate of profits during the first third of the nineteenth century.

On the whole, profits rose during this period after a catastrophic fall during the Revolution. In 1789, a very good year for Anzin, profits were estimated at about 1,000,000 francs.[123] At the nadir of its fortunes in 1794, Anzin actually suffered losses of about 100,000 francs.[124] By 1805 the company was again prospering, although the prerevolutionary level of output and profits had not been fully regained; profits in that year amounted to 720,000 francs, according to information imparted by Renard to Daubuisson.[125] The Etats de Redevance are almost useless as a source of information on profits for the years after 1815, but their figure of 770,000 francs profit for each of the years 1812–14 may be fairly accurate.[126]

Profits probably rose with production after 1817–18; they certainly rose greatly after 1821. For each of the years 1821–33, the Registres provide figures for the amount of profit distributed per denier of ownership. Total profits distributed for the 288 deniers into which ownership in the company was divided amounted to 1,584,000 francs in 1821. In 1822 profits distributed rose to 1,728,000 francs, in 1826 to 2,016,000, and in 1828 to 2,304,000, a level maintained through 1832.[127] These figures represent the minimum profit earned in any given year, since the company set aside sizable sums in undistributed profits during most of these years.[128] Only for 1822 is there certain evidence of the full amount of profit earned. Based on figures for sales and costs provided by the Perier Report, profits that year were 1,862,000 francs.[129] This sum was only 134,000 francs more than the amount distributed to owners, but in most years the difference between profits earned and profits distributed was certainly greater than in 1822. Investment in new facilities was especially heavy in 1822. And the sum of 2,304,000 francs distributed to owners in 1830 was identical with that distributed in 1828, despite the fact that output had risen almost 20 percent between the two dates. Presumably, much of the profit earned in 1830 found its way into the company bank accounts rather than the owners' pockets.[130] Hostile observers may not have been far wrong when they estimated Anzin profits for 1830 at 3,000,000 francs.[131]

Anzin's rate of profit as a percentage of sales revenue was remarkably high throughout the early nineteenth century.[132] Daubuisson's figures yield a rate of 20 percent, but he probably overestimated costs and may have underestimated sales revenue. According to the Etats de Redevance, which were used by the government to levy a profit tax and open to public scrutiny, Anzin's rate of profit in 1812 was about 33 percent, yet the engineer Migneron, who established the final figures for production, sales revenue, and costs from which the amount of profit was calculated, was on the whole friendly to Anzin and could hardly be accused of a desire to harm Anzin by increasing its taxes or providing ammunition for its enemies.[133] On the more certain basis of Perier's figures, Anzin profits in 1822 amounted to almost 40 percent of sales. This rate is certainly very close to that achieved throughout the period after about 1820, when a slight decline in prices was offset by an approximately equivalent decline in costs. Indeed, the rate of about 40 percent derived from Perier's global figures is about what one would expect from a comparison of the cost of production of a hectoliter of coal and the price at which it was sold during the 1820s. And since there is reason to believe that the rates of profit calculated for the earlier period from the Daubuisson and Etat figures are underestimates, it is not unlikely that Anzin achieved a rate of profit close to 40 percent throughout the entire period from about 1800 to 1833.

Since Anzin's cost of production and its sales prices at the riverside were both higher than those of Mons producers, it is clear that such a rate of profit did indeed reflect monopolistic advantages, as Anzin's enemies charged. Before 1814 Anzin's ability to compete with Mons and still make large profits was partly due to the difference in the cost of coal transportation from the Anzin and Mons loading areas to consumers in the French portion of the Anzin market area and partly due to Anzin's possession of hard-coal deposits of a kind essential for certain important uses and not found in the Mons area. After 1814 the importance of both differential transport costs and the hard-coal monopoly declined. The opening of the Mons-Condé Canal in 1814 lowered the transport costs of

Mons producers selling to French consumers. Anzin sales of hard coal declined proportionately as the consumption of soft coal increased with the growth of industry, and they probably declined absolutely after the prohibitive Belgian tariff of 1821. Anzin was able to maintain its monopolistic prices and profits, however, with the help of the tariff of 1816, which remained in force until 1835. The coal tariff did not "save Anzin from extinction" as company spokesmen were fond of claiming; Anzin's transport costs to its French markets remained lower than those of its Mons rivals, and even in the absence of any tariff protection these lower costs would have enabled Anzin to charge prices that would have more than covered its production costs and allowed it to earn a neat profit. But the tariff permitted Anzin to keep prices far higher than it would otherwise have been able to do. The tariff of 0.33 centimes per hectoliter represented an outright subsidy to Anzin and enabled the company to maintain and perhaps increase during the Restoration the monopolistic rate of profit it had earned under the Empire.[134]

About 1833 certain events, more or less fundamental or fortuitous, marked the end of an era for Anzin. One serious shock was the death in 1832 of Casimir Perier, the company's presiding genius in the twenties, but under his brother Joseph and the general agent Mark Jennings, Anzin continued to pursue Casimir's aggressive policies successfully. In 1833 Anzin's first great strike presaged the bitter class warfare that was to mar the history of the Nord coalfield in the later nineteenth century. Equally novel and disturbing to Anzin's equanimity was the formation of new companies outside its concessions. In contrast to the challengers of the eighteenth century, these rivals hit on workable seams and were not so easily destroyed by influence, purchase, or price-cutting. Anzin would have to face the breaching of its local monopoly and readjust its tactics to take account of the novel situation of local competition. On the whole, however, Anzin was able to dominate the newcomers and continued to set the pattern of prices and wages in the Nord until the richer Pas-de-Calais extension of the field opened after 1850. Another catastrophe that threatened also failed to develop. The clamor by industrialists for reduction or abolition of the coal tariff led to a government investigation in 1833 that produced a monumental collection of evidence and equivocal recommendations. The modest reductions that resulted did not destroy the profitability of the coal industry in the Nord as Anzin's lobbyists had predicted.

The death of Perier, the first signs of labor unrest, and the disquieting groundswell of sentiment for tariff reform all seemed so many ominous signs of an uncertain future. In fact, the future would be expansive far beyond reasonable calculation, and the apparent difficulties were destined to be more than offset by the rising demand for coal that accompanied the improvement of transport and the general progress of industry. Anzin in the thirties moved into an increasingly competitive but highly profitable position in the French coal industry.

Notes

Introduction

1. Quoted in *Mines et métallurgie* (Nov. 1959), p. 628, on the occasion of a descent into the pits during a visit to the Nord.
2. For the "original crisis," see E. A. Wrigley, "The Supply of Raw Materials in the Industrial Revolution," *Economic History Review*, 2d series, XV (1962), pp. 1–16. For the recent travails of coal, see Firmin Lentacker, *La révolution énergétique de notre temps.*
3. Claude Fohlen, "Charbon et Révolution industrielle en France (1815–1850)" in Louis Trenard, ed., *Charbon et les sciences humaines*, pp. 141–48.
4. Bertrand Gille, "Les plus grandes compagnies houillères françaises vers 1840: Essai sur la structure du capitalisme" in Trenard, p. 158.
5. Edouard Grar, *Histoire de la recherche, de la découverte et de l'exploitation de la houille dans le Hainaut français, dans la Flandre française et dans l'Artois, 1716–1791*; A. de Saint-Léger, ed., *Les mines d'Anzin et d'Aniche pendant la Révolution.* Volume III of Grar contains 271 pages of documents as well as extended genealogies for all the early families of Anzin, and volume II has the fullest account of the company from its origins to the Revolution. Volume I of Saint-Léger, published in two parts in 1938 and 1939, contains over 500 pages of documents, including the essential source for this study, the "Registres de déliberations" of the board of directors from the formation of the company to 1811, and a very useful introduction by the author concerning mainly the legal problems and vicissitudes Anzin underwent during the Revolution.
6. See especially the work of Marcel Gillet in Jean Bouvier, François Furet, and Marcel Gillet, *Le mouvement du profit en France au XIXe siècle*, pp. 71–163, and his "L'âge du charbon et l'essor du bassin houiller du Nord et du Pas-de-Calais (XIXe siècle–début du

XXe siècle)" in Trenard, pp. 25–50, reprinted in F. Crouzet, W. H. Chaloner, and W. M. Stern, eds., *Essays in European Economic History, 1789–1914,* pp. 179–202. Of the earlier, more descriptive works, Emile Vuillemin, *Le bassin houiller de Pas-de-Calais* is the most important, although development in the Nord is incidental to Vuillemin's main concern, the Pas-de-Calais. Less purely statistical are Georges Michel and Alfred Renouard, *Histoire d'un centre ouvrier*, and Louis Reybaud, *Le fer et la houille*, pp. 156–207, which first appeared in *La Revue des Deux-Mondes,* XCVI (11 Nov. 1871), pp. 134–65. All three works, so far as they are not simply technical accounts, strongly reflect the owners' point of view; Vuillemin and Michel were mining engineers, the former a manager of a coal-mining company in the area, and Renouard and Reybaud were popular publicists. Unfortunately, Marcel Gillet's magistral doctoral thesis on the Nord–Pas-de-Calais field in the later nineteenth century appeared too late to be consulted.

7. For a brief recent account stressing the activities of the Perier family and their business milieu, see Richard J. Barker, "Casimir Perier (1777–1832) and William Ternaux (1763–1833)," pp. 189–244, and "French Entrepreneurship during the Restoration: The Record of a Single Firm, The Anzin Mining Company," *Journal of Economic History,* XXI (June 1961), pp. 161–78.
8. For an earlier period, see Marcel Rouff, *Les mines de charbon en France au XVIIIe siècle, 1744–1791*, and for a slightly later period, see Pierre Guillaume, *La Compagnie des Mines de la Loire, 1846–1854.* A good general introduction can be found in Norman J. G. Pounds and William N. Parker, *Coal and Steel in Western Europe.* See also Eugene Schneider, *Le charbon,* and R. J. Lafitte-Laplace, *L'économie charbonnière de la France.*
9. See, for example, Rondo E. Cameron, "Economic Growth and Stagnation in France, 1815–1914," *Journal of Modern History,* XXX (Mar. 1958), p. 7.
10. See especially Bertrand Gille, *Recherches sur la formation de la grande entreprise capitaliste (1815–1848),* and *La banque et le crédit en France de 1815 à 1848*; Maurice Lévy-Leboyer, *Les banques européennes et l'industrialisation internationale dans la première moitié du XIXe siècle*; A. L. Dunham, *The Industrial Revolution in France, 1815–1848*; Pierre Léon, *La naissance de la grande industrie en Dauphiné (fin du XVIIIe siècle—1869)*; and Claude Fohlen, "The Industrial Revolution in France" in Rondo Cameron, ed., *Essays in French Economic History,* pp. 201–25, a revision of "La rivoluzione industriale in Francia," *Studi Storici,* II (1961), pp. 517–47. For a good summary view, see Tom Kemp, *Economic Forces in French History,* chapters 5, 6.
11. See the publications of the team working under the direction of J. Marczewski on the "Histoire quantitative de l'économie fran-

çaise," in *Cahiers de l'Institut de science économique appliquée*, series AF, nos. 1–9 and especially nos. 4–6 (1963–66). For a summary and the significance of these findings, see Maurice Lévy-Leboyer, "La croissance économique en France au XIXe siècle: Résultats préliminaire," *Annales: Economies, Sociétés, Civilisations*, XXIII (1968), pp. 788–807. The average compound annual growth rate of industrial gross product for 1803–12 to 1815–24 was 2.4 percent, for 1815–24 to 1825–34, 2.8 percent, and for 1825–34 to 1835–44, 2.9 percent. The only decade before the end of World War II with a higher growth rate was 1920–24 to 1925–34 (J. Marczewski, "Le produit physique de l'économie française de 1789 à 1913," *Cahiers de l'I.S.E.A.*, no. 4 [1963], p. xci).

12. T. Markovich, "L'industrie française de 1789 à 1964: Sources et méthodes," *Cahiers de l'I.S.E.A.*, no. 4, table 1. The data for this and the preceding note are reproduced in T. Markovitch, "The Dominant Sectors of French Industry" in Cameron, *Essays*, pp. 243, 241.
13. Charles Ballot, *L'introduction du machinisme dans l'industrie française*; Henri Sée, *Histoire économique de la France*, originally published in German in 1936.
14. See, for example, A. Chabert, *Essai sur les mouvements des prix et des revenues en France de 1798 à 1820,* and François Crouzet, "Wars, Blockade, and Economic Change in Europe, 1792–1815," *Journal of Economic History*, XXIV (Dec. 1964), pp. 567–88. For an example of the old, tendentious history glorifying the Emperor, see Odette Viennet, *Napoléon et l'industrie française.*
15. See Rondo Cameron, *France and the Economic Development of Europe, 1800–1914*, pp. 21–63.
16. See, for example, Jean Chaptal, *De l'industrie française*, II.
17. Charles Pouthas, "Les projets de réforme administrative sous la Restauration," *Revue d'histoire moderne*, I (Oct.–Nov. 1926), pp. 321–67. For industry in particular, see Hubert Bourgin and Georges Bourgin, eds., *Les patrons, les ouvriers et l'Etat*, I, Introduction.
18. The best general work on the Restoration is G. de Bertier de Sauvigny, *La Restauration.*
19. Markovich, table 2. The figures for coal are: 1803–12 to 1815–24, 2.2 percent; 1815–24 to 1825–34, 5.9 percent; 1825–34 to 1835–44, 5.6 percent (Cameron, *Essays*, p. 243).
20. Markovich, table 4a. By value added for all French industry, coal accounted for only 0.2 percent in 1802–12 and 0.3 percent in 1825–34 (Cameron, *Essays,* p. 235).
21. Marczewski, p. xcvi. An index of production, for all French mining, based on annual output rather than decadal averages, confirms the picture of rapid growth and low absolute output. François Crouzet,

"An Annual Index of French Industrial Production in the Nineteenth Century" in Cameron, *Essays*, p. 252, translated from "Essai de construction d'un indice annuel de la production industrielle française au XIXe siècle," *Annales: Economies, Sociétés, Civilisations*, XXV (1970), pp. 56–99.

22. Pounds and Parker, p. 140, n.1, and map, p. 141.

23. L. J. Gras, *Histoire économique générale des mines de la Loire*, I, 182, 256; "Notice sur les accidents arrivées dans les mines de houille du département de la Loire . . . ," *Annales des Mines*, no. 2 (1832), pp. 496–98. See also Lefebvre, "Aperçu générale des mines de houille exploitées en France . . . ," *Journal des Mines*, no. 2 (1802), pp. 371–75.

24. Figures for the Loire as cited above; for Anzin and France, see Emile Dormoy, *Topographie souterraine du bassin houiller de Valenciennes*, pp. 43, 59.

25. J. H. Clapham, *The Economic Development of France and Germany, 1815–1914*, p. 57.

26. Figures from AN, F14 1315–16, "Etat générale des substances combustibles," for the Jemmappes department (Mons). Jemmappes and the other areas which were to form Belgium after 1830 were, of course, a part of the French Empire to 1815 and then part of the Kingdom of the Netherlands until 1830, when Belgium established its independence.

27. F. Lentacker, "Les charbons belges sur le marché français au cours du XIXe siècle," *Bulletin des Séances* (1964–66), p. 1402.

28. A. Wibail, "L'évolution économique de l'industrie charbonnière belge depuis 1831," *Bulletin de l'Institut des Sciences Economiques*, VI (Nov. 1934), p. 3.

29. J. H. Clapham, *An Economic History of Modern Britain*, I, 431. For the Belgian Industrial Revolution before 1830, see Robert Demoulin, *Guillaume Ier et la transformation économique des provinces belges* (1815–1830), and Jan Craeybeckx, "The Beginnings of the Industrial Revolution in Belgium" in Cameron, *Essays*, pp. 187–200. For the insignificance of English coal exports to France before 1833, see François Crouzet, "Le charbon anglais en France au XIXe siècle" in Trenard, p. 173.

30. Maurice Baumont, *La grosse industrie allemande et le charbon*, p. 75.

31. Baumont, pp. 64–88; E. A. Wrigley, *Industrial Growth and Population Change*, p. 26; W. O. Henderson, *The State and the Industrial Revolution in Prussia, 1740–1870*, pp. 15–17, 56–69.

32. Pounds and Parker, p. 218.

33. Henderson, *Industrial Revolution in Prussia*, pp. 25–57; Baumont, pp. 40–49.

34. See David S. Landes, "French Entrepreneurship and Industrial Growth in the Nineteenth Century," *Journal of Economic History*, IX (May 1949), pp. 45–61; Guy Palmade, *French Capitalism in the Nineteenth Century*, pp. 101, 228–30, and "Translator's Introduction," pp. 9–27, by Graeme M. Holmes; Kemp, pp. 107–11; and below, chapter 1, under "The Decision Makers and the Managers."
35. For the phrase and the concept, see Alfred D. Chandler, Jr., *Strategy and Structure*.

1 Owners, Managers, and Miners: The Exercise of Control

1. See the evidence and works cited in Pounds and Parker, p. 95; Norman J. G. Pounds, *The Ruhr*, pp. 48–53; and John U. Nef, *The Rise of the British Coal Industry*, I, 7, n.1.
2. A. Lequeux, "Les prospections houillères dans la France du Nord et leurs conséquences humaines," *Revue du Nord*, XL (Apr.–June 1958), p. 324.
3. Ibid., p. 324.
4. Vuillemin, II, 249–53.
5. Saint-Léger, I, xii.
6. Where not otherwise indicated, the following account of the history of Anzin to the Revolution is drawn from Grar; Saint-Léger, I, xi–xxv; and Vuillemin, II, 321–32. The last two draw heavily on Grar, but Saint-Léger especially uses fresh sources as well. For the Desandrouin family and fortune, see also Douxchamps-Lefèvre, "L'exploitation houillère dans la région de Charleroi au début du XVIIIe siècle" in Trenard, pp. 109–10.
7. Christophe Mathieu, son of Jacques, claimed that prior to Desandrouin's search for coal in French Hainaut, "there were numberless attempts, not one succeeded, and the individual men of wealth successively involved were all ruined" (Michel and Renouard, p. 32).
8. Quoted from a contemporary account in Lequeux, p. 325.
9. Rouff, p. 193.
10. Germain Martin, *La grande industrie sous la règne de Louis XV*, pp. 150, 158.
11. Christophe Dieudonné, *Statistique du départment du Nord*, I, 165.
12. Rouff, pp. 65–81.
13. The nobility were among the founders of almost all the eighteenth-century French mining companies and especially of those bound for future greatness (ibid., pp. 173–208). For the important role of

the nobility in the Charleroi basin, where both the Desandrouin and Arenberg families operated mines directly, see Douxchamps-Lefèvre in Trenard, pp. 113–14.

14. Rouff, pp. 196–99.
15. It can be found in Saint-Léger, I, 129–34, in the first pages of the Registres, and in an eighteenth-century copy in AN, F14 1307.
16. Grar, II, 134.
17. Production at Aniche was only six thousand tons in 1791. For a full account of the company's difficulties, see Saint-Léger, II.
18. Dieudonné, I, 165. Among many other observers praising the management of Anzin were several government mining inspectors. See especially, AN, F14 4244, Pajot des Charmes, "Mémoire sur la manière d'exploiter la charbon de terre dans le Hainaut français."
19. Grar, II, 206–7.
20. Dieudonné, I, 165, 169, 175. Cf. Grar, II, 178–80.
21. Except where otherwise indicated, the account of Anzin during the Revolution given below is drawn from Saint-Léger, I, xxvi–xlv.
22. AN, ADxiii 14, "Discours de Mirabeau."
23. C. Richard, "La crise du charbon dans le Nord en 1793," pp. 325–37.
24. AN, F14 7822, "Renseignements sur la mine d'Anzin," 6 Dec. 1795.
25. Registres, 11 Aug. 1795.
26. The comments of the board of directors in passing on the new price rises are all cut from the same pattern, e.g., "The same high price of food that necessitates a rise in the daily wage of the worker forces us also to meet our engagements by raising the price of coal" (Registres, 17 June 1795). Cf. the similar price-wage spiral in the mines at Littry in western Normandy, as described by Georges Lefebvre, "Les mines de Littry (1774–an VIII)," in his *Etudes sur la Révolution française*, pp. 126–31.
27. Registres, 6 Nov. 1795.
28. Ibid., and passim to 1800.
29. AN, F14 7822.
30. Registres, 10 Feb. 1796.
31. Ibid., 14 June 1797, 17 Oct. 1797, 15 June 1799.
32. A. Cavrois, *Les sociétés houillères du Nord et du Pas-de-Calais*, p. 31.
33. AN, F14 1307, Renard to prefect, 23 May 1808, and minister of the interior to prefect, 28 June 1808.
34. See Cavrois, pp. 31–42, and Gille, *Recherches*, pp. 33–39.

35. For example, in 1821 when Arenberg replaced Desandrouin (Registres, 7 Sept. 1821).
36. Saint-Léger, I, 132. Acte de Société, article 9.
37. Saint-Léger, I, 133. Acte de Société, article 17.
38. Saint-Léger, I, 132. Acte de Société, article 12. For example, the board refused to authorize a sale to a Valenciennes merchant named Ewbank, who was a member of a rival company (Registres, 29 June 1828).
39. For the following paragraphs, see the Registres, where all transfers of ownership were recorded.
40. Anzin first listed its sols on the Lille stock exchange in 1875. By 1912 it had 5,550 owners (Gillet in Bouvier, Furet, and Gillet, pp. 82, 120.
41. Registres, 6 Oct. 1806, 7 Oct. 1807, 4 Feb. 1808, 5 Oct. 1809, 14 June 1810, 3 June 1812, 10 Sept. 1813, 13 May 1814, 6 Sept. 1821.
42. Pierre Barral, *Les Perier dans l'Isère au XIXe siècle*, p. 68; Registres, 28 Sept. 1822, 28 Sept. 1826, 29 June 1828.
43. In addition to note 41 above, see Registres, 1 June 1806, 5 Feb. 1807, 14 June and 4 Oct. 1810, 7 Feb. 1810, 21 June 1826.
44. For the biographical information in the following paragraphs where not otherwise cited, see, in addition to the Registres and standard collective biographies, Grar, III, and Saint-Léger, I, notes, passim.
45. *Annuaire Statistique du Département du Nord.*
46. Henri de Laplane, *Eloge de Louis-Alexandre-César Taffin de Givenchy*, p. 16.
47. Ibid., p. 29.
48. Registres, 4 Feb. 1808, 3 May 1821, 25 Sept. 1823, 29 June 1828, 4 Nov. 1832.
49. See David S. Landes, *The Unbound Prometheus*, pp. 301–7 on "logistical strangulation" and the inefficiencies of size at a later period.
50. Registres, passim.
51. For example, the names of the Boisseau, Mathieu, Quinet, Castiau, Postiau, and Gellé families appear throughout the Registres. It was, of course, not unusual for three generations of a family to be employed at the same time, together with numerous cousins, uncles, and nephews.
52. Saint-Léger, I, 30, n.1.
53. Registres, 10 Aug. 1795.
54. Ibid., 17 Oct. 1797.
55. Ibid., 17 June 1803.

56. See Dieudonné, I, 171, for the doubts of local opinion concerning Desandrouin's capabilities.
57. For the fullest account of the banking and business activities of Casimir and Scipion, see Barker, "Casimir Perier," pp. 57–61, 67–74. For their father and the family background, see ibid., chapters 1–3, and Barral, chapters 1–3.
58. Each director named his own assistant, who could attend board meetings only in the absence of the director but who usually moved up when the full director died or retired.
59. Scipion was twenty-five years old in 1801, Casimir a year younger.
60. Barral, p. 49.
61. E. Choulet, *La famille Casimir Perier*, pp. 112–13.
62. Barral, p. 50; Frank Manuel, *The New World of Henri Saint-Simon*, pp. 106, 182–83, 196.
63. "Le deuxième centenaire de la fondation (19 novembre 1757) de la Compagnie des Mines d'Anzin," *Mines*, no. 6 (1958), p. 578. This series of articles consists of selected letters from Joseph Perier, Casimir's brother and successor to Scipion in the Perier bank, to Christian Paulze d'Ivoy, general agent of Anzin from 11 April 1825 to 21 June 1827. Included is a useful running commentary by the anonymous editor, a descendent of Paulze.
64. Lachapelle was, however, married to Léonard Mathieu's niece and hence related by marriage to the several members of the Mathieu family who were among the most stubborn enemies of change. Registres, 25 Sept. 1823.
65. "Le deuxième centenaire d'Anzin," no. 6 (1958), p. 578.
66. Registres, 20 Sept. 1824. "Le deuxième centenaire d'Anzin," no. 6 (1958), p. 578.
67. Registres, 11 Apr. 1825.
68. "Le deuxième centenaire d'Anzin," no. 1 (1958), p. 26, n.4. He was the son of a tax farmer who was guillotined with his son-in-law Lavoisier in 1794.
69. Ibid., no. 1 (1959), p. 26.
70. Registres, 21 June 1827. Paulze thereupon resumed his career as a prefect. "Le deuxième centenaire d'Anzin," no. 1 (1958), p. 26, n.4.
71. Registres, 10 Sept. 1827.
72. For the character of this correspondence, see "Le deuxième centenaire d'Anzin," passim, and no. 6, pp. 575–77.
73. "Le deuxième centenaire d'Anzin," no. 6 (1958), pp. 578–79.
74. Registres, 15 June 1826. For the record-keeping system essentially as Perier found it in 1821, see Registres, 11 Aug. 1795.
75. Ibid., 15 June 1826.

76. Ibid., 28 Sept. 1822.
77. Ibid., 11 Jan. 1825. For the steps leading to this conclusion, see ibid., 22 Sept. 1822, 13 Mar. 1823, 25 Sept. 1823, 20 May 1824.
78. Ibid., 15 June 1826.
79. For the best introduction to the methods and significance of accounting practice in the early nineteenth century, see Sidney Pollard, *The Genesis of Modern Management*, chapter 6.
80. Joseph A. Schumpeter, *The Theory of Economic Development*, pp. 74–94.
81. See Landes, "French Entrepreneurship," pp. 45–61; David S. Landes, "French Business and the Businessman: A Social and Cultural Analysis," and John E. Sawyer, "Strains in the Social Structure of Modern France," in Edward Meade Earle, ed., *Modern France*, pp. 293–312, 334–53; the Landes-Sawyer thesis has been challenged, most notably by Alexander Gerschenkron in "Social Attitudes, Entrepreneurship, and Economic Development," *Explorations in Entrepreneurial History*, VI, no. 1 (1953–54), which has been reprinted in much the same form in his *Economic Backwardness in Historical Perspective* (Cambridge, Mass., 1962). A series of comments, rebuttals, and re-rebuttals by Landes, Sawyer, Gerschenkron, and Thomas Cochran appear in *Explorations in Entrepreneurial History*, VI (1953–54), pp. 181–83, 245–97, and VII (1954), pp. 111–19. For the same problem in the eighteenth century, see Paul W. Bamford, "Entrepreneurship in Seventeenth- and Eighteenth-Century France: Some General Conditions and a Case Study," *Explorations in Entrepreneurial History*, IX, no. 4 (1957), pp. 204–13, and Elinor G. Barber, *The Bourgeoisie in Eighteenth-Century France*. For a comparison contrasting English and French entrepreneurship, see Bert F. Hoselitz, "Entrepreneurship and Capital Formation in France and Britain since 1700," in National Bureau of Economic Research, *Capital Formation and Economic Growth*, pp. 291–338.
82. Chandler, p. 14.
83. See Introduction.
84. E. Beau de Loménie, *Les responsabilités des dynasties bourgeoises*, I, 12.
85. Ibid., I, 31.
86. Among the Anzin owners, Berryer helped draw up the bank's statutes, and Cambacérès was an early sol purchaser. So too was Cretet, a Perier relative by marriage and later governor of the bank and minister of the interior, while Henri Flory, an associate of Scipion and Casimir Perier in their private bank until 1807, was also a regent from 1805 to 1822 (Barker, "Casimir Perier," pp.49–50; Barral, p. 87).
87. Barral, pp. 86–87.

88. Registres, 4 Feb. 1808.
89. Barker, "Casimir Perier," pp. 286–326. All but one of the eight sons and two daughters of Claude Perier either sat as or were married to deputies at some time between 1815 and 1848. Casimir's son was minister of the interior, 1871–72 and 1873, and his grandson was president of the republic, 1894–95 (Barral, end table 2).
90. Barker, "Casimir Perier," pp. 169–73.
91. Ibid., p. 163.
92. Barral, pp. 115–18.
93. See, for example, Barker, "Casimir Perier," p. 67, on the lack of evidence of direct influence on the advantageous mining law of 1810.
94. See, for example, the threat posed by the former owners of Anzin, returned émigrés, who claimed that some of Anzin's concessions had expired and that the territory conceded exceeded the legal limits of size (Saint-Léger, I, xl–xvii; AN, F14 7823, Réplique pour MM de Villedeuil et consorts, 1838).
95. Dunham, p. 350.
96. Lentacker, p. 1400.
97. See, for example, Chaptal, I, 426–34; AN, F12 2534c, petitions from a variety of Nord industrialists and their professional organizations, 1821.
98. AN, F12 2534c, Becquey to Chaumont, 7 Mar. 1824.
99. Ibid., Avis de l'ingénieur des mines, 25 Oct. 1821; AN, F14 6674, Rapport sur la navigation de l'Escaut, 24 Sept. 1824.
100. AN, F14 4029, Procès-verbal, Comité d'évaluation, 5 July 1813.
101. Peter Stearns, "British Industry through the Eyes of French Industrialists (1820–1848)," *Journal of Modern History*, XXXVII (Mar. 1965), p. 59. For the general background of labor conditions and movements in the early stages of French industrialization, see Bourgin and Bourgin, I, Introduction; Ernest Labrousse, *Le mouvement ouvrier et les théories sociales en France de 1815 à 1848*; Georges Duveau, *La vie ouvrière en France sous la Second Empire*; Emile Levasseur, *Histoire des classes ouvrières et de l'industrie en France de 1789 à 1870.*
102. Registres, 27 June 1833, and passim.
103. AN, F14 7822, Renard to Council of Mines, 19 Dec. 1799.
104. Registres, 27 June 1833.
105. Ibid., 20 Sept. 1824.
106. See chapter 2, under "Haulage from Face to Shaft."
107. AD, Nord, M 626/1, Joseph Perier to prefect, 27 June 1833; Jean Pierre Aguet, *Contributions à l'histoire du mouvement ouvrier français*, pp. 244–59.

108. Registres, 27 June 1833.
109. For an interpretation of early French industrial strikes, see Peter N. Stearns, "Patterns of Industrial Strike Activity in France during the July Monarchy," *American Historical Review*, LXX (Jan. 1965), pp. 371–94.

2 The Mine and the Miners at Work

1. AN, F14 4244, Pajot des Charmes, "Mémoire."
2. AN, ADxiii 14, "Discours de Mirabeau."
3. Dieudonné, I, 170.
4. AN, F14 7822, Daubuisson, "Des houillères d'Anzin et de leur exploitation."
5. AN, F14 3848, Clère to prefect, 22 Mar. 1832; AN F14 4029, Clère, "Rapport sur l'assiette des redevances fixes et proportionnelles des mines du Département du Nord pour l'exercise 1818," 14 July 1818.
6. A. M. Héron de Villefosse, *De la richesse minérale*. See especially the last two volumes, published in 1819.
7. Michel Chevalier, "Observations sur les Mines de Mons et sur les autres mines de charbon qui approvisionnent Paris," *Annales des Mines*, nos. 2, 4 (1832), p. 433.
8. Grar, I, xi.
9. Rouff, pp. vii–viii.
10. On mining technique, generally, in the eighteenth and nineteenth centuries, see J. A. S. Ritson, "Metal and Coal Mining," in Charles Singer, E. J. Holmyard, and A. R. Hall, eds., *A History of Technology*, IV, chapter 3. For the technical conservatism of Anzin, cf. Grar, II, 203–24 in the eighteenth century; AN, F14 7822, Daubuisson, "Exploitation" in 1805; Héron de Villefosse, II, 184–95, 501–10 in 1819; Dormoy, pp. 61–113 in 1867. See also the accurate account by Emile Zola in the novel *Germinal*, first published in 1885. This example of the naturalistic technique par excellence was grounded in abundant research in contemporary mining literature in general and was based specifically on a guided visit to Anzin (Feb.–Mar. 1884), which included conversations with Emile Dormoy. Zola's technique has been analyzed, his sources tracked down, and part of his voluminous notes published in Ida Marie Frandon, *Autour de "Germinal,"* and Philippe Van Tieghem, *Introduction à l'étude d'Emile Zola.*
11. See below, under "Work at the Coal Face."
12. Elwood S. Moore, *Coal*, chapters 6, 8.
13. Wrigley, *Industrial Growth*, pp. 34–36.

14. George S. Rice and Irving Hartmann, *Coal Mining in Europe*, p. 14.
15. Ibid., pp. 8–21 and especially 14–15; Wrigley, *Industrial Growth*, pp. 31–32; Pounds, *The Ruhr*, pp. 24–26, 32; Dormoy, pp. 120–48.
16. AN, F14 7822, Daubuisson, "Exploitation"; Dormoy, pp. 137–38; Héron de Villefosse, II, 436–37; AN, F14 4244, Pajot des Charmes, "Mémoire."
17. Dormoy, pp. 131–32.
18. Ibid., pp. 148–51; A. Garcenot, *Les mines d'Anzin: Etude historique et technique,* end map; Rice and Hartmann, p. 15. On the classification of coals and their comparative nomenclature from country to country, see Rice and Hartmann, pp. 25–27.
19. Wrigley, *Industrial Growth*, pp. 34, 36.
20. Ibid., p. 34; Chevalier, pp. 432–33; J. F. Daubuisson, "Description des houillères d'Anzin," *Journal des Mines*, no. 2 (1805), pp. 133–34.
21. Rice and Hartmann, p. 14, 36, 42, 11, 19, 18.
22. Daubuisson, "Description," p. 132. Seams were often called by names describing their thickness. At Anzin, seams were first named in terms of so many hands thickness, later in terms of decimeters. By contrast, one seam in the Birmingham area was known as the "ten-yard seam." W. H. B. Court, *The Rise of the Midland Industries*, p. 166.
23. Chevalier, pp. 432–33.
24. Héron de Villefosse, II, 477–78; Dormoy, p. 77.
25. For the above two paragraphs, see the sources cited in note 15 as well as Daubuisson, "Description," p. 121.
26. Daubuisson, "Description," pp. 121–26; Dormoy, pp. 113–20, 63–64; Héron de Villefosse, II, 439–40; AN, F14 4244, Pajot des Charmes, "Mémoire."
27. Dormoy, p. 66.
28. Ibid., pp. 117–19. The Torrent measured 4,375 meters at its greatest width, 7,750 meters long, and an average of 9 meters thick. See also the good description in AN, F14 7822, Clère, "Procès-verbal d'enquête sur l'état actual et sur l'avenir probable des mines de houille d'Anzin . . . ," 4 May 1830. Hereafter cited as AN, F14 7822, Clère Report.
29. The sonde was first used in Hainaut in the late eighteenth century; previously, shallow pits were employed. E. Prost, *Historique de l'industrie houillère belge*, p. 11.
30. Héron de Villefosse, II, 109–22. At Anzin, a bore of some forty meters to the beds known as *bleus* was usually considered sufficient to establish the existence and depth of the coal measures beneath them (AN, F14 7822, Daubuisson, "Exploitation").

31. Dormoy, p. 61. The bit was prone to break off in the hole at considerable cost in time and money.

32. In most mining areas, equipment and crew could be hired by the day from a professional sondeur, and many departments maintained a sonde for the use of new companies willing to explore their surface for new mineral deposits. In 1821 the Nord prefect, on the request of private parties, bombarded the Ministry of the Interior with letters seeking an authorization to acquire a sonde, to be made by Anzin in its own workshops for 300 francs (AD, Nord, S 8285, correspondence, prefect and Ministry of the Interior, Mar.–Nov. 1821). A new and improved type of sonde was coming into use after 1830 (AN, F12 2534d, general director, Bureau of Mines, to minister of the interior, 1 Aug. 1834).

33. The mouth of the pit might be raised above ground level to prevent flash flooding (Héron de Villefosse, II, 156).

34. AN, F14 4244, Pajot des Charmes, "Mémoire"; AN, F14 7822, Daubuisson, "Exploitation." Pits had to be opened when the ground was not frozen and the weather was reasonably good. The usual season for this purpose, as for construction of all surface buildings and the emplacement of steam engines, was May to about the end of September when the fall rains began.

35. Daubuisson, "Description," pp. 163–64 n.1.

36. Registres, "Rapport fait à la Régie par Monsieur Casimir Perier," 15 June 1826. Hereafter cited as Perier Report.

37. Héron de Villefosse, II, 158–59, 190. A single pit might yield almost one million cubic feet of water in a twenty-four-hour period. The Newcomen engines were likely to be surplus machines available for irregular duties (Perier Report).

38. Perier Report. See below, chapter 4, under "The Attack on the Torrent."

39. Grar, II, 187–99. The system was not introduced in the mines of the Loire until 1838 and then only by "workers brought specially from the Nord" (Guillaume, p. 128).

40. Daubuisson, "Description," pp. 161–70; Héron de Villefosse, II, 190–94; Dormoy, p. 63.

41. The circular shafts adopted about 1850 were the natural end result (Dormoy, p. 63).

42. Registres, 28 Sept. 1822, 11 Apr. 1825; Michel and Renouard, pp. 83–85.

43. The apparent cause was the high cost of iron in France and wood in England. Héron de Villefosse, II, 194; Michel and Renouard, p. 85.

44. Registres, 26 Feb. 1804; AN, F14 7822, Daubuisson, "Exploitation."

45. Daubuisson, "Description," p. 119. The deepest pit in 1784 had been only 100 meters (AN, F14 4244, Pajot des Charmes, "Mémoire").
46. AN, F14 7822, Clère Report; AN, F14 7824, Mark Jennings, "Mémoire en défense pour la Compagnie d'Anzin," 30 Jan. 1830.
47. Héron de Villefosse, II, 160–61; AN, F14 7822, Daubuisson, "Exploitation."
48. AN, F14 7822, Daubuisson, "Exploitation." See also, in AN, F14 7822, the letter for 17 Nov. 1795 to the Directory from Mathieu and Renard, writing for Anzin to request draft exemptions for its miners on the grounds that only supple young men could negotiate galleries "two to three feet high" and two hundred to three hundred meters long.
49. Héron de Villefosse, II, 469–77.
50. For this and the following paragraphs on the system of extraction by *gradins renversés*, see AN, F14 7822, Daubuisson, "Exploitation"; Héron de Villefosse, II, 501–7; Dormoy, pp. 69–73. The word *height* (and *above, below*, etc.), as used here, does not refer to vertical distance, unless so specified, but rather to the distance between two points as measured along the slope of a seam. *Length* and related terms refer to horizontal distances along the strike of the seam. Thus, on a plat, the heights will be much greater than corresponding vertical distances, while on a droit the difference will be much less.
51. Michel and Renouard, pp. 96–98; AN, F14 7822, Daubuisson, "Exploitation."
52. Registres, 4 Nov. 1831; AN, F14 7824, Jennings, "Mémoire en défense"; AN, F14 7824, Council of Mines, extract from notes of meetings of 24 Jan., 7 Feb., and 21 Feb 1831. The pits at Fresnes and Vieux-Condé were much shallower and closer together until reorganized in the 1820s, and even at the newest pits near Denain the ideal layout was seldom obtained in practice.
53. AN, F14 7822, Daubuisson, "Exploitation"; Héron de Villefosse, II, 96–98, 478–79. Occasionally, the entire system of tailles was turned diagonally or even perpendicularly (*tailles montantes*) to the voie du fond, instead of parallel, in order to take advantage of the direction of the cleavage planes along which the coal broke most easily.
54. AN, F14 7822, Daubuisson, "Exploitation"; Dormoy, pp. 78–79.
55. AD, Nord, M 581 97, Charles Mathieu, "Les houillères anglaises et les houillères du département du Nord," 1860.
56. AN, F14 7822, Daubuisson, "Exploitation." The length of the individual relay might be longer if the working area involved was smaller and produced less coal, but the total meters covered and

weight drawn remained the same. Thus a hauler making 40-meter relays made only 130 round trips to fulfill his daily stint.

57. Registres, 4 June 1806, 7 Feb. 1811.
58. Ibid., 8 Oct. 1812, 4 May 1822, 28 Sept. 1822.
59. Ibid., 21 June 1826, 11 May 1833; Michel and Renouard, p. 95.
60. Registres, 7 Feb. 1811.
61. See, for example, AN, F14 4029, Etats de Redevance, passim.
62. Marcel Gillet, "Aspects industrial de la crise: Le département du Nord," in E. Labrousse, ed., *Aspects de la crise et de la dépression de l'économie française au milieu du XIXe siècle, 1846–1851*, p. 117.
63. AD, Nord, M 626 1, "Salaires: Délibérations des Associés Régisseurs des Mines d'Anzin," 27 June 1833.
64. T. S. Ashton and J. Sykes, *The Coal Industry of the Eighteenth Century*, pp. 62–65; Ritson, p. 84; "Sur plusieurs machines propre à transporter les minerais dans les galeries souterraines . . . ," *Journal des Mines*, no. 4 (1802), pp. 829–42; Héron de Villefosse, II, 243–44, 279. On the other hand, railed haulage remained the exception in the mines of the Loire until after 1844 (Guillaume, p. 125). R. A. Mott, "English Waggonways of the Eighteenth Century," *Transactions of the Newcomen Society*, XXXVII (1964–65), pp. 1–33, deals mainly with rails above ground, but also shows in passing the extensive use of iron wheels and rails underground in English coalfields where iron was cheap.
65. Perier Report.
66. For full details on Anzin's workshops, see chapter 3.
67. Registres, 4 May 1822.
68. Ibid., 16 Sept. 1825.
69. Perier Report.
70. AN, F14 4029, Etats de Redevance.
71. See, for example, Ashton and Sykes, p. 61. A visitor to Anzin in 1799 "explained" that sledges were used there in preference to sacks slung over the back, as in Auvergne, because the galleries at Anzin were too low to permit the haulers to stand upright! (AN, F14 7822, Busche to Council of Mines, 25 Nov. 1799).
72. For the miner psychology, see Rouff, pp. 314–24.
73. Registres, 20 Sept. 1824.
74. AD, Nord, M 626 1, "Salaires."
75. The use of horses would of course have further reduced the number of haulers. In fact, the attempt to employ horses for haulage at Anzin in 1847 led to a strike begun by the haulers (Gillet, "Aspects industriel de la crise," p. 117).

3 Winding, Drainage, and Ventilation

1. Héricart-Thury, "Mémoire sur les machines à vapeur de rotation . . . dans les houillères de Litry . . . ," *Journal des Mines*, no. 1 (1803), pp. 178–79; AN F14 7822, Daubuisson, "Exploitation."
2. Guenyveau, "Sur l'emploi des bœufs au service des machines à molettes," *Journal des Mines*, no. 1 (1812), p. 437.
3. AN, F14 7822, Daubuisson, "Exploitation"; AN, F14 4244, Pajot des Charmes, "Mémoire"; Vuillemin, I, 220, AN F14 4259, Laverrière, "Observations sur la machine à molettes," 14 May 1795.
4. AN, F14 4259, Perier to Lamillière, 13 July 1794; ibid., Mathieu to Lamillière, 14 Dec. 1787; Grar, II, 211–12; Prost, pp. 13–14; AN, F14 7822, Daubuisson, "Exploitation." Most of the horses were used on winches for extraction, and it was claimed, in 1802, just before the switch to steam, that Anzin's horses cost the company more than 200,000 francs annually to buy and keep ("Annonces concernant les mines," *Journal des Mines*, no. 1 [1802], p. 272). The exact number of horses used for extraction by Anzin is unknown, but it was probably about 160 out of a total of about 200 (Guenyveau, pp. 437–47).
5. J. R. Harris, "The Employment of Steam Power in the Eighteenth Century," *History*, LII (June 1967), p. 147.
6. Grar, II, 221–24. Cf. Prost, p. 17, and W. O. Henderson, *Britain and Industrial Europe, 1750–1870*, p. 105.
7. For the three versions, see Grar, II, 216–20. Cf. H. W. Dickinson, *A Short History of the Steam Engine*, p. 57.
8. About 1750 the mines of Littry in Normandy tried and abandoned an engine for reasons of economy. Aniche had two Newcomen engines in 1791. The Loire field had only one, and Anzin had twelve, all for pumping (Lefebvre, "Littry," p. 110; Gras, I, 185).
9. H. W. Dickinson, "The Steam-Engine to 1830" in Singer et al., IV, 185–86.
10. A. R. Griffin, *Coalmining*, pp. 33–37. In Belgium, Newcomen engines still predominated in 1838 (Prost, p. 17).
11. No relation to the Periers of Anzin.
12. Jacques Payen, *Capital et machine à vapeur au XVIIIe siècle*. In tracing the relations between the two sets of Perier brothers, I have followed the much fuller and more accurate account to be reconstructed from the sources for Anzin.
13. AN, F14 4259, Perier to Lamillière, 13 July 1794; ibid., Committee on Public Safety, 1794 [?].
14. Héricart-Thury, pp. 175, 187–88, and plate 5, p. 247.
15. Registres, 4 Nov. 1800.

16. Ibid., 19 Feb. 1801.
17. Héricart-Thury, p. 187.
18. AN, F14 7822, Renard to Council of Mines, 30 June 1803; Registres, 23 Sept. 1802, 17 June 1803, 2 June 1806. Only sixteen were finally delivered; the contract was fulfilled by substituting parts for new drainage engines for the other four winding engines (Registres, 4 Oct. 1810).
19. Héricart-Thury, p. 186.
20. AN, F14 7822, Renard to Council of Mines, 30 June 1803.
21. AN, F14 4029, Etats de Redevance, 1812–20.
22. Vuillemin, II, 342; Demoulin, p. 257; Gras, I, 214, 254, 258; Henderson, *Industrial Revolution in Prussia*, p. 14; Manes, "Mémoire sur les machines à vapeur . . . du Creusot," *Annales des Mines*, no. 1 (1840), p. 99. In some English fields the change was equally late where pits were not as deep as at Anzin (Court, *Midland Industries*, pp. 164, 167).
23. Registres, 5 Feb. 1806.
24. AN, F12 2467, Ministry of Commerce, "Note relative à divers manufacturiers," 1807 [?]; Chabert, II, 363; Payen, p. 202.
25. Registres, 7 Feb. 1810.
26. For bonuses and pay raises received, see Registres, 4 July 1804, 6 June 1811, 10 Sept. 1813, 8 Mar. 1816.
27. Perier Report.
28. Ibid.; Registres, 25 Sept. 1823.
29. AN, F14 4029, Etats de Redevance, passim.
30. AD, Nord, S 8285, Situations, 1811.
31. Registres, 23 Oct. 1802, 17 June 1803.
32. Gendarme, p. 55.
33. Registres, 4 Feb. 1808.
34. Ibid., 14 June 1810.
35. Ibid., 4 Oct. 1810.
36. Ibid., 7 Feb. 1811.
37. Ibid., 6 June 1811.
38. Ibid., 8 Oct. 1812, 6 Apr. 1815.
39. *Annuaire Statistique du Départment du Nord* (1814), p. 211.
40. Dickinson, "Steam-Engine," pp. 188–95; A. Mallet, *Compound Engines*, pp. 9–11; Rhys Jenkins, "A Cornish Engineer—Arthur Woolf, 1766–1837," *Transactions of the Newcomen Society*, XIII (1932–33), pp. 55–59. T. R. Harris, *Arthur Woolf, the Cornish Engineer, 1766–1837*, is of little use.
41. Jenkins, pp. 59–60; *Bulletin de la Société d'Encouragement pour l'Industrie Nationale*, XIV (1815), p. 297, and XVI (1817), p.

267; Annales des Artes et manufactures, 2d series, VI (1816–17), pp. 104–6.

42. It was only on the continent that "the merits of Mr. Woolf's engines are fully appreciated; for the cost of fuel being a considerable item of expense in manufactories abroad, there is a stronger inducement to adopt the most economical engines" (John Farey, *A Treatise on the Steam Engine*, II, 192).

43. Jenkins, p. 61; Mallet, p. 10; Charles Dollfus, remarks on W. H. Chaloner and W. O. Henderson, "The Manbys and the Industrial Revolution in France, 1819–84," *Transactions of the Newcomen Society*, XXX (1956–57), p. 75; Emile Koechlin, "Mémoire sur les machines à vapeur . . . ," *Bulletin de la Société Industrielle de Mulhouse*, IX (1835), pp. 79–182. For a description and engravings of the Woolf engine made by Edwards, see Héron de Villefosse, III, 84.

44. Registres, 8 Mar. 1816.

45. Ibid.

46. The purchase price of 40,000 francs was 5,000 francs under the list price given in *Annales des Artes et Manufactures*, 2d series, IV (1816–17), p. 105.

47. Registres, 4 Sept. 1817.

48. Ibid., 8 Mar. 1816, 4 Sept. 1817.

49. Ibid., 7 Apr. 1818.

50. Ibid., 9 Mar. 1820, 7 Sept. 1820.

51. Ibid., 7 Sept. 1821.

52. Ibid., 28 Sept. 1822.

53. AD, Nord, S 8285, "Réponse de la Compagnie d'Anzin . . . aux pétitions de quelques fabricants du département du Nord . . . ," 1822.

54. Registres, 25 Sept. 1823, 21 June 1827.

55. Registres, 4 Nov. 1832.

56. Cf. Gille, *Recherches*, p. 27.

57. AN, F14 4029, Etat de Redevance, 1820; Registres, 7 Sept. 1821; AD, Nord, M 623 12, "Questions annexés à la circulaire . . . sur la fabrication et l'usage des machines en France," 1825. Mr. Alexander, giving evidence for the "Third Report, Select Committee on Artizans and Machinery" in *British Parliamentary Papers*, 1824, V, 102–3, stated that "Mr. Edwards has put up in France about 100 [steam engines], that he made in England; and since the increase in duty, he has put up 200, which he has made at Chaillot" and added that some parts of the French engines, "particularly boilers," were imported from England. Quoted in Sidney Pollard and C. Holmes, eds., *Documents of European Economic History*, II, 286.

58. Payen, p. 204.
59. Barker, "Casimir Perier," p. 190. Cf. Cameron, *France*, p. 7, n.1.
60. Registres, 4 May 1822. The equipment for making improvements in the winding apparatus was also ordered from Chaillot, at a cost of 4,500 francs per engine, because Chaillot engaged to furnish it for the same price at which it could be made in Anzin's own workshops (Registres, 28 Sept. 1822).
61. Henderson, *Britain and Industrial Europe*, pp. 119–34; Gille, *Recherches*, p. 81.
62. AD, Nord, M 623 12, "Questions annexés," 1825; AN F14 4029, Etat de Redevance, 1826.
63. Perier Report.
64. AN, F14 4030, Clère to general director, Bureau of Mines, 7 Sept. 1830.
65. Registres, 11 Apr. 1825, 16 Mar. 1824, 16 Sept. 1825; Perier Report.
66. Perier Report; AN F14 4029, Etat de Redevance, 1826; AD, Nord, M 623 12, "Questions annexés," 1825.
67. Registres, 21 June 1826, 26 Sept. 1826, 16 May 1829. These were mostly sixteen- rather than ten-horsepower engines, in order to cope with deepening shafts.
68. In 1835, aside from two-horse winches of the old type used for handy extraction of the waste at new pits being dug, there were twenty-nine Edwards engines and eight Perier engines at Anzin for winding (AN, F14 4029, Etat de Redevance, 1836). One Perier engine set up in 1802 was still in regular operation in 1874 (Reybaud, p. 179).
69. AN, F14 7822, Daubuisson, "Exploitation." A new pump was to be installed in 1808 at Beaujardin to serve eleven pits, or almost all of those then active on the whole southern faisceau (AD, Nord, S 8285, Situation, 1808).
70. AN, F14 7822, Daubuisson, "Exploitation"; Héron de Villefosse, II, 188. Cf. "Mémoire sur les pompes employées dans les mines par John Taylor Esq.," *Annales des Mines*, no. 1 (1832), pp. 205–28.
71. See, for example, Registres, 5 Oct. 1809, 5 Sept. 1816.
72. Ibid., 11 Apr. 1825, 21 June 1827, 10 Sept. 1827.
73. Pumping engines at Anzin ranged from forty to eighty horsepower.
74. Thomas Tredgold, *The Steam Engine*, I, 275.
75. Compare the use of Trevithick's high-pressure pumping engines in the Cornish tin mines far from a cheap supply of coal (Dickinson, "Steam-Engine," pp. 191–92).
76. Registres, 19 Feb. 1802.
77. Ibid., 23 Oct. 1802, 6 Apr. 1815, 26 Apr. 1830.

78. Ibid., 17 June 1803.
79. AN, F14 4029, Etats de Redevance, 1812, 1818.
80. Registres, 4 Sept. 1817, 4 May 1822, 21 June 1827. It may, however, have been the first pumping engine whose piston was directly coupled to the pumps instead of being connected by the horizontal beam (Michel and Renouard, p. 93).
81. AN, F14 4029, Etats de Redevance, 1822–31; AD, Nord, M 623 12, "Questions annexés," 1825.
82. Registres, 16 Sept. 1825, 21 June 1826, 25 Sept. 1823; Perier Report.
83. In 1812 the company sent samples of the water to Paris to be analyzed by Scipion Perier, who was a chemist and a member of the board of directors (Registres, 4 Oct. 1812).
84. Ibid., 16 Sept. 1825, 21 June 1827.
85. Combes, "Aérage des mines," *Annales des Mines*, no. 1 (1839), p. 157; AN, F14 4244, Pajot des Charmes, "Mémoire." See also de Tréba, "Mémoire sur la température de l'intérieur des mines," *Annales des Mines*, no. 1 (1818), pp. 59–64.
86. AD, Nord, M 626 1, "Salaires."
87. AN, F14 7822, Daubuisson, "Exploitation."
88. Héron de Villefosse, II, 216.
89. AN, F14 4244, Pajot des Charmes, "Mémoire"; Grar, II, 207; Dieudonné, I, 180; AN, F14 7822, Daubuisson, "Exploitation."
90. AD, Nord, S 8111, Clère to prefect, 20 May 1816. Most European mine fields followed the same procedure. For the French Midi, see AN, ADxii 14, general director, Bureau of Mines, "Instruction sur l'emploi des lampes de surêté dans les mines . . . ," Apr. 1824; for England, see Ashton and Sykes, pp. 44–45.
91. AD, Nord, S 8111, Procès-verbal, 4 Jan. 1818.
92. AN, F14 4029, Etats de Redevance, 1820, 1821.
93. AD, Nord, S 8112, Brice to prefect, 17 Apr. 1823.
94. Quoted in Prost, p. 12, for Mons, 1637.
95. AN, F14 4259.
96. Héron de Villefosse, II, 218.
97. Combes, "Aérage des mines," passim.
98. AD, Nord, S 8098, Ministry of Public Works, "Instruction sur les mesures de surêté à prendre dans les mines à grisou . . . ," 1872.
99. Combes, "Aérage des mines," pp. 115–83. Anzin apparently never used chimneys in the eighteenth century because it had adopted furnaces so early. Therefore, when it decided to try them in 1810, Anzin was taking a momentary step backward from the point of view of the logical order of development in ventilation technique.

Chimneys continued to play a very minor part at Anzin (Registres, 7 Feb. 1810).

100. Combes, "Aérage des mines," pp. 291–92.

101. Ibid., pp. 183–200; AN, F14 7822, Daubuisson, "Exploitation."

102. AN, F14 7822, Daubuisson, "Exploitation."

103. Ibid.

104. AD, Nord, S 8111, Boudousquié, "Notice sur les accidents arrivés dans les mines de houille du département du Nord de 1822 à 1832," 13 Oct. 1833.

105. For various early projects for ventilating coal mines by machine in France, see AN, F14 4259.

106. AN, F14 1313–14, Council of Mines, 1 Feb. 1815; "Note sur une machine propre à renouveler l'air dans les mines," *Annales des Artes et Manufactures*, LV (1815), pp. 225–28; "Moyens de prévenir les effects funestes de l'inflammation du gas hydrogène," *Annales des Artes et Manufactures*, 2d series, II (1815), pp. 215–16.

107. Registres, 4 Feb. 1808. Presumably the device was steam-powered.

108. AD, Nord, M 557 7, Paulze d'Ivoy to prefect, 21 May 1827.

109. Michel and Renouard, pp. 89–92. Apparently the first mines anywhere to use mechanical ventilators successfully were a few smaller Belgian mines where furnaces had met with no success; the deeper the mine, the less effective both furnaces and ventilators were. A Mons mine led off in 1830 and others followed in the next ten years, but the technique was still in the experimental stage in 1839 (Combes, "Aérage des mines," pp. 201–45, 285–86; Combes, "Notes additionnelles . . . ," *Annales des Mines*, no. 2 [1839], pp. 187–91).

110. AN, F14 7822, Daubuisson, "Exploitation," and Combes, "Aérage des mines," pp. 190–200, 291–93 and passim, give excellent general descriptions of the ventilation system at Anzin at two different epochs, 1805 and 1839, respectively. Combes, "Notes additionnelles," pp. 177–200, gives a multitude of details for particular Anzin pits concerning temperatures, consumption of fuel by furnaces, etc.

111. AD, Nord, S 8112, Clère to prefect, 8 Dec. 1824.

112. AD, Nord, S 8112, general director, Bureau of Mines, to prefect, 10 Feb. 1825; AD, Nord, S 8112, prefect's order, 22 Feb. 1825.

113. AD, Nord, M 626 1, "Salaires."

114. AN, F14 3848, Lorieux, "Observations sur les procès-verbaux de visite des mines du département du Nord," 1839. Combes, "Aérage des mines," uses Anzin as an example of excellence again and again.

115. Prost, p. 13; Héron de Villefosse, II, 133–35.
116. Héron de Villefosse, II, 134.
117. AD, Nord, S 8112, Brice to prefect, 17 Apr. 1823. The chief engineer, Clère, was away from the Nord at the time of the explosion. Long after the fact, an entirely different suggestion associated the explosion with the partial flooding and interruption of ventilation at Chauffour due to the collapse of the cuvelage and flooding of a neighboring pit (AD, Nord, M 581 97, Mathieu, "Les houillères anglaises").
118. AD, Nord, S 8112, Brice to prefect, 17 Apr. 1823.
119. AN, F14 4259, Tournelle to Herchy, 23 Apr. 1811, and Renard to Herchy, 2 May 1811 (documents attached to Picard, "Notice sur les exhalaisons des mines," 13 June 1811).
120. Ritson, pp. 95–97; AD, Nord, S 8111, Brice to prefect, 17 Apr. 1823; AD, Nord, S 8111, general director, Bureau of Mines," "Envoi d'une instruction sur les lampes de surêté . . . ," 10 May 1824.
121. AD, Nord, S 8111, Renard to prefect, 25 May 1823; Brice to prefect, 17 Apr. 1823. Brice took the occasion to criticize a tariff system that worked against an entire class of French workers.
122. Registres, 22 Sept. 1822, 13 Mar. 1823.
123. AD, Nord, S 8112, Brice to prefect, 17 Apr. 1823.
124. Among the numerous despairing references to the workers' propensity to break open the lamps, see especially "Circulaire sur un nouveau mode de fermeture pour les lampes de surêté," *Annales des Mines*, no. 1 (1832), pp. 459–64.
125. AD, Nord, S 8112, prefect's order, 22 Apr.; Taffin de Sorel to prefect, 6 May; prefect to Taffin de Sorel, 10 May; director general, Bureau of Mines, to prefect, 12 May; prefect to Renard, 19 May 1823; AD, Nord, S 8111, Renard to prefect, 27 May 1823.
126. AD, Nord, S 8111, Brice to prefect, 16 July and 2 Oct. 1823.
127. Ibid., 16 July 1823.
128. AD, Nord, S 8111, general director, Bureau of Mines, "Envoi d'une instruction sur les lampes de surêté . . . ," 10 May 1824.
129. AN, ADxii 14, general director, Bureau of Mines, "Instruction sur l'emploi des lampes de surêté dans les mines . . . ," Apr. 1824; AD, Nord, S 8098, "Décret impérial contenant des dispositions de police relatif à l'exploitation des mines," 3 Jan. 1813.
130. AD, Nord, S 8112, prefect's order, 30 Sept. 1824.
131. Ibid., 22 Feb. 1825.
132. AD, Nord, S 8111, police report to mayor of Valenciennes, 28 June 1824.

133. Ibid., Valenciennes subprefect to prefect, 28 July 1826.
134. AD, Nord, S 8112, prefect to general director, Bureau of Mines, 20 Dec. 1827.
135. AD, Nord, S 8111, Boudousquié, "Notice sur les accidents."
136. Ibid.; Bourgin and Bourgin, III, 178–82, 191–95.
137. AD, Nord, S 8111, Boudousquié, "Notice sur les accidents."
138. See, for example, the remarks of Combes, "Aérage des mines," pp. 296–307.

4 Expansion of Workings and Concessions

1. Perier Report.
2. AN, F14 7822, Clère Report. For dating pit openings before 1791 and pit closings before 1847, see Grar, II, 154–55.
3. AN, F14 7822, Clère Report; Registres, 15 June 1799, 26 Oct. 1804.
4. AN, F14 7822, Clère Report; Registres, 16 Sept. 1825, 10 Sept. 1827.
5. Registres, 8 Mar. and 5 Sept. 1816; AN, F14 7822, Clère Report.
6. AN, F14 4029, Etat de Redevance, 1819.
7. Registres, 28 Sept. 1822, 16 Sept. 1825.
8. Ibid., 16 Sept. 1825, 28 Sept. 1827, 11 June 1830, 4 Nov. 1832; AN, F14 7822, Clère Report.
9. AN, F14 7822, Clère Report; Registres, 15 June 1799, 4 Nov. 1801, 23 Oct. 1802, 2 June 1806, 4 Nov. 1831.
10. AD, Nord, M 581 97, Mathieu, "Les houillères anglaises."
11. Chauffour was still operating in 1860, almost one hundred years after its opening (AD, Nord, M 581 97, Mathieu, "Les houillères anglaises").
12. AN, F14 4029, Etat de Redevance, 1812.
13. The process of closing down Barrière illustrates the difficulty of fixing the exact date to be assigned to the abandonment of a pit. Working operations were suspended there in 1809, and in 1811 the pit was ordered closed and filled in "as soon as possible," but according to Clère the actual filling was not completed until 1817 (Registres, 5 Oct. 1809, 6 June 1811; AN, F14 7822, Clère Report).
14. To allay any suspicion of subversive tunneling under the walls of Valenciennes' citadel, the permission of the minister of war was required before Saint Joseph could be sunk (Registres, 17 June 1803).

15. For the pits Saint Jean, Verger, Mouton Noir, Barrière, and Saint Joseph, see ibid., 17 June 1795, 15 June 1803, 5 Oct. 1809, 6 June 1811; AN, F14 7822, Clère Report.
16. Grar, II, 155; AN, F14 7822, Clère Report.
17. AN, F14 7822, Clère Report.
18. AN, F14 4029, Etat de Redevance, 1813.
19. AD, Nord, S 8285, Situations, 1800–1810; AN, F14 4029, Etats de Redevance, 1811–20.
20. AN, F14 7822, Clère Report.
21. Ibid., Council of Mines to Renard, 20 Oct. 1800, and Renard to Council of Mines, 15 Feb. 1801. For the renewal of work on the Torrent, see Registres, 15 June and 18 Oct. 1799.
22. AD, Nord, S 8285, Situations, 1801, 1802, 1803.
23. AN, F14 7822, Daubuisson, "Exploitation."
24. Grar, II, 155.
25. Daubuisson, "Description," p. 120.
26. AN, F14 7822, Clère Report; Registres, 17 June 1803, 27 Feb. 1804; AD, Nord, S 8285, Situations, 1804, 1805, 1806.
27. AN, F14 7822, Clère Report; AN, F14 4029, Etats de Redevance, 1811–33.
28. Registres, 8 Oct. 1812.
29. AN, F14 7822, Renard to Council of Mines, 30 June 1803.
30. AD, Nord, S 8285, Situation, 1811. A number of the pits, especially on Anzin, were not in production. This accounts for the discrepancy between Anzin and Raismes in the ratios of pits to production and workers.
31. Ibid.
32. Significantly, this period was also devoid of important technical improvements. It corresponds almost exactly to the years between the establishment of the Perier and Edwards winding engines.
33. AN, F14 7822, Clère Report.
34. Registres, 8 Mar. 1816.
35. Perier Report.
36. AN, F14 4029, Clère, "Rapport au Comité d'évaluation du département du Nord sur le travail des redevances pour l'exercise de 1819," 27 Dec. 1819.
37. Ibid., Renard to general director, Bureau of Mines, 13 Dec. 1820.
38. Registres, 4 Mar. 1819.
39. Ibid., 4 May 1822.
40. Ibid., 4 May and 28 Sept. 1822.
41. Ibid., 13 Mar. 1823.
42. Perier Report.

43. Ibid.
44. Registres, 16 Mar. and 20 Sept. 1824.
45. Perier Report.
46. AN, F14 4029, Etats de Redevance, 1820–30; Perier Report.
47. AN, F14 7824, Jennings, "Mémoire en défense."
48. AN, F14 7822, Clère Report.
49. Registres, 5 Aug. 1825. Joseph Perier and Hocquart undertook to get these extraordinary measures approved by the full board at the next regular meeting in September. They seem to have had no difficulty in doing so, so firmly were the Periers in the saddle (Ibid., 16 Sept. 1825).
50. Ibid., 20 Sept. 1824, 11 Apr., 5 Aug., and 16 Sept. 1825, 21 June 1826, 10 Sept. 1827; AN, F14 7822, Clère Report. A barrier of two hundred meters of intervening rocks was left between the pits of the Torrent and the older Anzin pits in order to prevent a possible flood in a Torrent pit from inundating all the pits of the southern faisceau. For the same reason, the workings at the new pits were begun thirty meters below the top of the coal measures in order to preserve this vertical distance from the Torrent above. Thus the Torrent pits were a self-contained system, linked together for drainage, but preserved from all contact with the pits to the east (AN, F14 7822, Clère Report).
51. AN, F14 4029, Etats de Redevance, 1829, 1830; Registres, 4 Nov. 1831.
52. AN, F14 7822, Mark Jennings, "Réplique pour la Compagnie des Mines d'Anzin du précis de la Compagnie Lacoste et Eubank," 10 July 1831; ibid., "Réponse de la Compagnie d'Anzin à l'avis de M. le Préfet du Département du Nord . . . ," 26 Apr. 1830.
53. AN, F14 4029, Etat de Redevance, 1831.
54. AN, F14 7822, Clère Report.
55. Ibid.
56. Registres, 4 May 1822, 20 Sept. 1824; AN, F14 7822, Clère Report; Perier Report.
57. Registres, 21 June 1826, 12 June and 10 Sept. 1827, 16 Oct. 1829; AN F14 7822, Clère Report; AD, Nord, M 581 97, Mathieu, "Les houillères anglaises."
58. Registres, 10 Sept. 1827; AN, F14 7822, Clère Report.
59. Registres, 16 Mar. 1824, 21 June 1826, 10 Sept. 1827, 16 May 1829, 26 Apr. 1830, 4 Nov. 1831, 18 July 1833; AN, F14 7822, Clère Report.
60. AN, F14 7822, Clère Report; Registres, 4 Nov. 1831.
61. AN, F14 7822, Clère Report; AD, Nord, M 581 97, Mathieu, "Les houillères anglaises."

62. AN, F14 4029, Etats de Redevance, 1830–36.
63. AD, Nord, M 581 97, Mathieu, "Les houillères anglaises."
64. Registres, 16 June 1800, 27 June 1803, 27 Feb. and 27 Oct. 1804, 2 June 1806, 7 Feb. 1810, 10 Oct. 1811; AD, Nord, S 8285, Situations, 1800–1810; AN, F14 7822, Clère Report; AN, F14 4029, Etat de Redevance, 1820.
65. AN, F14 4029, Etats de Redevance, 1820–24.
66. Registres, 4 May 1822.
67. Ibid., 25 Sept. 1823, 20 Sept. 1824, 11 Apr. 1825; Perier Report.
68. Registres, 11 Apr. and 16 Sept. 1825, 21 June 1826, 21 June and 10 Sept. 1827, 8 Aug. 1830.
69. AN, F14 7822, Clère Report; Registres, 4 Nov. 1832.
70. AN, F14 7822, Clère Report.
71. AN, F14 4029, Etat de Redevance, 1836.
72. J. Laloux, *Le rôle des banques locales et régionales du nord de la France,* pp. 14–16; "Travaux de la Compagnie départementale du Haut-Rhin pour la recherche de nouvelles mines de houille de 1822 à 1832," *Bulletin de la Société Industrielle de Mulhouse*, VII (1833), pp. 205–98; J. Levainville, "Les recherches de la houille dans la Seine-Inférieure pendant la Révolution," *Bulletin d'Histoire Economique de la Révolution*, VIII (1913), pp. 364–70; A. H. de Bonnard, "Notice sur diverses recherches de houille entreprises dans le département du Pas-de-Calais," *Journal des Mines*, no. 2 (1809), pp. 415–46; Vuillemin, III, 22–62.
73. See, for example, the efforts of Castiau, a former employee of Anzin, who offered his considerable skills and savings to three successive companies exploring in the Paris basin (1796–1827), Pas-de-Calais (1806–10), and Belgian Campine district (1806–?), respectively. For the Parisian company, at least, he hired his sondeurs "in the Nord," presumably at Anzin. All his enterprises failed completely (Levainville, "Recherches dans la Seine-Inférieure," pp. 365–70; de Bonnard, "Recherches dans le Pas-de-Calais," pp. 434–44; Guy Richard, "Un épisode peu connu de la recherche du charbon en Normandie, les mines de Meulers [1797–1806]," *Annales de Normandie*, no. 2 [1968], pp. 135–62). Anzin "temporarily retired" him in 1798 and suspended his pension in 1806, apparently amid considerable controversy over his "disloyalty" in undertaking outside ventures on his own. There is no suspicion in the secondary accounts or evidence in Anzin's records that would show any kind of concealed interest in his ventures on the part of the company, although in 1811 his pension was restored and in later years his son continued to rise in Anzin's managerial ranks (Registres, 11 Aug. 1795, 15 Feb. 1798, 2 June 1806, 6 June 1811, 28 Sept. 1822, 13 Mar. 1823, 16 Mar. 1824, 26 Apr. 1830).

74. See, for example, the "departmental company" in the Haut-Rhin, which included most of the members of the new cotton aristocracy from such families as Dollfus-Mieg, Koechlin, Schlumberger, and Heilmann ("Travaux de la Compagnie départementale du Haut-Rhin," pp. 205–6).

75. Castiau based his ventures both in the Paris basin and near Arras on his belief that the seams found at Mons, Valenciennes, and Aniche turned southwest beyond Aniche, whereas in fact they bent northwest toward the Channel. Castiau's view was generally held by the experts until the discoveries of the late 1840s. See, for example, de Bonnard, "Recherches dans le Pas-de-Calais," pp. 428, 434–36.

76. Ibid., pp. 424–26. A more surprising difficulty might be the fact that "the local inhabitants who would derive the greatest advantage from the success [of these efforts] show no haste to cooperate, that they even decry as thoughtless the expenses required and treat as chimerical a project whose purpose they cannot conceive. Their ignorant apathy is thus joined to the fears and ever active jealousy of the producers of nearby areas [i.e., Anzin] who fear new competition, with the result of discouraging the 'homme industrieux' and ruining his enterprise" (ibid., p. 426).

77. Most failed when the proposed initial capital levy was not paid in, but in some cases the whole affair seems to have been a speculative swindle. For example, governmental officials were concerned about the activities of two men who raised money for "mining" companies from the public on the strength of their announced "discoveries" with a divining rod. See the large correspondence in AN, F14 7916 and 7822 between the Nord prefect, the general director of the Bureau of Mines, and the mining engineer Miché, between 1806 and 1812.

78. The eleven soundings and two exploratory shafts attempted between 1822 and 1833 in the Haut-Rhin cost nearly 150,000 francs ("Travaux de la Compagnie départementale du Haut-Rhin," p. 210). Castiau claimed a loss of 600,000 francs in the Pas-de-Calais and 460,000 francs in Normandy (Vuillemin, III, 44; G. Richard, p. 162).

79. Aniche made a passing effort in the Pas-de-Calais in the first years of the century, as before the Revolution (Vuillemin, III, 36, 51–56).

80. In 1781 Anzin received permission to explore an area between Saint Pol and Arras and made at least eight soundings there before abandoning them in 1783. It also received several similar permissions for the areas around its concessions in the Nord, but again no encouraging results were obtained. All these efforts seem to have been somewhat half-hearted; sounding equipment in the

eighteenth century was often incapable of ascertaining the presence of coal by itself, and Anzin's motives were chiefly a desire to maintain its monopoly against its many ephemeral rivals rather than to find coal (Vuillemin, III, 38–39; Grar, II, 325–27).

81. Grar, II, chapters 1–2, and pp. 235–48, 252–60.
82. AD, Nord, S 8285, Anzin to prefect, 5 Mar. 1821.
83. *Annuaire Statistique* (1814), p. 105.
84. *Petites Affiches de Valenciennes*, 25 May 1822.
85. On the vicissitudes of this area in the eighteenth century, see the bundle of miscellaneous documents in AN, F14 7816, especially the widow Duquesne to prefect, 4 Feb. 1804. See also Grar, II, 252–60.
86. AN, F14 7816, Council of Mines to minister of the interior, 4 June 1804; Thieffries de Beauvoir to engineer Miché, 10 Oct. 1807; Thieffries de Beauvoir to Napoleon, 20 Sept. 1806.
87. AN, F14 7816, copies of acts of sale, 31 Oct. and 16 Nov. 1805, 10 June and 16 Oct. 1807.
88. Registres, 7 Oct. 1807.
89. AN, F14 7816, copies of acts of sale, 31 Oct. 1807, 8 Feb. 1808.
90. Ibid., Renard to prefect, 27 Apr. 1808. In the act of sale of 8 Feb. 1808, by which Thieffries transferred his interests to Anzin, it was specified that Anzin would bend every effort to effect the transfer of the concession, but if it were unsuccessful, the sale was nullified and the costs of negotiation between the parties and with the government were to be borne by Anzin.
91. Ibid., Miché to prefect, 16 May 1808.
92. AN, F14 7822, Renard to Council of Mines, 18 Jan, 4 Aug., and 19 Nov. 1810; ibid., Clère Report; Registres, 4 Oct. 1810, 10 Sept. 1813.
93. AN, F14 4029, Clère, "Rapports sur l'assiette des redevances fixes et proportionnelles," 1818–20.
94. Ibid., Etats de Redevance, 1820–30.
95. AN, F14 7822, Clère Report.
96. Registres, 11 June 1831; AN, F14 7816, J. Perier and Hocquart to general director, Bureau of Mines, 30 June 1831.
97. See the series of government warnings and recommendations, 1835 to 1845, in AN, F14 7816.
98. Registres, 2 June 1806, 5 Feb. 1807; AN, F14 7822, Renard to prefect, 13 May 1806; Council of Mines to minister of the interior, 1 Dec. 1809; Renard to Council of Mines, 4 Aug. 1810.
99. AN, F14 7822, Renard to Council of Mines, 19 Nov. 1810; Council of Mines to Renard, 13 Dec. 1810.
100. My italics. Registres, 4 Oct. 1810.

101. For a complete record of research permits granted, renewed, and lapsed in the Nord and Pas-de-Calais during this period, see *Annuaire Statistique*, 1803–17.
102. Registres, 7 Oct. 1807.
103. Ibid., 8 Feb. and 5 Oct. 1809; Grar, III, 30–31.
104. Registres, 16 Sept. 1819.
105. Ibid., 9 Sept. 1819, 9 Mar. 1820; Vuillemin, II, 345.
106. For a large number of *mémoires, répliques, pétitions, réponses,* etc., reviewing the history of the dispute from the point of view of all three parties, see AN, F14 7823, and AD, Nord, S 8285.
107. Vuillemin, II, 350; Dormoy, p. 44.
108. The material on Denain-Douchy is similar to that for Bruille-Odomez (see AN, F14 7822, 7823, 7824).
109. Gillet in Trenard, p. 29.
110. Dormoy, pp. 47–48.
111. Gillet in Trenard, p. 31.
112. Dormoy, pp. 47–58; Vuillemin, II, 349–58, III, 62–63, 97–108.
113. AN, F12 2534d, A. Divoy, "Réplique à la Compagnie d'Anzin sur la question des houilles," Mar. 1834.
114. H. Watelet, "Le rôle de l'individu dans l'industrie houillère belge au début du XIXe siècle: L'exemple de De Gorge-Legrand," in Trenard, pp. 127–40; Lévy-Leboyer, *L'industrialisation*, pp. 631–32.
115. Lévy-Leboyer, *L'industrialisation*, p. 131, n.176.
116. Registres, 4 Nov. 1831, 4 Nov. 1832.
117. Watelet in Trenard, p. 127, n.5; Lévy-Leboyer, *L'industrialisation*, p. 611, n.57.

5 Output, Costs, and Productivity

1. For an evaluation of the Michel and Renouard series used here and a comparison with other available series, see my Ph.D. dissertation, "The Anzin Company: The History of a French Coal Mining Firm, 1800–1833" (University of Minnesota, 1965), pp. 258–81.
2. Markovitch, "Sources et methodes," table 2.
3. These statements were the basis for a 5 percent tax on annual net income. For an evaluation of their accuracy and usefulness, see Geiger, "The Anzin Company," pp. 265–67.
4. The local government engineer, Boudousquié, reported in 1833 that many of the detailed cost figures supplied by Anzin seemed too high. This was undoubtedly because he was using much too

low an output base for his calculations. Boudousquié's successor, Blavier, commented in 1841 that the reported costs were probably only half as large as real costs (AN, F14 4030).

5. Françoise Duchastelle, "Le commerce du charbon des Mines d'Aniche de 1855 à 1873," p. 88; Baumont, p. 602; Pounds and Parker, p. 7; "Amélioration de la situation financière des houillères du Nord et du Pas-de-Calais," *Mines et Metallurgie* (Aug. 1959), p. 443.
6. Pierre Léon, I, 297. See also, for the middle of the nineteenth century, the figures in Alfred Marc, *L'évolution des prix depuis cent ans,* p. 24.
7. P. N. Berryer, *Observations présentées par les propriétaires et exploitants des mines de houille de France . . .* (Paris, 1826[?]), p. 10.
8. Perier Report.
9. Ibid.
10. AN, F14 7822, Renard to Council of Mines, 19 Dec. 1799. The regulatory system had, however, been indifferently successful in Hainaut even before the Revolution. Paul W. Bamford, *Forests and French Sea Power, 1660–1789,* pp. 73–74.
11. Dieudonné, I, 32, 271, 280–81, 538.
12. AN, F14 7822, Renard to Council of Mines, 19 Dec. 1799.
13. Registres, 27 Feb. 1804; AN, F14 7822, Daubuisson, "Exploitation."
14. Registres, 4 Sept. 1817.
15. Ibid., 17 Apr. 1818.
16. Dunham, p. 86.
17. Berryer, p. 11.
18. Registres, 10 Apr. 1825.
19. AN, F14 7111, J. Cordier, "Mémoire sur le projet de canal de la Sambre à l'Escaut," 1838.
20. AN, F14 7823, "Rapport de l'inspecteur de la division du Nord sur la réclamation de la Compagnie d'Anzin ayant pour objet de défendre l'exportation des perches servant au boisage," 10 June 1848.
21. Registres, 28 Sept. 1822, 13 Mar. 1823, 20 Sept. 1824, 11 Apr. and 16 Sept. 1825.
22. Ibid., 16 Sept. 1825.
23. Perier Report. The saving was partly due to the replacement of wooden planks by iron rails for underground haulage, but it was mainly due to better prices and the use of props of lesser dimensions.
24. Registres, 10 Sept. 1827.

25. Dunham, p. 86; AN, F14 7823, "L'exportation des perches."
26. See, for example, Berryer, p. 10.
27. Perier Report.
28. Registres, 17 Oct. 1797; AD, Nord, S 8285, Situation, 1811.
29. Perier Report.
30. Registres, 17 Oct. 1797, 18 Oct. 1798.
31. Ibid., 18 Oct. 1798.
32. Ibid., 28 Sept. 1822, 13 Nov. 1823.
33. Anzin even constructed a "horse infirmary" in 1797 (Registres, 17 Oct. 1797).
34. Perier Report.
35. Registres, 4 May 1822.
36. Perier Report.
37. Registres, 13 Mar., 25 Sept., and 3 Nov. 1823; Perier Report. The company's former Lille candle dealer protested in vain a decision which left 34,000 kilograms of unmarketable candles on his hands.
38. Registres, 11 Apr. 1825; AN, F14 4029, Etats de Redevance, 1829–31.
39. Registres, 13 Mar. 1823; Perier Report.
40. Perier Report.
41. Ibid.
42. Ibid.
43. Registres, 4 May 1822, 25 Sept. 1823.
44. Perier Report.
45. Registres, 28 Sept. 1823.
46. For the changes in the Anzin accounting system, see ibid., 11 Jan. 1825.
47. Ibid., 20 Sept. and 16 Mar. 1824, 16 Sept. 1825, 21 June 1826; Perier Report.
48. Registres, 16 Sept. 1825; Perier Report.
49. Registres, 25 Sept. 1823; Perier Report.
50. Perier Report.
51. This confusion of functional and raw material categories probably resulted in some double counting. See AN, F14 4030, Boudousquié, comments on Etat de Redevance, 1833.
52. Daubuisson, "Description," p. 170.
53. AN, F14 4029, Clère to prefect, 31 Oct. 1821.
54. Renard Report, p. 21; Chevalier, p. 220.
55. AN, F14 4029, Renard to general director, Bureau of Mines, 13 Dec. 1820.

56. Examples of this policy appear throughout the Registres. Indemnities may have amounted to an average of 30,000 francs per year. See table 2.
57. Registres, 17 Apr. 1818.
58. Ibid., 4 Sept. 1817, 21 June 1827.
59. Ibid., 28 Sept. 1822.
60. Perier Report.
61. AN, F14 4029, Etat de Redevance, 1831.
62. The largest and most expensive Watt-type engine was the eighty-horsepower one made by Edwards for 100,000 francs. See above, chapter 3, under "Winding."
63. Information on the cost of pumping and winding machinery itself is extremely slight. Most such apparatus was made in the company's own shops, and its cost is hidden in the Etat figures for labor, metals, wood, etc.
64. Profits and gross cost estimates are based on figures given in the Perier Report and the Etats. See below, chapter 8, under "Profits."
65. The figures for the various items are based on a variety of sources and calculations. New pits cannot easily be assigned to a specific year; the figure of 400,000 francs represents the cost of one new pit, since slightly less than one pit a year was opened between 1819 and 1833. The figures for steam engines and railed haulage represent 20 percent of the total spent for these items during the years 1821–25, as indicated in the Perier Report. The costs of the last three items are based on the Etats, as discussed in the text.
66. Perier Report.
67. AD, Nord, S 8252, Renard to prefect, 7 Jan. 1815.
68. Perier Report.
69. Chevalier, p. 434.
70. AN, F12, 2534a, Mark Jennings, "Réponse aux questions à faire résoudre pour l'enquête sur les houilles," 8 Dec. 1832. Hereafter cited as AN, F12 2534a, Jennings Report.
71. AN, F14, 4030, Boudousquié, comments on Etat de Redevance, 1833. As might be expected, figures for costs per ton as calculated from production and cost figures given in the Etats de Redevance are on the high side and fluctuate widely in no discernable pattern.
72. In areas like the Loire and Mons, where many mining companies of greatly varying size and efficiency operated, Chevalier's figures are for the largest and most efficient producers.
73. AD, Nord, S 8111, Boudousquié, "Notice sur les accidents," and Clère, "Notice supplémentaire."
74. Héron de Villefosse, II, 504.

75. Chevalier, pp. 433–34.
76. AN, ADxiii 14, "Observations des propriétaires des mines de houille de Littry," 1822; AN, F14 4253, "Description du gîte houiller d'Epinac," 15 Nov. 1839.
77. AN, F14, 4253, "Description du gîte houiller de Blanzy," 24 Nov. 1839, and "Description du gîte houiller de Creusot," 15 Dec. 1839.
78. Baumont, p. 566, n.22.
79. This amount would not, however, be as much as the 50 percent difference at Anzin, where a larger proportion of all workers were employed at the surface than elsewhere.
80. Garcenot, p. 99; Baumont, p. 566.

6 Transportation: From Pit Head to Market

1. Dunham, pp. 14–22; J. Letaconnoux, "Les transports en France au XVIIIe siècle," *Revue d'Histoire Moderne et Contemporaine,* X (1908–9), pp. 105–14. For the clearest contemporary account of the problems faced, by the leading government exponent of vigorous cooperation by government and private capital, see F. L. Becquey, *Rapport au Roi sur la navigation intérieure de la France.* For Becquey's life, policies, and achievements, see A. A. Beugnot, *Vie de Becquey . . . directeur général des Ponts et Chaussées et des Mines sous la Restauration.* See also AN, F14 7076, Commission des routes et canaux, 1828.
2. Yves Casail, "Transports et voies de communication en Flandre Wallonne . . . ," pp. 21–23, 29–33, 52–53, 64–69, 77–78; P. Caron, "Une enquête sur l'état des routes, rivières et canaux an début de l'an II," *Bulletin d'Histoire Economique de la Révolution,* X (1917–19), pp. 228–44; P. Dion, "Géographie de la circulation dans la partie centrale de la région du Nord," *Annales de la Société Geologique du Nord,* III (1939), pp. 97–129; R. Blanchard, *La Flandre: Etude géographique de la plaine flamande,* pp. 444–47; M. Blanchard, "Enquête administrative sur la roulage en 1811," *Revue de Géographie Alpine,* VIII (1920), pp. 585–96; AN, F14 1270, Enquête administrative sur la roulage en 1811, Nord.
3. AN, F14 1270, Enquête sur la roulage, Nord; AN, F14 6673, Cordier, Bureau of Bridges and Roads, and Baudrand, colonel of engineers, Procès-verbal de conférence, 27 June 1817; AD, Nord, M 653 2, Statistique industrielle et manufacturière, 1808.
4. For the condition of and improvements in water transport in the Nord during this period, see J. Dutens, *Histoire de la navigation intérieure de la France,* I, 489–592; Caron, pp. 244–46; J. L. E. Cordier, *De la navigation intérieure du département du*

Nord, pp. 24–146; J. L. E. Cordier, *De la navigation du département du Nord.*

5. Registres, 16 Oct. 1829, 12 Oct. 1833; AN, F14 4030, Etat de Redevance, Aniche, 1833.
6. Registres, 17 Oct. 1797, 15 June 1799, 17 Feb. 1800, 11 Apr. and 28 Sept. 1826, 21 June 1827, 29 June 1828, 16 May and 16 Oct. 1829, 26 Apr. 1830, 4 Nov. 1831, 11 May and 18 July 1833; Perier Report.
7. AN, F14 7111, Cordier, "Mémoire sur le canal de la Sambre à l'Escaut," 1838.
8. AN, ADxiii 60, [Anzin], "Observations," 15 Jan. 1822; *L'écho du Nord,* 21 Jan. 1820, 24 Jan. 1824.
9. Chevalier, p. 482.
10. AN, F14 1270, Enquête sur le roulage, Nord, Condé, 1812.
11. AN, F14 4030, Etat de Redevance, Aniche, 1833.
12. Registres, 26 Apr. 1830; AN, F12 2534c, "Renseignements divers sur les principales mines de charbon de terre existant en France," [1833].
13. Registres, 7 Feb. 1810.
14. De Gallois, "Des chemins de fer en Angleterre," *Annales des Mines,* no. 3 (1818), pp. 129–44. De Gallois also pointed out the British use of stationary steam engines to haul wagons on long inclines. He estimated that tramways were more economical than either roads or canals for moving large volumes of bulky goods, regardless of the distances involved, and he foresaw the time when steam-powered railroads would replace waterways for long-distance transportation. See also Mott, passim.
15. One, the Lyon-Saint Etienne (opened 1832), carried general traffic as well as coal and used locomotives on one level stretch. The other (opened 1834) merely improved the rail connection of the Saint Etienne mines to the Loire (Georges Lefranc, "The French Railroads, 1823–43," *Journal of Economic and Business History,* II [Feb. 1930], pp. 298–331; Dunham, pp. 49–57).
16. Prost, p. 21.
17. Registres, 26 Apr. 1830.
18. Ibid., 18 July and 12 Oct. 1833.
19. Michel and Renouard, p. 76; Société de hauts fourneaux, forges et aciéries de Denain et d'Anzin, *Livre d'Or,* pp. 15–16; AD, Nord, S 8097, Jennings to prefect, 16 Jan. 1836.
20. Registres, 8 Oct. 1812; Perier Report.
21. Registres, 17 June and 27 Oct. 1803, 27 Oct. 1804, 2 June 1806, 4 Feb. 1808, 4 Oct. 1810.

22. AN, ADxiii 60, "Observations sur l'approvisionnement de Paris en combustible," [1816?].
23. Perier Report.
24. Registres, 11 Apr. 1825.
25. Ibid., 25 Nov. 1824.
26. There is no evidence that the loaders at Anzin had tried to formally organize to raise their wages as did the carters and boatmen. But cf. Bourgin and Bourgin, III, 288, on the attempt of the loaders at the river port of Cambrai to set up in 1817 a hereditary "corporation" after the style of the Old Regime and, they claimed, in imitation of one at Saint Quentin.
27. Perier Report; Registres, 25 Sept. 1823, 16 Mar. 1824, 11 Apr. and 16 Sept. 1825; AN, F14 6674, "Rapport sur la navigation de l'Escaut," 1824.
28. AN, F14 7823, Piet, "Mémoire pour les sieurs Saugrain, Charpentiers et Compagnie contre la Compagnie d'Anzin," [1830?]; AN, F14 7824, Dumas Company to Council of Mines, 16 July 1830.
29. Registres, 18 July 1833.
30. Ibid., 26 Apr. 1830; Reybaud, pp. 182–83.
31. AN, F12 2534c, Coupin, "Examin critique d'un mémoire," 1832.
32. AN, ADxiii 15, Condé boatmen to National Assembly, 12 June 1791.
33. AD, Nord, M 653 2, Statistique industrielle, 1801; AN, F14 1270, Enquête sur la roulage, Nord, Condé, 1812.
34. Casail, pp. 102–6, 110–11; Letaconnoux, pp. 275–80; AN, F14 1270, Enquête sur le roulage, Nord, Condé, 1812; AN, F14 6968, "Pétition adressée à . . . la Chambre des Députés par les Bateliers du Canal de Condé (Nord)," 1832.
35. Casail, pp. 120–28.
36. Ibid., pp. 122–25; Grar, II, 170–73.
37. Grar, II, 175–76.
38. See various documents, including copies of laws, administrative rulings, etc., in AN, ADxiii 15.
39. AN, ADxiii 15, Legislative Corps, Council of Five Hundred, "Motion d'ordre fait par Foncez (de Jemmappes) sur la navigation intérieure," 30 Sept. 1798.
40. AN, ADxiii 12, Arrêté concernant la navigation de l'Escaut, 2 June 1803.
41. Registres, 4 Nov. 1801, 7 Oct. 1807, 7 Feb. 1811.
42. Ibid., 4 Feb. 1808, 4 Oct. 1814; *L'écho du Nord,* 2 Mar. 1824.
43. Registres, 28 Sept. 1822.

44. Ibid., 20 May 1824.
45. Ibid.
46. Demoulin, p. 303, n.1.
47. AN, F14 1270, Enquête sur le roulage, Nord, Condé, 1812; AD, Nord, M 653 2, Statistique industrielle, 1808. The reason given in 1808 was a decline in river traffic due to increased water tolls and decreased land tolls, with the result that much coal was being transported by wagon. This undoubtedly applied mainly to Mons coal.
48. AN, F14 1270, Enquête sur le roulage, Nord, Condé, 1812; AN, F14 6968, "Pétition . . . par les Bateliers de Condé."
49. Registres, 15 July 1832. Cf. AN, F12 2534c, Coupin, "Examin critique d'un mémoire," 1832.
50. The Loire boatmen carrying coal from the Saint Etienne and Rive-de-Gier mines reacted similarly to pressure by mine owners on shipping rates. There the government took the boatmen's side since it was easy to prove that their association had been formed in response to equally illegal collective action on the part of owners. Anzin was, of course, immune to the law of coalitions since it stood virtually alone in the Nord (Bourgin and Bourgin, I, 187–207).
51. The information in this section, where not otherwise noted, is drawn from the mass of documents found in AN, F14 6673. Mostly government reports on the improvements planned or completed on the Escaut at various times between 1817 and 1842, these documents give a detailed analysis of the problems faced, solutions proposed, and results achieved, as well as considerable historical background on developments from the mid-eighteenth century to the Restoration. They are especially important for technology and government policy. The most valuable are the reports of Louis Cordier, divisional inspector of the Bureau of Bridges and Roads in the Nord during the 1820s.
52. [Desfontaines de Preux], *Précis historique et statistique sur la ville de Valenciennes,* pp. 23–25.
53. See, for example, *Petites Affiches de Valenciennes*, 18 Aug. 1824. In 1812 it was claimed that the Escaut was "normally" closed from December to February or even later (AN, F14 1270, Enquête sur la roulage, Nord, Valenciennes, 1812).
54. *Archives de la Chambre de Commerce de Lille, 1832–40,* p. 1.
55. For the seasonality of the coal traffic, see Lévy-Leboyer, *L'industrialistation,* p. 499.
56. Bourgin and Bourgin, III, 149.
57. Casail, pp. 106–8.

58. This was especially true between Bouchain and Cambrai, where the river was well controlled by seven chamber locks dating from before the Revolution (Caron, "Enquête de l'an II," p. 244).

59. Cordier, *Navigation intérieure* (1820), pp. 26–33, 47–79.

60. Ibid., pp. 24–26; Dutens, I, 517–19; AN, ADxiii 18, "Rapport et projets sur les ponts et chaussées," 24 Feb. 1806.

61. AN, F14 6673, Cordier and Baudrand, Procès-verbal de conférence.

62. AN, F14 9685, prefect's report, 14 July 1828.

63. Dutens, I, 451–82.

64. *Annuaire Statistique* (1811), p. 140.

65. AN, F14 7822, Renard to Council of Mines, 7 Nov. 1798, 28 March 1800, 18 Jan. 1810; Registres, 4 Feb. 1806.

66. See below, Chapter 8, under "The Distribution of Sales and the Search for New Markets."

67. AN, F14 7111, Bureau of Bridges and Roads, "Rapport d'une Commission . . . sur le projet d'un canal de la Sambre à l'Escaut," 1846.

68. Registres, 4 Nov. 1801.

69. Ibid., 21 June 1827, 4 Nov. 1831, 15 July 1832.

70. Ibid., 4 Nov. 1801, 28 June 1802.

71. Ibid., 4 Sept. 1817, 7 Sept. 1820.

72. Ibid., 7 Sept. 1820.

73. Ibid., 25 Oct. 1828.

74. Presumably, it intended to lease the falls to millowners rather than use them as a source of power for its own workshops, which were already supplied with steam power.

75. Registres, 4 Sept. 1817, 17 Apr. 1818.

76. AN, F14 7076, Commission des routes et canaux, 1828. See also, A. Lesieur, ed., *Opinions et discours de M. Casimir Périer*, I, 318–20; Gille, *La Banque et le crédit,* p. 204.

77. For Honnorey's widespread activities, see AN, ADxixN 192, Ministry of Public Works, "Actes législatifs et dépenses concernant les travaux de navigation intérieure et maritime, 1814–1900," 1902, pp. 13, 77–83.

78. AN, F14 7111, Bureau of Bridges and Roads, "Mémoire sur le projet d'un canal de jonction de la Sambre à l'Escaut par le Rhonelle," 1844.

79. Dutens, I, 501–9.

80. *Archives de la Chambre de Commerce de Lille*, pp. 25–34.

81. Ibid., pp. 134–41; Cordier, *Navigation du département* (1826), pp. 96–97, 122–23. The cost of transport from Mons to Lille by the Scarpe and Deule was 0.80 francs per hectoliter in 1833; the cost to Dunkerque by Belgian waterways was only 0.85 francs, although the distance was nearly twice as great (*Archives de la Chambre de Commerce de Lille*, p. 52).
82. *Archives de la Chambre de Commerce de Lille*, pp. 49–53; Dutens, I, 492–99; Cordier, *Navigation intérieure* (1820), pp. 35–36; Cordier, *Navigation du département* (1826), pp. 93–125.
83. Registres, 3 Feb. 1813.
84. Dutens, I, 497; *Archives de la Chambre de Commerce de Lille*, pp. 397–400.
85. Dutens, I, 501–4, 509–10.
86. About the same time, the Lille Chamber of Commerce pointed out that sometimes it took longer for freight to make the voyage from Dunkerque to Lille than from America to Dunkerque (*Archives de la Chambre de Commerce de Lille*, pp. 112–13, 127–30).
87. Cordier, *Navigation intérieure* (1820), pp. 36–38, 149; Cordier, *Navigation du département* (1826), pp. 1–36; AN, ADxixN 192, Ministry of Public Works, "Les travaux de navigation intérieure et maritime," p. 83.

7 Anzin Markets: The Consumers of Coal

1. Jean Lestocquoy, *Histoire de la Flandre et de l'Artois*, p. 100.
2. Georges Lefebvre, *Les paysans du Nord pendant la Révolution française*, I, 734–35.
3. AN, F14 7824, Clère to Bureau of Mines, 9 Oct. 1926. See also, ibid., Dumas Company, circular to local chambers of commerce, 1829; Dumas Company to general director, Bureau of Mines, 20 Feb. 1829; and Brochant de Villiers to Council of Mines, 7 Feb. 1830, printed in Jennings, "Mémoire en defense," 1830, p. 21; AN, F14 7111, Cordier "Mémoire sur le canal de la Sambre à l'Escaut," 1838; A. de Villeneuve-Bargemont, *Economie politique chrétienne*, II, 51.
4. Renard Report.
5. Baumont, p. 8; Lefebvre, *Les paysans du Nord*, I, 314. The process of householders converting to coal was still continuing as late as 1865 (L. Simonin, "De l'extraction croissante et de l'épuisement de la houille," *Revue des Deux-Mondes* [6 June 1865], p. 267). In England, the same process occurred but much earlier (Nef, I, 196, 198).
6. In London the transport of "sea coal" by the vast fleet of colliers that plied the coastal waters between Newcastle and the Thames

had induced Londoners of all classes to adopt coal fuel for household use as early as 1700 (Nef, I, 196–97).

7. Prost, p. 9.
8. [P. Caron], "Un mémoire sur l'approvisionnement de Paris en charbon en 1793," *Bulletin d'Histoire Economique de la Révolution,* VIII (1913), pp. 210–12.
9. Bertier de Sauvigny, p. 348. See also AN, F12 2467, Costaz to minister of the interior, 30 July 1811.
10. AN, ADxiii 60, "Observations sur l'approvisionnement de Paris en combustible," [1816?].
11. Chevalier, pp. 218, 488. See also *Journal du Commerce,* 4 Feb. and 1 Nov. 1830.
12. See under "Houille" in *Encyclopédie des gens du monde,* 22 vols. (Paris, 1833–44), XIV, 287.
13. Caron, "L'approvisionnement de Paris," pp. 210–12; Charles Dupin, *Forces productives et commerciales de la France,* II, 259; Chevalier, passim.
14. Lequeux, p. 323.
15. Rouff, p. 415.
16. AN, F14 1307, "Avis des députés du commerce sur la requête des marchands de charbon de terre de la ville de la Bassée et de Lille," 1778; AN, F14 7816, "Mines de charbon de terre en Flandres," [1777].
17. Renard Report.
18. Registres, 4 Oct. 1810.
19. Schodet, *Almanach du commerce des arts et métiers des villes de Lille, Armentières, Roubaix et Tourcoing,* pp. 26, 39–40.
20. *Archives de la Chambre de Commerce de Lille,* p. 281.
21. Lefebvre, "Aperçu générale," pp. 425–29; Dieudonné, I, 181–85, 407–9.
22. See, for example, *Annales des Arts et Manufactures,* 2d series, III (1816), pp. 182 ff.; *Annuaire Statistique* (1811), p. 222; AN, F14 7822, Lefebvre to minister of the interior, 2 Dec. 1798.
23. The population statistics in the following paragraphs are drawn from Charles Pouthas, *La population française pendant la première moitié du XIX*[e] *siècle.* Pouthas anticipated the publication of *La population française* so far as it concerned the Nord in an article in the *Revue du Nord,* XXXVI (Apr.–June 1954), pp. 331–38.
24. Lefebvre, "Aperçu général," p. 439.
25. Chaptal, II, 112–13; but compare Louis Cordier, p. 344.
26. AN, F14 7822, Lefebvre to minister of the interior, 2 Dec. 1798; Chevalier, p. 435.

27. AN, F12 2534c, "Tableau de la répartition et des prix de vente des combustibles minéraux en France dan l'état actuel de nos communications intérieures," [1833].

28. The others employed 16, 19, 45, and 120 workers, respectively (AD, Nord, M653 6, Statistique industrielle, 1813).

29. Registres, 4 Oct. 1810; Lévy-Leboyer, *L'industrialisation*, p. 310, n.50.

30. Dormoy, p. 59.

31. AN, F14 4029, Etats de Redevance, passim.

32. AN, F12 2534c, "Tableau de la répartition et des prix de vente," [1833].

33. It was not a fertilizer in the strict sense, but it lightened heavy clay soils and thus aided drainage and cultivation by retarding clodding; it also neutralized excess soil acidity and stimulated the activity of soil microorganisms that freed the phosphorus, potassium, and nitrogen deriving from other additives. For similar usages elsewhere, see Léon, II, 48; Paul Leuilliot, *L'Alsace au début du XIX*[e] *siècle,* II 81; AN, F12 638, "Pétition des propriétaires des mines de charbon de Litry," 15 Sept. 1814. Littry coal was used almost exclusively for limeburning when English coal reappeared on the Norman coast after 1815.

34. E. Beauzot, "Le département du Nord sous la Restauration (un rapport du préfet de Villeneuve-Bargemont en 1826)," *Revue du Nord,* XXV (1939–42, 1943), p. 32. Yields for all major crops were far higher in the departments of the Anzin market area than in any other region in France (William H. Newell, "The Agricultural Revolution in Nineteenth Century France," *Journal of Economic History*, XXXIII [Dec. 1973], pp. 707, 712–21).

35. Dieudonné, I, 407–9.

36. Pouthas, *La population française*, p. 70. No figures are available for any date earlier than 1836.

37. Lévy-Leboyer, *L'industrialisation*, p. 310.

38. Dieudonné, I, 504–5. Even in 1954, the Nord manufactured over 20 percent of the total French production of bricks (Gendarme, p. 32).

39. AN, F12 2534c, "Tableau de la répartition et des prix de vente," [1833].

40. AD, Nord, M653 6, Statistique industrielle, 1813.

41. Bertrand Gille, "Investissements immobiliers urbains et conjoncture," *Histoire des Entreprises*, I (Nov. 1958), pp. 5–6; AN, F14 7111, Cordier, "Mémoire sur le canal de la Sambre à l'Escaut," 1838.

42. P. M. Bondois, "Les centres sucriers français au XIXe siècle," *Revue d'Histoire Economique et Sociale*, XIX, nos. 1, 3 (1931), pp. 321–33, and especially 323–25.
43. AN, ADxiii 60, [Anzin], "Observations," 1822.
44. In the eighteenth century, partly refined cane from the West Indies was further processed in most of the great French ports like Dunkerque and in some inland towns like Paris and Lille. The fuel most often used was coal of either French or English origin (Bondois, "Les centres sucriers," pp. 27–76; Rouff, pp. 34, 418; Caron, "L'approvisionnement de Paris," p. 212).
45. Viennet, pp. 247–54; Firman Lentacker, "Sucre et betterave dans le département du Nord à la fin du premier Empire," *Revue du Nord*, XXXVI (Apr.–June 1954), pp. 325–27; Lévy-Leboyer, *L'industrialisation*, pp. 260–61.
46. Sée, pp. 146–47, 162; *Archives de la Chambre de Commerce de Lille,* pp. 110–11, 160.
47. *L'indicateur valenciennois*, p. 55.
48. Gille, *La banque et la crédit*, p. 64.
49. AN, F12 2534c, "Tableau de la répartition et des prix de vente," [1833].
50. AD, Nord, M557 7, Chamber of Commerce, Valenciennes, "Rapports spéciaux sur les diverses industries de l'arrondissement," 1831.
51. *Archives de la Chambre de Commerce de Lille*, p. 281.
52. AN, ADxiii 60, [Anzin], "Observations," 1822. *L'indicateur valenciennois*, p. 55, lists two such establishments in or near Valenciennes in 1827.
53. AD, Nord, M653 2, Statistique industrielle, 1808.
54. Viennet, p. 84; Demoulin, p. 304, n.4. Of four oil presses in the immediate vicinity of Valenciennes in 1827, one used a steam engine (*L'indicateur valenciennois*, p. 55).
55. *Annuaire Statistique* (1829), p. 336; Beauzot, p. 34. There were four hundred refineries in the five departments of the Oise, Aisne, Somme, Pas-de-Calais, and Nord, all part of the Anzin market area (Lévy-Leboyer, *L'industrialisation*, p. 300).
56. Baumont, p. 8; Caron, "L'approvisionnement de Paris," p. 210; Lefebvre, "Aperçu géneralé," p. 439; Cordier, "Sur les mines de houille de France," p. 344; Rouff, p. 418.
57. AD, Nord, M653 2, 11, Statistique industrielle, 1808, 1824; *L'indicateur valenciennois*, p. 55.
58. Dupin, II, 168.

59. AN, F12 2534c, "Tableau de la répartition et des prix de vente," [1833].
60. Rouff, p. 418.
61. Cordier, "Sur les mines de houille de France," p. 344; Demoulin, p. 304.
62. Lefebvre, *Les paysans du Nord*, I, 735.
63. Dupin, II, 168.
64. AD, Nord, M653 11, Statistique industrielle, 1824; *L'indicateur valenciennois*, p. 55.
65. AN, F12 2534c, "Tableau de la répartition et des prix de vente," [1833].
66. P. M. Bondois, "Le développement de la verrerie française au XVIIIe siècle," *Revue d'Histoire Economique et Sociale*, XXIII, nos. 3, 4 (1936–37), pp. 337–38, 353–61; Warren C. Scoville, *Capitalism and French Glass Making, 1640–1789*. Glass was produced by the fusion in a furnace of siliceous sand and other materials. The raw glass was then refined by a second heating and cast, blown, or rolled into various finished shapes. Wood had traditionally been the source of heat for both fusions until the late seventeenth century. In order to prevent the discoloration of the finished glass, wood continued to be used for the second fusion until the middle of the nineteenth century when the glass industry underwent a second technical revolution that placed it on an essentially modern footing.
67. Bondois, "La verrerie française," pp. 245–46; Laloux, pp. 37–38.
68. Registres, 4 Oct. 1810.
69. Ministry of Commerce, *Enquête relative à diverses prohibitions*, III, 230–31.
70. AD, Nord, M653 8, 10, Statistique industrielle, 1815, 1822; Renard Report, p. 13; AN, ADxiii 60, [Anzin], "Observations," 1822; AN, F14 6856, Bureau of Mines, registration and disposition of correspondence received, 1816–23.
71. The Bruai firm, when it incorporated in 1829, justified its location by citing the proximity of the Escaut and "especially" the mines of Anzin and Mons (AN, F12 6738, "Société anonyme pour l'exploitation de la verrerie de Bruai," 1829).
72. AD, Nord, M653 11–13, Statistique industrielle, 1824–26; *L'indicateur valenciennois*, p. 55; Registres, 29 June and 25 Oct. 1828; "Notice sur les houillères d'Epinac," *Bulletin de la Société Industrielle de Mulhouse*, V (1831), p. 253. Cf. Demoulin, p. 304, n.4; Cordier, "Sur les mines de houille de France," p. 354.
73. Ministry of Commerce, *Enquête relative à diverses prohibitions*, II, 166–67, and III, 230–31. Demand had been somewhat overestimated, and with the commercial downswing that occurred between

1826 and 1832, the industry found itself with excess capacity by the early 1830s. Four firms, including the one at Saint Saulve and one in the Avesnes region which had evidently been experimenting with Anzin coal, failed or suspended operations.

74. AN, F14 4253, "Tableau des verreries . . . dans le département du Nord," 1840.
75. Duchastelle, p. 17.
76. AN, F12 6738, "Statutes de la manufacture de Commentry," 1826.
77. Cameron, *France*, pp. 352–53. The chief firm there also owned Desandrouin's old Anzin works in 1840 (AN, F14 4253, "Tableau des verreries," 1840).
78. AN, F14 4253, "Tableau des verreries," 1840.
79. "These devils who have made their fortunes in coal and glass lead a gay life" (Gras, I, 233).
80. For the French iron industry in the early nineteenth century, see especially Jean Vial, *L'industrialisation de la sidérurgie française, 1814–1864*; Pounds and Parker, pp. 19–75, 105–15, 151–54; Dunham, chapter 6; J. Levainville, *L'industrie du fer en France*, pp. 29–109; Henderson, *Britain and Industrial Europe*, pp. 53–62; Gille, *Recherches*, pp. 47–56; and Bertrand Gille, *La sidérurgie française au XIXe siècle*.
81. For an indication of French forge masters' technical backwardness, see AN, F12 633, "Réclamation des maîtres de forges," 1814, in which a footnote remarks that "the English seem to have finally found a way to obtain wrought iron by using only mineral fuel." For the subtle impediments to transfer of English techniques, see J. R. Harris, "Industry and Technology in the Eighteenth Century: Britain and France," inaugural lecture, University of Birmingham, May 1971.
82. For the contrary point of view, see Fohlen in Trenard, pp. 146–47.
83. L. Compte, "Notice sur l'histoire, progrès, avenir de l'industrie du fer dans le département du Nord," *Annuaire Statistique* (1814), pp. 102, 108–14.
84. AN, F14 4263, de Gallois, "Mémoire sur les minerais de fer des houillères"; Gille, *La banque et le crédit*, pp. 186–88.
85. Daubuisson, "Description," p. 141.
86. Registres, 6 Mar. 1817.
87. AN, F14 6856, Bureau of Mines, registration and disposition of correspondence received, 24 June 1817.
88. AN, ADxiii 14, Rambourg, "De la surabondance et excellence des mines et usines de fer en France, pour tous les divers besoins," [n. d.].
89. Compte, pp. 115–16; Société de Denain et d'Anzin, pp. 16–20.

90. Compte, pp. 122–25; *Annuaire Statistique* (1814), p. 215.
91. Compte, pp. 135–36; *Petites affiches de Valenciennes*, 12 Mar. 1824.
92. AD, Nord, M653 11, Statistique industrielle, 1824; *L'indicateur valenciennois*, p. 55.
93. Henderson, *Britain and Industrial Europe*, p. 54.
94. Compte, p. 137. Only the key workers and foremen may have been English.
95. AD, Nord, M653 11, Statistique industrielle, 1824; *L'indicateur valenciennois*, p. 55.
96. Compte, p. 137. After 1831, Raismes used Maubeuge coked pig iron, and after 1840, Trith used Belgian coked pig iron.
97. AD, Nord, M653 11, Statistique industrielle, 1824; *L'indicateur valenciennois*, p. 55; Compte, p. 143.
98. Société de Denain et d'Anzin, pp. 16–20.
99. Unlike smelting, founding, and refining, all of which were influenced in the 1820s and 1830s by new, mainly English methods, most of the tertiary processes in the iron industry remained largely wedded to traditional techniques and organization. Many were of little importance for the coal industry, especially the host of craftsmen who worked by themselves or with a single helper to produce a wide variety of small fine wares from iron or steel. For example, in the Valenciennes-Douai area in 1815 there were 102 establishments producing edge-tools, knives, guns, locks, clocks, and plated tinware; their total workforce, including owners, was only 162. Most of the makers of edge-tools, knives, tinware, and perhaps locks were undoubtedly part of a putting-out organization. The others were probably independent craftsmen. But only one among all these metal-workers, an edge-tool maker at Vieux-Condé, was found in any of the mining villages near the Anzin pits. In any case, their total need for coal must have been quite small. Indeed, from Anzin's point of view, their needs were not much different from those of thousands of other householders and shopkeepers (AD, Nord, M653 8, Statistique industrielle, 1815).
100. The only nailmaking company in the Nord above the valley of the Sambre, at Marly just south of Valenciennes, had gone out of business during the Revolution (Compte, p. 162). It may have been operating again by 1813 (AD, Nord, M653 6, Statistique industrielle, 1813).
101. AD, Nord, M653 8, 10, 11, Statistique industrielle, 1815, 1822, 1824.
102. Renard Report.
103. AD, Nord, M653 11, Statistique industrielle, 1824.

104. Like the others, he concentrated his production on nails for shoeing horses but also made some cold-process nails for construction work and a few iron chains (AD, Nord, M653 11, Statistique industrielle, 1824; *L'indicateur valenciennois,* p. 55). The French armed forces were also minor users of Anzin coal for metalworking operations. The army ordnance works at Douai and the Royal Navy shipyards at Dunkerque, Brest, and Rochefort were among Anzin's customers (Renard Report; Registres, 11 Apr. 1825, 21 June 1826, 15 July 1832.

105. Of the forty largest metallurgical firms in France around 1845, the three in the Anzin market area ranked only sixteenth, twenty-third, and thirty-seventh (Bertrand Gille, "La sidérurgie française du XIXe siècle, avant l'acier," *Revue d'Histoire de la Sidérurgie,* VII [1966], pp. 158–64).

106. For the distribution of the Nord textile industry by fiber and process in the opening years of the nineteenth century, see the map in Gendarme, p. 64.

107. M. Failly, "L'industrie du départment du Nord," in Aimé Leroy and Arthur Dinaux, *Archives historiques et littéraires du nord de la France et du midi de la Belgique,* X, 189–99; *Annuaire Statistique* (1829), p. 336; Lévy-Leboyer, *L'industrialisation,* p. 106.

108. AD, Nord, M653 2, Statistique industrielle, 1801, 1808. The implication in these figures is confirmed by direct testimony in Berryer, p. 10.

109. AD, Nord, M653 11–13, Statistique industrielle, 1824–26; *L'indicateur valenciennois,* p. 55.

110. AD, Nord, M653 11, Statistique industrielle, 1824.

111. Gendarme, pp. 63, 66; *Annuaire Statistique* (1829), p. 334.

112. Charles Schmidt, "Les débuts de l'industrie cotonnière en France (1760–1806)," *Revue d'Histoire Economique et Sociale,* VII (1914–19), pp. 26–55; Chaptal, II, 356–57, 371, 374, 380, 400, 406, 409; P. Bruyelle, "L'industrie cotonnière à Lille-Roubaix-Tourcoing," *Revue du Nord,* XXXVI (July–Sept. 1954), pp. 21–24; *Annuaire Statistique* (1829), p. 334.

113. AN, ADxiii 60, [Anzin], "Observations," 1822; *Archives de la Chambre de Commerce de Lille,* p. 86.

114. AD, Nord, M653 12, Statistique industrielle, 1825. Aniche and Anzin together then had 38 engines.

115. They ranked second, third, seventh, and twelfth, respectively.

116. AN, F14 4233, "Notice statistique sur les machines, chaudières et bateaux à vapeur," 19 Feb. 1838. There were also between one and four engines employed by such diverse industries as flour milling, chemicals and printing, brewing, rope making, water supply, and tobacco and chicory processing.

117. It may be noted that, without counting the many engines used on the nearby coalfields, the borough of Birmingham alone had 169 engines in 1835, most of them used in the metal trades (Court, p. 257). See also Fohlen in Trenard, pp. 142–46, for the insignificant role of the steam engine in France before 1850.

118. AN, F14 4233, "Les machines, chaudières et bateaux à vapeur," 1838.

119. For modern grading nomenclature, see Rice and Hartmann, p. 27; for the early nineteenth century, see Chevalier, pp. 208–21.

120. AD, Nord, S8285, Situation, 1811; AN, F14 4029, Etats de Redevance, 1820, 1823; AN, F14 4822, Daubuisson, "Exploitation," end note; AN, F14 7822, Jennings, "Réplique au précis de Lacoste et Eubank," 1831. Anzin slack coal accounted for only about one to two-fifteenths of production. Most was sold to the very poor in the area directly around the pits at a price about two-thirds that of Anzin's standard grade. The rest was doled out to company workers for domestic heating as a supplement to wages or was used on company premises. The gaillettes were carefully separated out by raking or hand-picking any lumps three to four inches or more in diameter and sold at a premium of about 50 to 100 percent over ordinary coal because they were relatively clean, set up well on grates, and were easy to light and stoke, but they amounted to no more than one-fifteenth of all coal sold, despite numerous attempts or at least admonitions on the part of management to increase their quantity by more careful culling or handling. Anzin coal was relatively friable, a fact which unfortunately increased the amount of slack and reduced the number of gaillettes. It also meant that Anzin's middle grade tended to break up when stored exposed to the weather for long periods and when loaded and unloaded. Since open storage was the regular practice at Anzin and since between the cutting face and final consumption coal might be loaded and unloaded as many as six times, it was often partly reduced to slack before it was finally burned.

121. Cordier, "Sur les mines de houille de France," pp. 360–68, 332.

122. See above, under "The Limeburning and Construction Industries." For the qualities and uses of each of the types of coal mined by the Anzin Company as described in the following paragraphs, see especially Chevalier, pp. 431–35. See also Grar, II, 158, 168; AD, Nord, S8285, Situations, 1796, 1797; AD, Nord, S8285, Renard, "Renseignements sur la nature et le commerce des mines de . . . Anzin," 27 Dec. 1797; AN ADxiii 60, "L'approvisionnement de Paris en combustible," [1816?]; AN, F14 7823, Piet, "Mémoire pour les sieurs Saugrain, Charpentiers et Compagnie . . .," [1830?]; AN, F14 7822, Jennings, "Réplique au précis de Lacoste et Eubank," 1831; AN, F14 7824, "Observations pour

la Compagnie Dumas," 1830; AD, Nord, M581 97, Mathieu, "Les houillères anglaises"; AN, F12 2534c, Ministry of Commerce, note, Oct. 1833; AN, F12 2534c, High Council of Commerce, 25 Sept. 1832; AN, F12 2534d, Divoy, "Réplique à la Compagnie d'Anzin," Mar. 1834.

123. Although of essentially the same quality, the coal from these pits might vary considerably in the amount of gaillettes, slack, and stone, they furnished. In these respects, the newer pits of the Torrent were generally superior.

124. Renard Report; Chevalier, pp. 485–90; AN, F12 2534, Saint Cricq, Bureau of Commerce, to minister of the marine, 13 Mar. 1828; AN, F12 2534d, Clement-Dorsormer, Guenyveau and Lefroy, "Rapport . . . sur les houilles . . . employées pour le service des chaudières à vapeur," 19 Mar. 1834.

125. It ranked seventh in both 1801 and 1831. The ranking of the ten largest French towns remained completely unchanged over these thirty years (Pouthas, *La population française,* p. 98).

126. In 1801, 54,756; in 1831, 67,073 (ibid.).

127. Ibid., p. 107.

128. Jacqueline Beaujeu-Garnier, "Le dynamisme urbain dans le Nord," *Revue du Nord,* XXXVI (Apr.–June 1954), pp. 411–18; Casail, pp. 37–40, 46–49; *L'indicateur valenciennois,* pp. 95–103.

129. *Annuaire Statistique,* 1810, 1822, 1828, 1832.

8 Commercial Organization: Prices, Markets, and Profits

1. AN, F14 4029, Etat de Redevance, 1830.

2. Registres, 6 June 1811, 16 May 1829; Renard Report; AN, F14 4029, Etats de Redevance, 1812–33. See also AN, F14 7824, Jennings, "Mémoire en défense"; AN, F14 7822, Jennings, "Réplique au précis de Lacoste et Eubank," 1831; Registres, 11 Oct. 1811, 6 Feb. 1812, 29 June and 25 Oct. 1828, 16 Oct. 1829, 8 Oct. 1830, 4 Nov. 1831. By weight, eleven heaping hectoliters equaled ten metric quintals or one metric ton.

3. The task of establishing a price series for Anzin is not easy. The sources are all fragmentary, difficult to interpret, and often biased. The data found in the Registres are above suspicion, but they are complete (or nearly so) in only two instances, 1811 and 1829. The Etats de Redevance contain some outright statements of prices, and further calculations are possible on the basis of the dubious figures given there for quantity of output and value of gross sales. Statements and calculations, however, do not always agree. The Renard Report contains still another unclear and partial series of prices for 1800 to 1822. Another problem stems from

the fact that Anzin added to the basic price a charge for transport to loading areas, measurement, and loading, a charge which was really part of the standard sales price. Since this charge changed over the years and is not always known, and since Anzin preferred to regard it as an extra charge for services performed rather than as a part of the sales price, it is not always certain whether it is or is not included in the available price data or, if not, by how much the basic price ought to be increased to include it.

4. See, for example, the charges in quotations from a statement printed by the Lacoste Company and Anzin's replies in AN, F14 7822, Jennings, "Réplique au précis de Lacoste et Eubank," 1831. See also AN, F14 7823, Piet, "Mémoire pour les sieurs Saugrain, Charpentiers et Compagnie," [1830?]; Chevalier, p. 436; Renard Report; Berryer, p. 12.
5. The actual cost of these "services" was as little as one-half the charge (AN, F14 4029 and 4030, Etats de Redevance, engineer's remarks, 1820, 1833).
6. Changes in the prices of the other types of coal sold by Anzin can be summarized briefly by comparison with the primary series given for middle-sized soft coal sold at wholesale. The price of large-sized coal was about double that of the middle-sized to about 1820 and then fell to about 50 percent higher by the end of the 1820s. During the same decade, prices for hard coal, which had always been somewhat higher than those for soft, fell to the same level. Throughout the period, retail prices tended to average about 10 percent higher than wholesale, and slack coal, sold only at retail, maintained a price level betwen 0.90 and 1.00 francs. After Denain coal came on the market about 1830, it may have sold at a slight premium of about 10 percent over similar soft coal from the older Anzin pits (AN, F14 7823, Piet, "Mémoire pour les sieurs Saugrain, Charpentiers et Compagnie," [1830?]).
7. AN, F14 7824, Jennings to general director, Bureau of Mines, 4 Jan. 1831; ibid., Council of Mines, extract from reports of meetings for 7 Jan., 24 Jan., and 21 Feb. 1831; Lentacker, p. 404.
8. Chevalier, p. 224.
9. At times, some dealers were forced to operate at a loss ("Le deuxième centenaire d'Anzin," no. 1 [1959], p. 27).
10. Chevalier, p. 224.
11. The general analysis of rebates in the following paragraphs follows the Registres.
12. Registres, 4 June 1807.
13. Ibid., 4 Oct. 1810.
14. Ibid., 4 Feb. 1808, 7 Feb. 1810, 6 June and 10 Oct. 1811, 6 Feb. 1812.

15. AN, F14 4029, Etats de Redevance, 1826–29; Registres, 16 May 1829.
16. Registres, 15 July 1832.
17. As in the rebates, the information on which the general analysis of credit in the following paragraphs is based can be found scattered throughout the Registres.
18. Registres, 4 Oct. 1810.
19. Ibid., 13 June 1797, 4 Nov. 1801, 10 Sept. 1813, 26 April 1830.
20. The security demanded was ordinarily a real estate mortgage (ibid., 28 June 1802, 6 Apr. 1815, 29 June 1828). Cf. AN, F14 7824, Douchy Company to Council of Mines, [1830].
21. He eventually paid off all but 34,000 francs by liquidating his assets, mostly land holdings, and in 1827 Anzin granted him the lifetime use of a dwelling and a pension of 1750 francs (Registres, 4 Oct. 1810, 5 Sept. 1816, 9 Sept. 1819, 21 June and 10 Sept. 1827).
22. "Le deuxième centenaire d'Anzin," no. 1 (1959), p. 26.
23. Registres, 20 Sept. 1824, 29 June 1828, 16 Oct. 1829, 26 Apr. and 8 Oct. 1830, 11 June 1831, 15 July 1832, 11 May and 12 Oct. 1833.
24. Ibid., 26 Apr. 1830.
25. See the copies of actual letters from Anzin billing customers in AN, F12 2534c, for 1824 and AN, F12 2534b, for 1837. The French banking and credit system was still rudimentary in the nineteenth century. There was no public (i.e., incorporated) bank in the Nord until 1836 when the Bank of France opened a branch at Lille (Gille, *La banque et le crédit*, pp. 94–95).
26. Registres, 26 Jan. 1828.
27. Such evidence may exist in the notarial archives, which have not been examined. The system was used by Aniche by the middle of the century (Duchastelle, p. 33).
28. "Le deuxième centenaire d'Anzin," no. 2 (1958), p. 178.
29. See Gille in Trenard, pp. 166–67. In fact, the Periers were forced to defend the interest rate they allowed Anzin against complaints from other owners who thought the company might do better with another banking firm ("Le deuxième centenaire d'Anzin," no. 2 [1958], p. 178).
30. Such contracts may exist for the earlier period in the notarial archives; these have not been examined.
31. Duchastelle, p. 25; Michel and Renouard, pp. 128–30.
32. With the exception of the firm of Carlier and Mirland, discussed below.
33. Registres, 7 Feb. and 4 Oct. 1810.

34. In making the conversions, a single price has been used throughout, the 1.30 francs charged for a heaping hectoliter of soft coal of middle-sized grade bought wholesale. Almost all the listed firms bought at wholesale prices, but the slightly higher prices charged for hard coal and the much higher prices charged for the large-sized grade have been ignored. The quantities given in the text are thus somewhat larger than they should be. Conversion of heaping hectoliters to tons is at the standard ratio of 11:1.

35. Renard Report. A slight reduction in the price was also made for Valenciennes hospitals (Registres, 16 Sept. 1825).

36. Renard Report; Registres, 8 Oct. 1812.

37. Registres, 4 Oct. 1810. The company general agent suggested in 1825 that Anzin set up a central retail outlet in Valenciennes to supply better service at lower cost than that provided by the existing system of petty dealerships. The Periers received his suggestion well, but the outcome is not known ("Le deuxième centenaire d'Anzin," no. 1 [1959], p. 26). For local sales during the Revolution, see Saint-Léger, I, 296–97.

38. Registres, 4 Oct. 1810.

39. Ibid., 6 Mar. 1817. On Mons competition at Valenciennes, see AN, F14 6674, "Rapport sur la navigation de l'Escaut," 1824, and the testimonial letter of Madame Renaux in AN, F12 2534c.

40. Registres, 7 Sept. 1820, 4 May 1822, 3 Mar. 1823, 16 Mar. and 20 Sept. 1824, 24 June 1826, 29 June and 25 Oct. 1829. Attempts to deal directly with large industrial firms outside the local area where they could be served by wagon were of little significance, although the Periers put considerable emphasis on the development of such outlets. There is only one certain example of Anzin's succeeding in by-passing the wholesale dealers in an area where they regarded themselves as the normal intermediaries between the mines and industry. In 1828 a rebate of 5 percent was extended to the Royal Glassworks at Saint Gobain near Saint Quentin "in order to activate relations," which lapsed in the early 1820s ("Le deuxième centenaire d'Anzin, no. 1 [1959], p. 25; Registres, 29 June 1828).

41. AN, F14 4029, Etats de Redevance, 1830, 1831.

42. AN, F12 633, Isoard, "Informations recueillies à Valenciennes," 5 Jan. 1835.

43. AN, F12 2534c, "Tableau de la répartition et des prix de vente," [1833].

44. The société en commandite was essentially a partnership that was permitted to have sleeping partners liable only for the amount of capital they invested.

45. Registres, 4 Oct. 1810, 6 Feb. 1812; Schodet, *Almanach du commerce,* pp. 39–41, 167; Bourgin and Bourgin, III, 294, 356.

In the eighteenth century, Anzin had attempted to circumvent the dealers and play the role of middleman by setting up its own depots for direct sales to consumers in the major towns of its market area (Rouff, p. 400; AN, F14 1307, "Requête présentée au Conseil d'Etat par les marchands de Lille et de la Bassée," 1778). The dealers opposed this scheme of "monopoly" so vigorously that the company never dared repeat the attempt, except in new markets where there were no established dealers and possibly at Valenciennes where the many dealers were extremely small and weak.

46. "Le deuxième centenaire d'Anzin," no. 1 (1959).
47. The phrase Perier used was "dorer la pilule" ("Le deuxième centenaire d'Anzin," no. 1 [1959], p. 27).
48. Ibid.
49. Ibid., passim.
50. Registres, 6 Feb. 1812.
51. de Bonnard, pp. 444–45.
52. Registres, 26 Apr. 1830.
53. The fact that Jean, one of the two dealers, was bankrupt by 1816 may point to inherent difficulties in wholesaling to customers located at a distance and help explain why the method was so rarely used.
54. Registres, 4 June 1810.
55. Ibid., 4 Oct. 1810, 9 Sept. 1819.
56. Ibid., 7 Feb. 1810.
57. "Le deuxième centenaire d'Anzin," no. 6 (1959), p. 577. More examples of the use of demonstrations could undoubtedly be cited if a fuller knowledge of Anzin's commercial practices were available; the position taken here is that they would nevertheless remain exceptions to the general policy that left the burden of selling to the wholesale dealers.
58. AD, Nord, S 8285, Situation, 1798.
59. Registres, 4 Nov. 1801, 19 Feb. and 17 Apr. 1802.
60. AN, F12 1958, Council of Commerce and Agriculture of Dunkerque to prefect, 9 Aug. 1802.
61. Ibid.
62. AN, F14 7822, Renard to Council of Mines, 23 Mar. 1802, 15 July 1802.
63. Ibid., 15 July 1802.
64. Registres, 28 June 1802.
65. AN, F12 1748, Dunkerque port records, 1813.
66. Registres, 4 Oct. 1810.
67. Demoulin, pp. 301–2.

68. *Archives de la Chambre de Commerce de Lille,* pp. 286, 291, 293; Michel and Renouard, p. 154; *Journal du Commerce,* 1830.
69. Registres, 25 Sept. 1823, 11 Apr. 1825, 21 June 1826, 21 June 1827.
70. Ibid., 15 July 1832.
71. AN, F12 2534b, Isoard, "Information recueillies," 1835. For Belgian coal sold coastwise from Dunkerque and amounting to about six times the Anzin total, see Lentacker, p. 1402, and Lévy-Leboyer, *L'industrialisation,* p. 307.
72. Registres, 4 Feb., 8 June, and 6 Oct. 1808, 4 Oct. 1810.
73. Chevalier, p. 228; AN, F14 1270, Enquête sur la roulage, Nord, Valenciennes, 1812; AN, ADxiii 60, "L'approvisionnement de Paris en combustible," [1816?].
74. Registres, 10 Oct. 1811, 3 Feb. 1813.
75. Ibid., 10 Sept. 1813.
76. Ibid., 6 Apr. 1815.
77. AN, ADxiii 60, "L'approvisionnement de Paris en combustible," [1816?].
78. Registres, 3 May 1821.
79. Ibid., 20 Sept. 1824.
80. Gille, *Recherches,* p. 58; Chevalier, p. 228.
81. Chevalier, p. 228.
82. AN, F12 2534a, Girard to Ministry of Commerce and Public Works, 23 Dec. 1832.
83. Registres, 26 Oct. 1828, 26 Apr. 1830.
84. Chevalier, p. 437.
85. Registres, 21 June 1827, 29 June 1828, 15 July 1832.
86. AN, F12 2534d, A. Pichault de la Martinière, "Mémoire sur la nécessité de modifier la législation des douanes . . .," [1830].
87. Registres, 26 Apr. 1830.
88. AN, F12 2534d, Pichault de la Martinière, "Mémoire," [1830].
89. Levainville, "Recherches dans la Seine-Inférieure," p. 365.
90. AD, Nord, S 8285, "Renseignements sur la nature et le commerce des mines de charbon de terre d'Anzin," 1798[?]; AN, F14 1313–14, Delacroix, "Mémoire d'observations relative à l'emploi de charbon de terre pour les fabriques et usines," 1802[?]; AN, F12 1958, Delacroix to Council of Mines, 3 Mar. 1802.
91. Registres, 10 Oct. 1811, 6 Feb. 1812, 3 Feb. 1813.
92. AN, F12 2534a, Ministry of Commerce, "Charbon de terre arrivé à Rouen pendant les années 1834, 1835, et 1836" [1837?]; AN, F12 2534b, Isoard, "Information recueillies," 1835.

93. AD, Nord, M 557 7, Chamber of Commerce, Valenciennes, "Rapports spéciaux sur les diverses industries," 1831; Registres, 29 June and 25 Oct. 1828, 11 June 1831.

94. AN, F12 2534b, Isoard, "Information recueillies," 1835.

95. AN, F12 2534c, "Tableau de la répartition et des prix de vente," [1833]. See also Registres, 25 Oct. 1828, 16 Oct. 1829, 15 July 1832.

96. Chabert, II, 422–23; Gille, *La Banque et le crédit,* pp. 293–337. Chabert analyzes the cycle from 1798 to 1820, Gille from 1815 to 1848. See also Viennet, pp. 107–33 and passim; Labrousse, *Le mouvement ouvrier,* pp. 90–92; and Guy Thuillier, "Pour une histoire monétaire du XIXe siecle: La crise monétaire de l'automne 1810," *Revue Historique,* CCXXXVIII (July–Sept. 1967), pp. 51–84.

97. There are some differences of opinion about the exact dates of initial slump and final recovery for some of these periods, notably during the years 1814–22. The dates given here are based on a conflation of the sources cited in the previous note.

98. The three major schools are represented by first, Labrousse and followers like Chabert who have worked out the theory of an initial agricultural crisis in a "traditional" economy; second, Gille and others who stress a financial crisis, a crisis of credit; and third, many others, then and now, who have viewed slumps as the result of industrial overproduction or gluts. Both Chabert and Gille are open to causes differing from their chief choice, and no single cause seems to cover all the panics, recessions, and depressions of the early nineteenth century.

99. See Fernand Bezy, "Les phases de la conjoncture au sein de la firme: Les agencements cycliques de coûts de production dans un charbonnage belge 1836–1939," *Bulletin de l'Institute de Recherches Economiques et Sociales* (1951), pp. 29–74, for a very elegant example of what can be done by the application of economic theory to a mass of accurate data. Most of the conclusions reached by his study of a Belgian coal firm over a period of one century cannot be tested for applicability to Anzin because of lack of data.

100. In theory, and to a certain extent in practice, production in the coal industry tends to fall later than sales in an economic downswing, with the result that unsold stocks pile up; it also tends to recover later on the upswing, because the stocks must first be absorbed by increased sales before output can expand again. The reason for this effect is the "increasing returns" obtained in the coal industry by "spreading overhead" (fixed costs) over more units of output and thus lowering per unit costs. Increasing

returns offer a motive for keeping output near capacity at all times (ibid., pp. 51–54). At least until the late 1820s, however, Anzin could not permit stocks to pile up indefinitely because of lack of stockage space and the deterioration suffered by coal exposed to the weather on storage dumps. Anzin certainly tried to keep its work force occupied fully at all times in order to lower unit costs and for other reasons, but it was not always successful, e.g., at Fresnes and Vieux-Condé during the 1820s (An, F12 2534c, Renard to general director of customs, 9 May 1824). In fact, there was usually excess productive capacity at Anzin in the form of inactive pits or working areas which could be quickly reopened in the event of sudden rises in demand. The major exception was the lack of sufficient good soft-coal pits in the early 1820s and perhaps since 1817–18. Coal supply at Anzin, then, was fairly "elastic," i.e., it adjusted rather easily to changes in demand as reflected by sales, and thus in practice the output and sales curves probably moved fairly closely together.

101. AD, Nord, S 8285, Situations, 1800–1810 is the sole quantitative source for unsold stocks, and it is not fully reliable for a number of reasons.
102. See, for example, the conclusions of Gillet in Bouvier, Furet, and Gillet, *Le mouvement du profit,* pp. 121–23.
103. Registres, 4 Nov. 1801.
104. AD, Nord, S 8285, Situations, 1807–8; Registres, 4 June 1807.
105. AD, Nord, S 8285, Situations, 1808–10; Registres, 4 Oct. 1810.
106. Registres, 6 June 1811, 6 Feb. 1812.
107. Ibid., 6 Feb. 1812.
108. Chabert, II, 422.
109. AN, F14 7822, Renard to prefect, 28 Apr. 1814.
110. For the frontier question, 1814–15, see Bertier de Sauvigny, pp. 88–89, 166, 173–74.
111. Registres, 6 Mar. 1817.
112. Dormoy, p. 59.
113. Labrousse, *Le mouvement ouvrier,* p. 90.
114. AD, Nord, M 557 7, Chamber of Commerce, Valenciennes, "Rapports spéciaux sur les diverses industries," 1831.
155. AN, F7 9685, perfect's report, 14 July 1828.
116. AD, Nord, M 557 7, Chamber of Commerce, Valenciennes, "Rapports spéciaux sur les diverses industries," 1831.
117. Dormoy, p. 59.
118. Demoulin, pp. 304–5; Chevalier, p. 224; Registres, 20 May 1824, 16 May 1829.
119. Registres, 1824–30, passim.

120. Marcel Gillet, "Aspects de la crise de 1845–1851 dans le bassin houillère du Nord," *Revue du Nord,* XXXVIII (Jan.–Mar. 1956), p. 15.
121. Labrousse, *Aspects de la crise,* p. xiv.
122. Unless otherwise specified, the word profits is used in this section to refer to net profits, the difference between gross revenues (essentially, income from sale of coal) and total costs. Anzin treated depreciation costs, investments in new productive facilities, and taxes as current operating costs, which become, therefore, the equivalent of total costs. For an example of the elegant variety of methods of estimating profits that can be obtained with more adequate sources, see Gillet in Bouvier, Furet, and Gillet, *Le mouvement du profit,* pp. 72–91. For the difficulties in calculating the profits of French businesses in the early nineteenth century, see Vial, I, 187–89.
123. Grar, II, 179.
124. Registres, 1 Apr. 1795.
125. Daubuisson, "Description," end page.
126. AN, F14 4029, Etats de Redevance, 1812–14. Although output had risen by about one-sixth since 1805, profits had probably not risen proportionately because of price cuts.
127. Registres, 17 Apr. 1818, 4 May 1822, 21 June 1826, 21 June 1827, 29 June 1828, 16 Oct. 1829, 8 Oct. 1830, 4 Nov. 1831, 4 Nov. 1832. Profits distributed per denier amounted to 5,500 francs in 1821, 6,000 in 1822–25, 7,000 in 1826–27, and 8,000 in 1828–32.
128. Undistributed profits were mostly held in various accounts with the Perier Bank. In the years immediately after 1800, Anzin overdrew its account with the Perier Bank; the overdraft constituted a sort of running loan on which Anzin paid 6 percent interest. The company soon succeeded in wiping out its indebtedness (which was evidently never very large) and built up a favorable current account on which it in turn drew 6 percent interest. At some point it began to set aside funds from its swelling current account and placed them in a separate account with the Periers as a reserve fund. In 1826 its total assets with the Perier Bank stood at 4,329,597 francs, about evenly divided between current account and reserve ("Le deuxième centenaire d'Anzin," no. 2 [1958], p. 178; no. 1 [1959], p. 33). See also, Registres, 21 June 1827. Anzin's "reserves" were not intended to cover any specific anticipated future expenses but were merely a pool of undistributed assets from which it drew revenue and a sense of security against possible catastrophes such as revolution, invasion, or a major flood. In this light, Anzin's high rate of profit is all the more remarkable.

129. Perier Report.

130. The maintenance of distributed profits at a given level over a period of several years, although output and hence total profits were in any particular year usually rising, makes it evident that the proportion of profits distributed to total profits was probably considerably lower at the end of the period than at the beginning. Thus, increases in the amount of profits distributed normally lagged behind increases in total profits.

131. *Archives de la Chambre de Commerce de Lille,* p. 71. By far the greater part of Anzin revenues were derived from coal sales, but the company also received income from interest on its current and reserve accounts with the Perier Bank (the "reserve" funds probably being invested by the Periers in government bonds, mortgages, or other safe long-term securities), from rents on its considerable land holdings in the area of the mines, and in the late twenties from its holdings in canal companies, especially those of the Saint Quentin Canal after 1828. Although these non-mining revenues tended to increase in the twenties, as late as 1830 they probably did not exceed 200,000 francs a year, well under 10 percent of total revenues (AN, F14 7823, Piet, "Mémoire pour les sieurs Saugrain, Charpentiers et Compagnie," [1830?]). Cf. Michel and Renouard, pp. 136–43, and Reybaud, p. 176, for the much greater importance of these sources of income after 1833.

132. It is not possible to speak of the rate of profit as a proportion of the capital value of the company's "stock" or assets. Since the deniers that represented the fractions into which Anzin's ownership was divided had never had a par value and almost never were sold on the market, their value, and hence the value of the capital assets of Anzin as a whole, was not determinable. The company sometimes proceeded as if its deniers had a determinable value, but it calculated their value at any given time as a fixed multiple of profits distributed. Since capital value was defined on the basis of profits, it would obviously be unrealistic to attempt to reverse the procedure and calculate a rate of profit on capital value. While the increase in profits distributed per denier could be used to calculate capital gains to the owners, they almost never "took" their gains by selling their deniers.

133. Migneron estimated that his production figures, and hence those for sales, were underestimated by one-sixth (AN, F14 4029, Comité d'évaluation, procès-verbal, 1813).

134. Though the effects of the tariff for Anzin seem clear, "surprisingly little historical work" exists "on the effect of tariffs on industrial structure" or firms (J. D. Gould, *Economic Growth in History,* p. 289).

Bibliography

Archival Materials

Most of the material for this study has been drawn from manuscripts and pamphlets deposited either in the Archives Nationales at Paris or the Archives Départementales of the Nord at Lille. Some sources are found in both depositories. This duplication is notably true for the most important single source used, the "Registres des délibérations de la Régie de la Compagnie des Mines d'Anzin, 1758–1873." The Nord archives contain a manuscript copy in the series J 316. At Paris a microfilm of the Nord copy is located in the series AQ 49 1. The Registres are an unimpeachable source containing information on every aspect of the company's activities. They are the indispensable starting point for any study of the Anzin Company.

The Registres are all the more essential because no other private records of the company survive. Whatever financial records or commercial correspondence had endured the ravages of time and neglect until the twentieth century were lost to fire and bombing before the company's archives were opened to the public by nationalization after World War II. Nor are any of the papers of the leading figures associated with the company in the early nineteenth century still extant, with the exception of the valuable letters of Casimir and Joseph Perier to the general agent Paulze d'Ivoy, 1825–27. Some of these letters have been published with a helpful anonymous commentary in "Le deuxième centenaire de la fondation (19 novembre 1757) de la Compagnie des Mines d'Anzin," *Mines*, nos. 1, 2, 6 (1958); no. 1 (1959). Other material on the business activities of the Periers and other company figures might be found in the notorial archives, which were not used for this study. Richard Barker, however, has consulted the notorial archives for his Ph.D. dissertation, "Casimir Perier (1777–1832) and William Ternaux (1763–1833): Two French Capitalists" (Duke University, 1958), without turning up any material of much significance for the history of Anzin.

Aside from the Registres, most of the archival material relating to Anzin falls into two broad classifications. One consists of the records of the government bodies charged with enforcing the network of laws and administrative regulations pertaining to coal mines. Mines operated under a government concession and were subject to inspection, regulation, advice, and taxation by various official agents. Some of the official records concerning Anzin are the product of the Nord department's prefectural administration and others of the Ministry of Commerce, but most stem from the activities of local and national representatives of the Bureau of Mines in the Ministry of the Interior. The second major type of archival record is the mass of manuscripts and pamphlets addressed to various government bodies by Anzin and its opponents. Most such material concerns directly new concessions and tariff policy, but all aspects of Anzin's history and operations are covered. These manuscripts and pamphlets exist in stupifying abundance, they are often repetitious, and their point of view is not disinterested. Nevertheless, these products of heated controversy are often the sources of facts and opinions found in the apparently more reliable official records, and, used with care, the conflicting testimony they offer is extremely valuable. For a listing and analysis of "Les archives de l'industrie houillère" by Bertrand Gille, see *Histoire des Entreprises,* I (Nov. 1958), pp. 78–103.

Archives Nationales (AN)

The most important series relating to coal mining in the Archives Nationales is F14, Travaux Publics, which contains the records of both the Bureau of Mines and the Bureau of Bridges and Roads. The manuscript inventory of Charles Schmidt, "Inventaire des documents versés en 1909" (24 May 1910), Inventaire 665, is invaluable as a guide to the important documents found in cartons 6501–8665. Elsewhere in the series F14 (as for the cartons used in the other series discussed below) one must depend on the much less detailed inventory provided by the *Etat sommaire des versements des ministres* (Paris, 1924).

The most important cartons in series F14 are 7820–24, which deal with the concessions held by Anzin and those proposed or obtained by its rivals in the Nord. They contain both official records of the Bureau of Mines and a large collection of *mémoires, pétitions, rapports, observations*, etc., by Anzin and rival concessionaires. The "Etats de Redevance" for Anzin are found in the carton 4029, and although they must be used with great care, these Etats, drawn up for tax purposes, yield a wide variety of statistical information on equipment, manpower, production, sales, costs, and profits. Particularly to be noted are the comments on the Etats by the Bureau of Mines local engineers; these comments are found with the Etats or in carton 4030. The records of the Bureau of Bridges and Roads concerning French waterways fill over five hundred cartons, and even of those dealing with the rivers and canals carrying Anzin coal, only a small number could be examined in detail. Particular use has been made of cartons 6673 (Escaut River), 6968 (Mons-Condé Canal), 7036–42 and 7079 (Saint Quentin Canal), 7076 (reports of Molé, Becquey, etc.), and 7111 (Sambre-Escaut project). Useful maps and drawings of the works, concessions, geology, and topography of the Nord coalfield are found in cartons 10300 and 10301. The other cartons in series F14 in which documents cited in this study are found include: 1270, 1302, 1313–14, 1315–16, 3848, 4233, 4244, 4253, 4259, 4263, and 7816. Dozens of other cartons were examined, and much material found in them proved useful, although they have not been specifically cited.

Series F12 contains two major types of documents relating to Anzin. One concerns the movement of interior and foreign trade; soundings in such cartons as 1748 and 1810 proved of secondary significance, but a more thorough investigation would undoubtedly turn up useful information. Of prime value, however, is the massive testimony by chambers of commerce, industrialists, and coal companies, including Anzin, in cartons 633–38 and especially 2534 and 2534a–d (four cartons) in connection with various official investigations of tariff policy.

Series AD is specifically the repository for printed sources. It consists mainly of copies of laws and administrative decrees and orders, volumes by government officials (such as Emile Dormoy's *Topographie souterraine*) which can usually be found elsewhere, and government reports, but it also includes a few privately printed pamphlets. The documents from this series that have been specifically cited can be found in ADxiii 14, 15, 18, 60, and in ADxix 192.

Minor use has been made of three other series. Series F7, Police Générale, contains information on labor troubles, especially in carton 3796. Series F2, Administration Départementale, carton 1220–21, contains some material relating to Nord mines. The registers of correspondence received by the Ministry of the Interior and the Bureau of Mines in series F, Registres d'Enregistrement, cartons 6855 and 6856–71, are useful as a check on chronology.

Archives Départementales (AD)

Aside from the "Registres" in series J 316, the two series M and S contain much material relating to Anzin which cannot be found in the Archives Nationales. A complete and detailed inventory exists for each series. Unfortunately, many of the most interesting entries in the inventories proved upon request to have "disappeared."

The information on Nord industry and commerce found in series M provides the necessary background for an understanding of the Anzin market. Of the large number of bundles examined, those numbered 557 7, 581 97, 623 12, 626 1, and 653 2, 6, 8, 11, and 13 proved most useful.

The series S contains roughly 250 numbered bundles or single items relating to the Nord mines. Nearly 100 concern the granting of concessions. A vast supply of information of all kinds has been preserved by the detailed nature of the government requirements for applying for, winning, and keeping a concession. The bundles concerning Anzin's four original concessions are numbered 8284–85, 8327, and 8353. Other bundles consulted include those for Aniche (8272–73), Bruille (8299), Denain (8311), Douchy (8312), Odomez (8340) and Saint Saulve (8345). The important official reports on individual mine accidents, including the summaries by the engineers Clère and Boudousquié, are found in bundles 8111–13.

Printed Works Cited

Works to 1850

Annales des Artes et Manufactures. Paris, 1800–1817.

Annales des Mines. Paris, 1816–48.

"Annonces concernant les mines." *Journal des Mines,* no. 1 (1802).

Annuaire Statistique du Département du Nord. Lille, 1803–35.

Archives de la Chambre de Commerce de Lille, 1832–40. Lille, 1850.

Becquey, F. L. *Rapport au Roi sur la navigation intérieure de la France.* Paris, 1820.

Berryer, P. N. *Observations présentées par les propriétaires et exploitants des mines de houille de France. . . .* Paris, 1826 (?).

Bonnard, A. H. de. "Notice sur diverses recherches de houille entreprises dans le département du Pas-de-Calais. . . ." *Journal des Mines,* no. 2 (1809).

Bulletin de la Société d'Encouragement pour l'Industrie Nationale. Paris, 1815–20.

Chaptal, Jean. *De l'industrie française.* 2 vols. Paris, 1819.

Chevalier, Michel. "Observations sur les Mines de Mons et sur les autres mines de charbon qui approvisionnent Paris." *Annales des Mines,* nos. 2, 4 (1832).

"Circulaire sur un nouveau mode de fermeture pour les lamps de sûreté." *Annales des Mines,* no. 1 (1832).

Combes. "Aérage des mines." *Annales des Mines,* no. 1 (1839).

———. "Notes additionnelles. . . ." *Annales des Mines,* no. 2 (1839).

Compte, L. "Notice sur l'histoire, progrès, avenir de l'industrie du fer dans le département du Nord." *Annuaire Statistique,* 1844.

Cordier, J. L. E. *De la navigation du département du Nord. . . .* Paris, 1826.

———. *De la navigation intérieure du département du Nord. . . .* Paris, 1820.

Cordier, Louis. "Sur les mines de houille de France." *Journal des Mines,* no. 1 (1814).

Daubuisson, J. F. "Description des houillères d'Anzin." *Journal des Mines,* no. 2 (1805).

[Desfontaines de Preux]. *Précis historique et statistique sur la ville de Valenciennes. . . .* Valenciennes, 1825.

Dieudonné, Christophe. *Statistique du département du Nord.* 3 vols. Douai, 1804.

Dupin, Charles. *Forces productives et commerciales de la France.* 2 vols. Paris, 1827.

Dutens, J. *Histoire de la navigation intérieure de la France.* 2 vols. Paris, 1829.

L'écho du Nord. Lille, 1821–24.

Encyclopédie des gens du monde. 22 vols. Paris, 1833–44.

Gallois, de. "Des chemins de fer en Angleterre." *Annales des Mines,* no. 3 (1818).

Garcenot, A. *Les mines d'Anzin: Etude historique et technique.* Paris, 1844.

Gérando, de. "Mémoire sur le progrès de l'industrie. . . ." *Bulletin de la Société Industrielle de Mulhouse,* XIV (1840).

Guenyveau. "Sur l'emploi des boeufs au service des machines à molettes." *Journal des Mines,* no. 1 (1812).

Héricart-Thury. "Mémoire sur les machines à vapeur de rotation . . . dans les houillères de Litry. . . ." *Journal des Mines,* no. 1 (1803).

Héron de Villefosse, A. M. *De la richesse minérale.* 3 vols. Paris, 1810–19.

L'indicateur valenciennois: Almanach historique, statistique, administratif, judicaire et commercial de l'arrondissement de Valenciennes. Valenciennes, 1827.

Journal des Mines. Paris, 1792–1815.

Journal du Commerce. Paris, 1820–30.

Koechlin, Emile. "Mémoire sur les machines à vapeur. . . ." *Bulletin de la Société Industrielle de Mulhouse,* IX (1835).

Lefebvre. "Aperçu générale des mines de houille exploitées en France. . . ." *Journal des Mines,* no. 2 (1802).

Leroy, Aimé, and Dinaux, Arthur. *Archives historiques et littéraires du nord de la France et du midi de la Belgique.* 18 vols. Valenciennes, 1829–57.

Lesieur, A., ed. *Opinions et discours de M. Casimir Périer.* 4 vols. Paris, 1838.

Manes. "Mémoire sur les machines à vapeur . . . du Creusot." *Annales des Mines,* no. 1 (1840).

"Mémoire sur les pompes employées dans les mines par John Taylor, Esq." *Annales des Mines,* no. 1 (1832).

Ministry of Commerce. *Enquête relative à diverses prohibitions.* 3 vols. Paris, 1835.

Ministry of Commerce and Public Works. *Enquête . . . sur les houilles.* Paris, 1833.

Molinari, Gustave de. *Histoire du tarif.* 2 vols. Paris, 1847.

"Moyens de prévenir les effects funestes de l'inflammation du gas hydrogène." *Annales des Artes et Manufactures,* 2d series, II (1815).

"Note sur une machine propre à renouveler l'air dans les mines." *Annales des Artes et Manufactures,* LV (1815).

"Notice sur les accidents arrivées dans les mines de houille du département de la Loire. . . ." *Annales des Mines,* no. 2 (1832).

"Notice sur les houillères d'Epinac." *Bulletin de la Société Industrielle de Mulhouse,* V (1831).

Petites affiches de Valenciennes. Valenciennes, 1821–25.

Schodet, ed. *Almanach du commerce des arts et métiers des villes de Lille, Armentières, Roubaix et Tourcoing.* Lille, 1829.

"Sur plusieurs machines propre à transporter les minerais dans les galeries souterraines. . . ." *Journal des Mines,* no. 4 (1802).

"Travaux de la Compagnie départementale du Haut-Rhin pour la recherche de nouvelles mines de houille de 1822 à 1832." *Bulletin de la Société Industrielle de Mulhouse,* VII (1833).

Tréba, de. "Mémoire sur la température de l'intérieur des mines." *Annales des Mines,* no. 1 (1818).

Tredgold, Thomas. *The Steam Engine.* 2d ed. 2 vols. London, 1838.

Villeneuve-Bargemont, A. de. *Economie politique chrétienne; ou, Recherches sur la nature et les causes du paupérisme en France et en Europe.* 3 vols. Paris, 1834.

Works after 1850

Aguet, Jean Pierre. *Contributions à l'histoire du mouvement ouvrier français: Les grèves sous la Monarchie de Juillet (1830–1847).* Geneva, 1954.

Ashton, T. S., and Sykes, J. *The Coal Industry of the Eighteenth Century.* Manchester, 1929.

Ballot, Charles. *L'introduction du machinisme dans l'industrie française.* Paris, 1923.

Bamford, Paul W. "Entrepreneurship in Seventeenth and Eighteenth Century France: Some General Conditions and a Case Study." *Explorations in Entrepreneurial History*, IX, no. 4 (1957).

———. *Forests and French Sea Power, 1660–1789.* Toronto, 1956.

Barber, Elinor G. *The Bourgeoisie in Eighteenth-Century France.* Princeton, 1955.

Barker, Richard J. "Casimir Périer (1777–1832) and William Ternaux (1763–1833): Two French Capitalists." Ph.D. dissertation, Duke University, 1958.

———. "French Entrepreneurship during the Restoration: The Record of a Single Firm, the Anzin Mining Company." *Journal of Economic History*, XXI (June 1961).

Barral, Pierre. *Les Perier dans l'Isère au XIXe siècle.* Paris, 1964.

Baumont, Maurice. *La grosse industrie allemande et le charbon.* Paris, 1928.

Beau de Loménie, E. *Les responsabilités des dynasties bourgeoises.* 4 vols. Paris, 1943–63.

Beaujeu-Garnier, Jacqueline. "Le dynamisme urbain dans le Nord." *Revue du Nord*, XXXVI (Apr.–June 1954).

Beauzot, E. "Le département du Nord sous la Restauration (un rapport du préfet de Villeneuve-Bargemont en 1828)." *Revue du Nord*, XXV (1939–42, 1943).

Bertier de Sauvigny, G. de. *La Restauration.* Paris, 1955.

Beugnot, A. A. *Vie de Becquey . . . directeur général des Ponts et Chaussées et des Mines sous la Restauration.* Paris, 1852.

Bezy, Fernand. "Les phases de la conjoncture au sein de la firme: Les agencements cycliques des coûts de production dans un charbonnage belge 1836–1939." *Bulletin de l'Institute de Recherches Economiques et Sociales*, Louvain University, 1951.

Blanchard, M. "Enquête administrative sur la roulage en 1811," *Revue de Géographie Alpine*, VIII (1920).

Blanchard, R. *La Flandre: Etude géographique de la plaine flamande. . . .* Dunkerque, 1906.

Bondois, P. M. "Les centres sucriers français au XIXe siècle." *Revue d'Histoire Economique et Sociale*, XIX, nos. 1, 3 (1931).

———. "Le développement de la verrerie française au XVIIIe siècle." *Revue d'Histoire Economique et Sociale*, XXIII, nos. 3, 4 (1936–37).

Bourgin, Hubert, and Bourgin, Georges, eds. *Les patrons, les ouvriers et l'Etat: La régime de l'industrie en France de 1814 à 1830.* 3 vols. Paris, 1912–41.

Bouvier, Jean; Furet, François; and Gillet, Marcel. *Le mouvement du profit en France au XIXe siècle.* Paris and The Hague, 1965.

Bruyelle, P. "L'industrie cotonnière à Lille-Roubaix-Tourcoing." *Revue du Nord*, XXXVI (July–Sept. 1954).

Cameron, Rondo E. "Economic Growth and Stagnation in France, 1815–1914." *Journal of Modern History*, XXX (Mar. 1958).

———, ed. *Essays in French Economic History.* Homewood, Ill., 1970.

———. *France and the Economic Development of Europe, 1800–1914.* Princeton, 1961.

Caron, P. "Une enquête sur l'état des routes, rivières et canaux au début de l'an II." *Bulletin d'Histoire Economique de la Révolution,* X (1917–19).

[Caron, P.]. "Un mémoire sur l'approvisionnement de Paris en charbon en 1793." *Bulletin d'Histoire Economique de la Révolution*, VIII (1913).

Casail, Yves. "Transports et voies de communication en Flandre Wallonne. . . ." Thesis, University of Lille, 1955.

Cavrois, A. *Les sociétés houillères du Nord et du Pas-de-Calais.* Paris, 1896.

Chabert, A. *Essai sur les mouvements des prix et des revenues en France de 1798 à 1820.* 2 vols. Paris. 1945–49.

Chandler, Alfred D. Jr. *Strategy and Structure: Chapters in the History of the Industrial Enterprise.* Cambridge, Mass., 1962.

Choulet, E. *La famille Casimir Perier.* Grenoble, 1894.

Clapham, J. H. *The Economic Development of France and Germany, 1815–1914.* 3d ed. Cambridge, 1928.

———. *An Economic History of Modern Britain.* 2d ed. 3 vols. Cambridge, 1930–38.

Cochran, Thomas C. "Cultural Factors in Economic Growth." *Journal of Economic History*, XX (Dec. 1960).

Court, W. H. B. *The Rise of the Midland Industries, 1600–1838.* London, 1938.

Crouzet, François. "Wars, Blockade, and Economic Change in Europe, 1792–1815." *Journal of Economic History*, XXIV (Dec. 1964).

Crouzet, F.; Chaloner, W. H.; and Stern, W. M., eds. *Essays in European Economic History, 1789–1914.* London, 1969.

Demoulin, Robert. *Guillaume Ier et la transformation économique des provinces belges (1815–1830).* Liège and Paris, 1938.

"Le deuxième centenaire de la fondation (19 novembre 1757) de la Compagnie des Mines d'Anzin." *Mines*, nos. 1, 2, 6 (1958); no. 1 (1959).

Dickinson, H. W. *A Short History of the Steam Engine.* New York, 1939.

Dion, P. "Géographie de la circulation dans la partie centrale de la région du Nord." *Annales de la Société Géologique du Nord*, III (1939).

Dollfus, Charles. Remarks on W. H. Chaloner and W. O. Henderson, "The Manbys and the Industrial Revolution in France, 1819–84," *Transactions of the Newcomen Society*, XXX (1956–57).

Dormoy, Emile. *Topographie souterraine du bassin houiller de Valenciennes.* Paris, 1867.

Duchastelle, Françoise. "Le commerce du charbon des Mines d'Aniche de 1855 à 1873." Thesis, University of Lille, 1956.

Dunham, A. L. *The Industrial Revolution in France, 1815–1848.* New York, 1955.

Duveau, Georges. *La vie ouvrière en France sous la Second Empire.* Paris, 1946.

Earle, Edward Meade, ed. *Modern France.* Princeton, 1951.

Farey, John. *A Treatise on the Steam Engine: Historical, Practical, and Descriptive.* 2 vols. Vol. I, London, 1827. Vol. II was published in 1971 in Newton Abbot, Devon, as a reprint of corrected page proof left at the author's death in 1851.

Frandon, Ida Marie. *Autour de "Germinal": La mine et les mineurs.* Geneva and Lille, 1955.

Gendarme, René. *La région du Nord: Essai d'analyse économique.* Paris, 1954.

Gerschenkron, Alexander. "Social Attitudes, Entrepreneurship, and Economic Development." *Explorations in Entrepreneurial History*, VI, no. 1 (1953–54).

Gille, Bertrand. "Les archives de l'industrie houillère." *Histoire des Entreprises*, I (Nov. 1958).

———. *La banque et le crédit en France de 1815 à 1848.* Paris, 1959.

———. "Investissements immobiliers urbains et conjoncture." *Histoire des Entreprises*, I (Nov. 1958).

———. *Recherches sur la formation de la grande entreprise capitaliste (1815–1848).* Paris, 1959.

———. *La sidérurgie française au XIXe siècle: Recherches historiques.* Geneva, 1968.

———. "La sidérurgie française du XIXe siècle, avant l'acier." *Revue d'Histoire de la Sidérurgie*, VII (1966).

Gillet, Marcel. "Aspects de la crise de 1845–1851 dans le bassin houillère du Nord." *Revue du Nord*, XXXVIII (Jan.–Mar. 1956).

Gould, J. D. *Economic Growth in History: Survey and Analysis.* London, 1972.

Grar, Edouard. *Histoire de la recherche, de la découverte et de l'exploitation de la houille dans le Hainaut français, dans la Flandre française et dans l'Artois, 1716–1791.* 3 vols. Valenciennes, 1847–50.

Gras, L. J. *Histoire économique générale des mines de la Loire.* 2 vols. Saint Etienne, 1922.

Griffin, A. R. *Coalmining.* London, 1971.

Guillaume, Pierre. *La Compagnie des Mines de la Loire, 1846–1854: Essai sur l'apparition de la grande industrie capitaliste en France.* Paris, 1966.

Harris, J. R. "The Employment of Steam Power in the Eighteenth Century." *History*, LII (June 1967).

———. "Industry and Technology in the Eighteenth Century: Britain and France." Inaugural lecture, University of Birmingham, May 1971.

Harris, T. R. *Arthur Woolf, the Cornish Engineer, 1766–1837.* Truro, n.d.

Henderson, W. O. *Britain and Industrial Europe, 1750–1870.* Liverpool, 1954.

———. *The State and the Industrial Revolution in Prussia, 1740–1870.* Liverpool, 1958.

Jenkins, Rhys. "A Cornish Engineer—Arthur Woolf, 1766–1837." *Transactions of the Newcomen Society*, XIII (1932–33).

Kemp, Tom. *Economic Forces in French History.* London, 1971.

Labrousse, Ernest, ed. *Aspects de la crise et de la dépression de l'économie française au milieu du XIXe siècle, 1846–1851.* Paris, 1956.

———. *Le mouvement ouvrier et les théories sociales en France de 1815 à 1848.* Les Cours de Sorbonne. Paris, 1948.

Lafitte-Laplace, R. J. *L'économie charbonnière de la France.* Paris, 1933.

Laloux, J. *Le rôle des banques locales et régionales du nord de la France.* Paris, 1924.

Landes, David S. "French Entrepreneurship and Industrial Growth in the Nineteenth Century." *Journal of Economic History,* IX (May 1949).

———. *The Unbound Prometheus: Technological Change and Industrial Development in Western Europe from 1750 to the Present.* Cambridge, Mass., 1969.

Laplane, Henri de. *Eloge de Louis-Alexandre-César Taffin de Givenchy.* Saint Omer, 1859.

Lefebvre, Georges. *Etudes sur la Révolution française. Paris,* 1954.

———. *Les paysans du Nord pendant la Révolution française.* 2 vols. Paris and Lille, 1924.

Lefranc, Georges. "The French Railroads, 1823–42." *Journal of Economic and Business History,* II (Feb. 1930).

Lentacker, Firman. "Les charbons belges sur le marché français au cours du XIXe siècle." *Bulletin des Séances.* Brussels: Académie Royale des Sciences d'Outre-mer, 1964–66.

———. *La révolution énergétique de notre temps.* Paris, 1969.

———. "Sucre et betterave dans le département du Nord à la fin du premier Empire." *Revue du Nord,* XXXVI (Apr.–June 1954).

Léon, Pierre. *La naissance de la grande industrie en Dauphiné (fin du XVIIIe siècle—1869).* 2 vols. Paris, 1954.

Lequeux, A. "Les prospections houillères dans la France du Nord et leurs conséquences humaines." *Revue du Nord,* XL (Apr.–June 1958).

Lestocquoy, Jean. *Histoire de la Flandre et de l'Artois.* Paris, 1949.

Letaconnoux, J. "Les transports en France au XVIIIe siècle." *Revue d'Histoire Moderne et Contemporaine,* X, XI (1908–9).

Leuilliot, Paul. *L'Alsace au début du XIXe siècle.* 3 vols. Paris, 1959–60.

Levainville, J. *L'industrie du fer en France.* Paris, 1922.

———. "Les recherches de la houille dans la Seine-Inférieure pendant la Révolution." *Bulletin d'Histoire Economique de la Révolution,* VIII (1913).

Levasseur, Emile. *Histoire des classes ouvrières et de l'industrie en France de 1789 à 1870.* 2d ed. 2 vols. Paris, 1903–4.

Lévy-Leboyer, Maurice. "La croissance économique en France au XIXe siècle: Résultats préliminaires." *Annales: Economies, Sociétés, Civilisations,* XXIII (July–Aug. 1968).

———. *Les banques européennes et l'industrialisation internationale dans la première moitié du XIXe siècle.* Paris, 1964.

Lhomme, Jean. *La grande bourgeoisie au pouvoir, 1830–1880: Essai sur l'histoire sociale de la France.* Paris, 1960.

Mallet, A. *Compound Engines.* English ed. New York, 1884.

Manuel, Frank. *The New World of Henri Saint-Simon.* Cambridge, Mass., 1956.

Marc, Alfred. *L'évolution des prix depuis cent ans.* Paris, 1958.

Marczewski, J. "Histoire quantitative de l'économie française," *Cahiers de l'Institute de science économique appliquée*. Series AF, nos. 4–6 (1963–66).

Martin, Germain. *La grande industrie sous la règne de Louis XV*. Paris, 1900.

Michel, Georges, and Renouard, Alfred. *Histoire d'un centre ouvrier: Anzin*. Paris, 1891.

Mines et Métallurgie. Nov. 1959.

Moore, Elwood S. *Coal*. New York, 1922.

Mott, R. A. "English Waggonways of the Eighteenth Century." *Transactions of the Newcomen Society*, XXXVII (1964–65).

National Bureau of Economic Research. *Capital Formation and Economic Growth*. Princeton, 1955.

Nef, John U. *The Rise of the British Coal Industry*. 2 vols. London, 1932.

Newell, William H. "The Agricultural Revolution in Nineteenth Century France." *Journal of Economic History*, XXXIII (Dec. 1973).

Palmade, Guy. *French Capitalism in the Nineteenth Century*. English ed. New York, 1972.

Payen, Jacques. *Capital et machine à vapeur au XVIIIe siècle: Les frères Perier et l'introduction en France de la machine à vapeur de Watt*. Paris, 1969.

Pollard, Sidney. *The Genesis of Modern Management: A Study of the Industrial Revolution in Great Britain*. London, 1965.

Pollard, Sidney, and Holmes, C., eds. *Documents of European Economic History*. 2 vols. New York, 1968–.

Pounds, Norman J. G. *The Ruhr: A Study in Historical and Economic Geography*. London, 1952.

Pounds, Norman J. G., and Parker, William N. *Coal and Steel in Western Europe*. London, 1957.

Pouthas, Charles. "L'évolution démographique du département du Nord dans la première moitié du XIXe siècle." *Revue du Nord*, XXXVI (Apr.–June 1954).

———. *La population française pendant la première moitié du XIXe siècle*. Paris, 1956.

———. "Les projets de réforme administrative sous la Restauration." *Revue d'Histoire Moderne*, I (Oct.–Nov. 1926).

Prost, E. *Historique de l'industrie houillère belge*. Renaix, 1927.

Reybaud, Louis. *Le fer et la houille*. Paris, 1874.

Rice, George S., and Hartmann, Irving. *Coal Mining in Europe*. U.S. Department of the Interior, Bureau of Mines, Bulletin 414. 1939.

Richard, C. "La crise du charbon dans le Nord en 1793." *Revue du Nord*, V (1914–19).

Richard, Guy. "Un épisode peu connu de la recherche du charbon en Normandie, les mines de Meulers (1797–1806)." *Annales de Normandie*, no. 2 (1968).

Rouff, Marcel. *Les mines de charbon en France au XVIIIe siècle, 1744–1791*. Paris, 1922.

Saint-Léger, A. de, ed. *Les mines d'Anzin et d'Aniche pendant la Révolution*. 2 vols. Paris, 1935–36, 1938–39.

Schmidt, Charles. "Les débuts de l'industrie cotonnière en France (1760–1806)." *Revue d'Histoire Economique et Sociale*, VII (1914–19).

Schneider, Eugène. *Le charbon: Son histoire, son destin*. Paris, 1945.

Schumpeter, Joseph A. *The Theory of Economic Development*. Cambridge, Mass., 1934.

Scoville, Warren C. *Capitalism and French Glass Making, 1640–1789*. Berkeley, 1950.

Sée, Henri. *Histoire économique de la France: Les temps moderne, 1789–1914*. Paris, 1951.

Simonin, L. "De l'extraction croissante et de l'épuisement de la houille." *Revue des Deux-Mondes,* 6 June 1865.

Singer, Charles; Holmyard, E. J.; and Hall, A. R., eds. *A History of Technology*. 5 vols. Oxford, 1953–58.

Société de hauts fourneaux, forges et aciéries de Denain et d'Anzin. *Livre d'Or*. Paris, 1950.

Stearns, Peter N. "British Industry through the Eyes of French Industrialists (1820–1848)." *Journal of Modern History,* XXXVII (Mar. 1965).

———. "Patterns of Industrial Strike Activity in France during the July Monarchy." *American Historical Review,* LXX (Jan. 1965).

Thuillier, Guy. "Pour une histoire monétaire du XIXe siècle: La crise monétaire de l'automne 1810. *"Revue Historique,* CCXXXVIII July–Sept. 1967.

Trenard, Louis, ed. *Charbon et les sciences humaines*. Paris and the Hague, 1966.

Van Tieghem, Philippe. *Introduction à l'étude d'Emile Zola: "Germinal" (Documents inédits de la Bibliothèque Nationale)*. Les Cours de Sorbonne. Paris, n.d.

Vial, Jean. *L'industrialisation de la sidérurgie française, 1814–1864*. 2 vols. Paris and the Hague, 1967.

Viennet, Odette. *Napoléon et l'industrie française: La crise de 1810–1811*. Paris, 1947.

Vuillemin, Emile. *Le bassin houiller du Pas-de-Calais: Histoire de la recherche, de la découverte, et de l'exploitation de la houille dans ce nouveau basin.* 3 vols. Lille, 1880–83.

Wibail, A. "L'évolution économique de l'industrie charbonnière belge depuis 1831." *Bulletin de l'Institut des Sciences Economiques,* Louvain University, VI (Nov. 1934).

Wrigley, E. A. *Industrial Growth and Population Change: A Regional Study of the Coalfield Areas of Northwest Europe in the Later Nineteenth Century.* Cambridge, 1961.

———. "The Supply of Raw Materials in the Industrial Revolution." *Economic History Review.* 2d series, XV (Aug. 1962).

Zola, Emile. *Germinal.* Paris, 1885.

Index